Manual for Improving
Boiler and Furnace Performance

Thomas Garcia-Borras

Gulf Publishing Company
Book Division
Houston, London, Paris, Tokyo

Manual for Improving
Boiler and Furnace Performance

Library of Congress Cataloging in Publication Data

Garcia-Borras, Thomas.
Manual for improving boiler and furnace performance

Bibliography: p.
Includes index.
1. Furnaces—Efficiency. 2. Steam-boilers—Efficiency.
I. Title.
TH7140.G37 1983 621.1'8 82-21024
ISBN 0-87201-243-3

Contents

Preface

This book shows plant managers, chief engineers, and foremen how to improve the efficiency of boilers, furnaces, and dryers while saving up to 30% in fuel costs. The 20 methods presented are simple and direct. Ten of the methods require no expenses or capital investment; ten require some expenses and capital investment and have relatively short payback periods.

This is a practical manual. Only a minimum amount of theory is used. The analytical instruments recommended for energy audits are practical, relatively inexpensive, and can easily be carried in the trunk of a car or even on a motorcycle.

Tabulated and graphical data are provided here, and the use of specially designed slide rules is encouraged whenever possible. With these simple tools, many efficiency and heat-loss calculations can be performed in a matter of minutes without significantly affecting the accuracy of the results. Checklists for evaluating potential heat losses and analytical data for calculating efficiencies are also provided, along with glossaries of technical terms.

This book is based, in part, on several studies produced under contract for the Federal Energy Administration, the U.S. Department of Energy, and the U.S. Environmental Protection Agency.

Thomas Garcia-Borras
Fullerton, California

1

Introduction

If you are operating your steam boiler, furnace, or dryer as most people do, you should be able to increase its efficiency 10%–30% or even more. Said differently, you unknowingly may be wasting 10%–30% of your fuel.

If you have a 100-hp boiler producing 3450 lbs/hr steam, you could save $31,000–$93,000 per year in fuel cost. Or, if you have a larger boiler producing 100,000 lbs/hr steam, your savings could amount to an almost unbelievable $1 million to $3 million per year! Dyer and Maples estimated a fuel savings of $50 million per year when the efficiencies of 18 boilers were increased 6% or more.[1]

This book discusses 20 different methods to save fuel and gives examples of how these methods are used.

Rising Fuel Costs

Not too long ago, the job of the engineer or foreman responsible for boiler operation was to keep the combustion equipment operating safely, with no visible smoke in the stack. The boiler-room man and service technicians used their eyes and ears as their primary instruments, eyeballing the flame and listening to the operation of the equipment. Those were the days when fuel was inexpensive. Since 1973, fuel costs have gone up drastically, with no end in sight (see Figure 1-1). Comparative costs of different fuels are shown in Table 1-1. Today, one of the chief engineer's primary goals is to cut fuel costs.

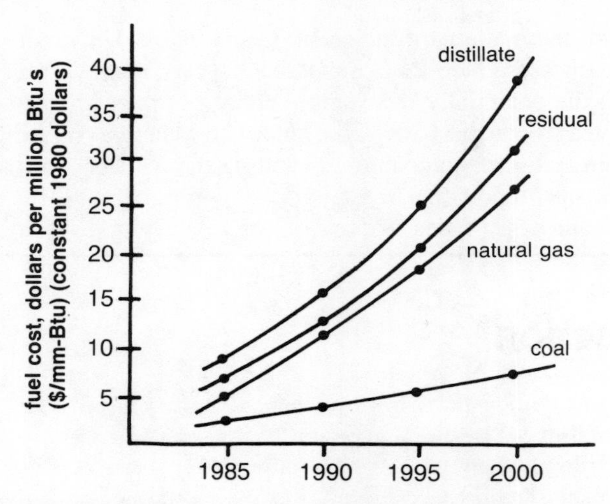

Figure 1-1. Hydrocarbon fuel-cost projections in the U.S.A. (Source: Dow Chemical Hydrocarbon and Energy Forecast, August 1980.)

Table 1-1
Comparative Costs of Fuels, First-Quarter 1983

Fuel	Price/Unit January 1983	Cost per Million Btu
Coal	$ 50/ton	$ 2.08
Natural gas	4.50/mm Btu	4.50
No. 6 oil	0.88/gallon	5.80
No. 2 oil	1.20/gallon	8.58
Propane	0.88/gallon	9.79
Gasoline	1.38/gallon	11.00
Electricity	0.045/kW-hr	13.20

Energy Consumption

The total energy output in the U.S. is in the range of 80 quadrillion Btu per year.* This is about one-third of the world's output in one year. Industry consumes some 41% of the 80 quads used in this country—17% in industrial boilers and the remaining 24% in glass and steel furnaces, kilns, foundry ovens, and for other industrial applications.[2,3,4] According to the Center for

*One quadrillion, or one quad, is equivalent to 1×10^{15} Btu. It takes 170 supertankers to haul one quad of oil.

Strategic and International Studies at Georgetown University, electrical power companies produce 24.3 quads each year, mostly from coal. It is reported that the potentially recoverable energy in the U.S. adds up to 30–40 quads a year, or about half of the total U.S. energy consumption.[3]

Boiler manufacturing companies, including the well-known Babcock and Wilcox company, report that most boilers operate below 70% efficiency.[5] Steel, glass, and other types of furnaces are even worse: their efficiencies usually run below 60%.

How to Save Fuel

For efficient combustion, the mechanical parts of the boiler or furnace must be operating properly. If the control devices such as valves and dampers are not doing their jobs, good control is impossible. The boiler and furnace themselves can be problems. For example, air flow cannot be minimized if large amounts of air leak into the boiler. Therefore, before anything else, the equipment must be in top operating condition. There are two principal ways to save fuel and increase boiler or furnace efficiency:

1. Analyze operations and improve efficiencies with *no additional expense* or capital investment (10 such methods are included in this manual).
2. Improve efficiencies *with some expenses* and capital investment.

The first option is essentially a do-it-yourself analysis of operations using recommended techniques and instruments. The second involves some engineering calculations that, although fairly simple, require the time of in-house engineers or outside consultants. All of these methods are covered in detail in this book.

It has been estimated that a boiler producing more than 30,000 lbs/hr of steam justifies substantial capital investment to recover heat losses.[5] Keep in mind that a boiler normally burns more than four times its original cost in fuel per year. Therefore, fuel saved on a large boiler usually has a quick payback on the additional capital investment.

Before installing sophisticated, expensive, automatic controls on your equipment, use the methods shown here to upgrade and optimize your operations with as little capital investment as possible. Only after this should you consider combustion efficiency controls. *Compare the best* you *can do with the best the automatic controls (or additional equipment) can do.*

It is not unusual to spend $150,000 or more installing new equipment to increase boiler efficiency by 20%, yet the simple techniques shown here can increase efficiency by 25%, without spending a penny.

Boiler Efficiency

A boiler's design features and the way it is operated affect its efficiency. Typically, hot-water boilers are the most efficient, followed by low-pressure and high-pressure steam boilers. Firetube boiler manufacturers claim a minimum of 80% efficiency at full load. Watertube boilers are several percentage points less efficient than firetube models. However, the larger exhaust-gas volumes and higher stack temperatures in watertube boilers may make the addition of an economizer worthwhile.

Manufacturers rate the efficiency of their boilers on the basis of fuel-to-steam efficiency. Boilers can also be rated by combustion efficiency, which is usually several percentage points higher because it does not consider radiation, convection, and other heat losses. Thus:

$$\begin{aligned}\text{combustion efficiency} + (&\text{radiation} + \text{convection} \\ &+ \text{blowdown heat losses}) \\ &= \text{true boiler efficiency}\end{aligned}$$

The combustion efficiency is really the effectiveness of the guts of the boiler itself, and relates to its ability to completely burn the fuel.

Some manufacturers report ratings as ''thermal efficiency,'' meaning the effectiveness of fuel-to-water heat transfer through the boiler tubes. It does not take into account boiler radiation and other heat losses.

Fuel-to-steam efficiency (the ratio of Btu output to Btu input) is the correct figure to use when determining fuel costs. It includes radiation, convection, and other heat losses. In this book, boiler efficiency is used interchangeably with fuel-to-steam efficiency. However, the *change* in overall efficiency will always be approximately the same for both combustion and boiler efficiencies. Therefore, an economic gain in combustion efficiency is the same as that based on overall boiler efficiency (keeping heat losses constant).

Furnace Efficiency

The efficiency of a glass or steel furnace can also be defined in the same manner:

$$\text{furnace efficiency} = \frac{\text{Btu output of the hot gases}}{\text{Btu input of the fuel}}$$

The bulk of the discussion in this book is on boilers, as there are more of them (1.8 million in the U.S. alone[6]). However, the same principles apply to other types of heating systems, whether they produce hot gases to melt glass and steel or to dry coffee beans in South America.

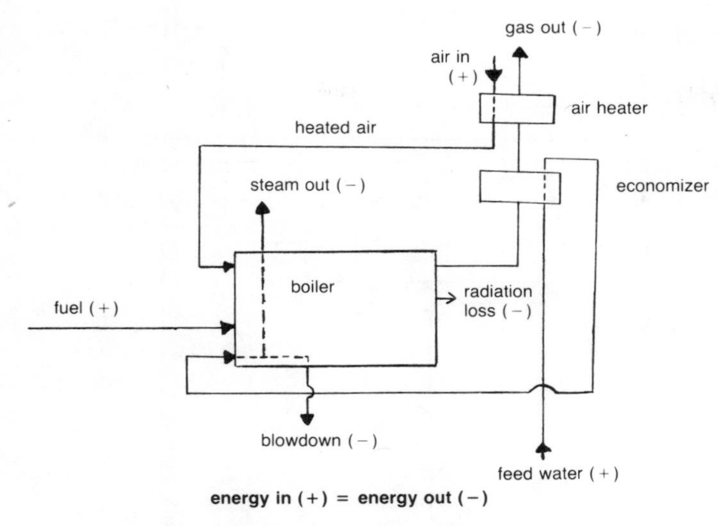

Figure 1-2. Schematic of a steam boiler.

Where to Save Fuel

Figure 1-2 is a schematic of a steam boiler. The areas where fuel can be saved include:

1. The boiler proper (improve fuel combustion; minimize hot-water blowdown; use better controls).
2. Fuel (keep at the proper viscosity; treat with additives).
3. Water (treat to minimize scale and deposits; improve deaerator operations).
4. Hot flue gases (extract waste heat; preheat water and air).
5. Steam (use optimum piping insulation; recover condensate; optimize steam trap operation).

Figure 1-3 is a schematic of a steel or glass furnace. All areas are important, and all can save a substantial amount of fuel.

References

1. Dyer, David F., and Glennon Maples, "Boiler Efficiency Improvement," Boiler Efficiency Institute, Auburn, Alabama, 1981.
2. Balzhiser, R. E., "A Clearer View of U.S. Energy Prospects," *Chemical Engineering*, p. 74, January 11, 1982.

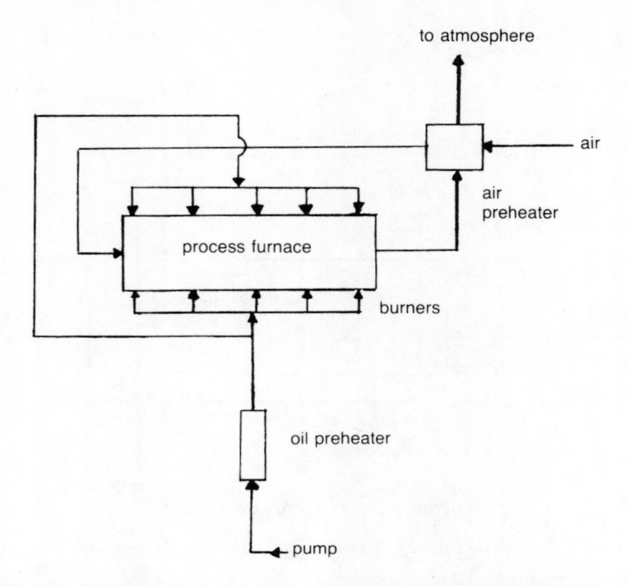

Figure 1-3. Schematic of a steel furnace and auxiliary equipment.

3. Miller, James Nathan, "The Energy Crisis: There *is* an Easy Answer," *Reader's Digest,* p. 73, June 1980.
4. *The Directory of Industrial Heat Processing and Combustion Equipment, U.S. Manufacturers, 1981–1982,* The Energy Edition, Information Clearing House, Inc.
5. Wilcox, J. C., "Improving Boiler Efficiency," *Chemical Engineering,* p. 122, October 9, 1978.
6. EPA-600/7-79-178a, August 1979, "Population and Characteristics of Industrial/Commercial Boilers in the U.S.," Interagency Energy/Environment R&D Program Report, National Technical Information Service, Springfield, Virginia 22161.

2

Boilers and Combustion

A boiler is basically a cast-iron or steel pressure vessel designed to transfer heat produced by combustion to a fluid. It is used to produce hot water, saturated steam (steam at its saturated temperature), or superheated steam (steam heated above the saturation temperature). Figure 2-1 is a typical packaged boiler used in industry, with capacities up to about 120,000 lbs/hr steam.

Boilers, except the electrical ones, have six basic parts:

1. Burner, or nozzle, which is the heart of the boiler.
2. Combustion space, or firebox.
3. Convection section.
4. Stack.
5. Air fans.
6. Instrumentation and controls.

Boilers can be classified in different ways according to the fuel used or type of construction. These are explained on the pages that follow.

Boilers Classified by Type of Fuel

Boilers can be classified according to the type of fuel they use:

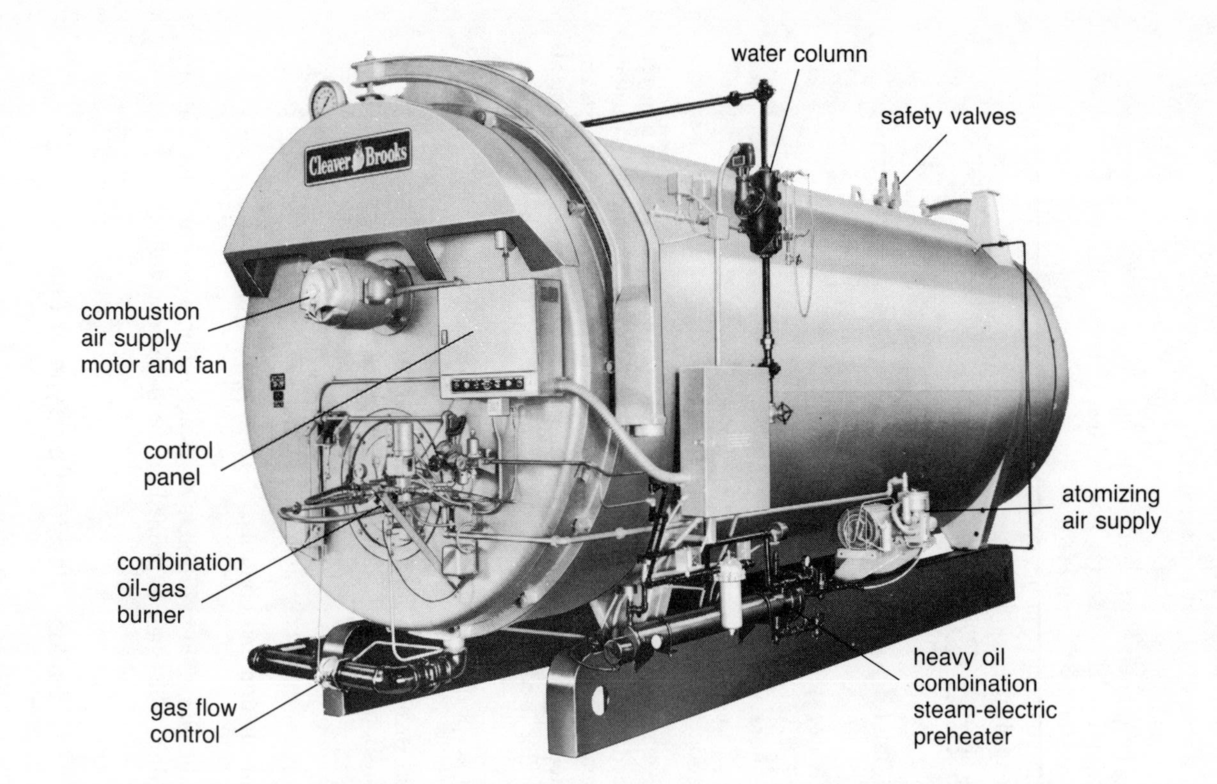

Figure 2-1. Typical packaged boiler. Packaged boilers include a pressure vessel, burner, all the controls, air fans, and insulation. The boiler is tested at the manufacturer's plant and shipped to the customer, ready for use, after installing the fuel lines and piping and electrical connections. (Courtesy Cleaver-Brooks.)

1. Liquid fuels, including No. 2 and No. 6 fuel oils. ASTM classifies liquid fuels using a scale from 1 to 6 according to certain characteristics.[1] Table 2-1 briefly describes these fuel oils.[2]
2. Natural gas.
3. Coal. These boilers are used when high steam production is required. They require large capital investments. Figure 2-2 shows a typical coal-fired boiler. Compare it to the typical oil-fired boiler in Figure 2-3 (the auxiliary equipment common to both systems is excluded).[3]
4. Other solid fuels: solid waste, bagasse, etc.
5. Combinations of the preceding. Liquid and natural gas boilers are common.

Boilers Classified by Type of Combustion and Heat Transfer

There are three general types of boilers.

Firetube boilers. In this type of boiler, hot gases flow inside tubes that are submerged in water within a shell. Design operating pressures are about 150 psig, producing up to 29,000 lbs of steam per hour, from 10 hp to 850 hp. This type of boiler is the most popular and is used in small industrial plants. The advantages of firetube boilers are that:

1. They require a low initial investment and are less expensive than watertube boilers.
2. They have high efficiencies—80% or higher.
3. They can meet wide and sudden load fluctuations with only slight pressure changes because of the large volume of water stored in the shell.
4. They are ready for use after being delivered on site (see Figure 2-4).

Table 2-2 gives the general characteristics of packaged firetube boilers. Firetube boilers are usually rated by horsepower. Table 2-2 includes the average fuel consumption rate of firetube boilers.

Watertube boilers. In these boilers, water flows through tubes that are surrounded by hot combustion gases in a shell. They are usually rated by pounds of steam per hour, and they range from about 2000 lbs/hr to 10,000,000 lbs/hr steam. Other characteristics are as follows:

1. They are used for high-pressure steam.
2. They require more instrumentation and more controls.
3. They are classified as A, D, O, or one of several other configurations according to their tube-and-drum arrangement. (The steam drum is at the top; the water drum(s) near the bottom.[4] See Figure 2-5.)

(text continued on page 15)

Table 2-1
Fuel Oil Preheating Requirements

Grade[a]	Description and Application	Preheating Requirements		Viscosity Range	Typical API Gravity
		For Pumping and Handling	For Burning	Saybolt Universal at 100°F[e]	Sec. at 60°F[f]
No. 1	Light distillate oil intended for vaporizing pot-type burners. Seldom used for pressure-type oil burners or commercial burners.	no	no	—	42
No. 2	Medium distillate oil for general-purpose domestic and commercial heating equipment.	no	no	33–38 SSU [35]	35
No. 3	Obsolete grade designation.	—	—	—	—
No. 4	Heavy distillate oil that may contain some residual oil. Suitable for firing most commercial burners.	usually no[c]	usually no[c,d]	45–125 SSU [80]	19
No. 5 light	Light residual oil for commercial-industrial burners. Generally contains a larger blended portion of distillate oil than No. 5 heavy. Usually requires preheat for burning but not for handling.	usually no[c]	usually yes[c]	>125–300 SSU [200]	18
No. 5 heavy	Medium-viscosity residual fuel oil for commercial-industrial burners. Usually requires preheat for burning.	usually no[c]	yes	>300–900 SSU [550]	16
No. 6	High-viscosity grade of residual fuel oil for the largest commercial-industrial burners with full preheating. Sometimes referred to as ''Bunker C.''	yes	yes	>900–9000 SSU [5000]	13
Low-sulfur resid[b]	Residual fuel oil for commercial-industrial burners that is refined or blended to meet local sulfur regulations.	usually no[c]	usually yes[c]	45–9000 SSU	—

Table 2-1 continued

[a]Grade numbers No. 1, No. 2, No. 4, No. 5 light, No. 5 heavy, and No. 6 are ASTM designations.*

[b]"Low-sulfur resid" is a recent term used to describe residual oil grades recently shipped to meet local regulations; it is essentially replacing No. 5 and No. 6 where sulfur regulations are in effect (for example, along the East Coast). (The sulfur content of this grade of fuel oil is generally 1%, or less.) The viscosity of present low-sulfur resid is in the range of No. 5.† (It is not clear what the viscosity of these fuels may be in the future.)

[c]Preheating requirement depends on pour point and viscosity in relation to climate.

[d]May require heating for burning when using mechanical atomization.

[e]Viscosity limits specified by ASTM D396-75 for number grade shown. Range for low-sulfur resid is estimated. Average viscosity for U.S. refined fuels from ERDA Heating Oils Survey, 1975,** is shown in brackets and is presented as a typical value.

[f]Average API gravity for U.S. refined fuels from ERDA Heating Oils Survey, 1975.**

*"Specifications for Fuel Oils," ASTM D-396-75. *ASTM Standards for Petroleum Products* (Part 17), American Society for Testing and Materials, 1975.

†"Low-Sulfur Fuels are Different," Siegmund, C. W., *Hydrocarbon Processing,* February 1970, pp. 89–95.

**"Burner Fuel Oils, 1975," Shelton, E. M., Bartlesville Energy Research Center, BERC/PPS-75/2 (August 1975), available from U.S. Energy Research and Development Administration, Bartlesville, Oklahoma.

Source: "Guidelines for Burner Adjustments of Commercial Oil-Fired Boilers," EPA-600/2-76-088, March 1976.

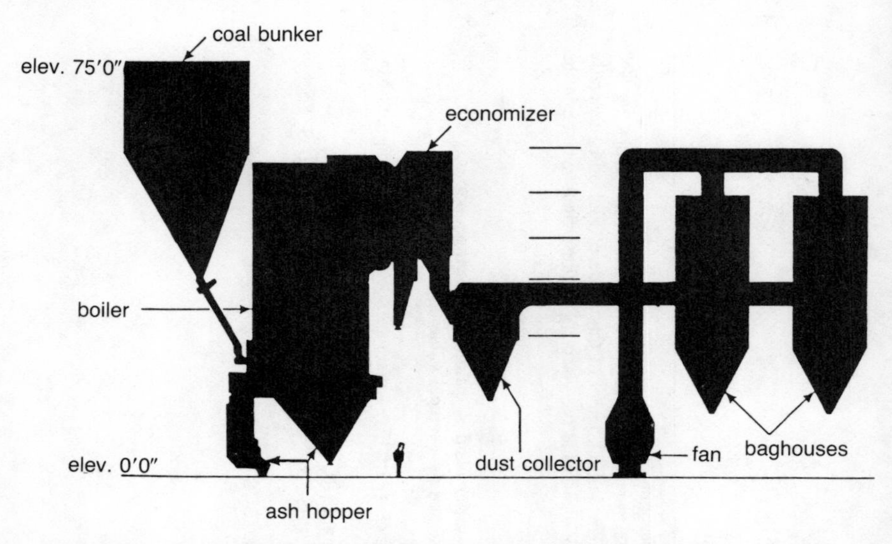

Figure 2-2. Typical coal-fired boiler. (Source: Quillman, B., "Present and Future Coal Utilization at DuPont," *Chemical Engineering Progress*, March 1980, p. 43.)

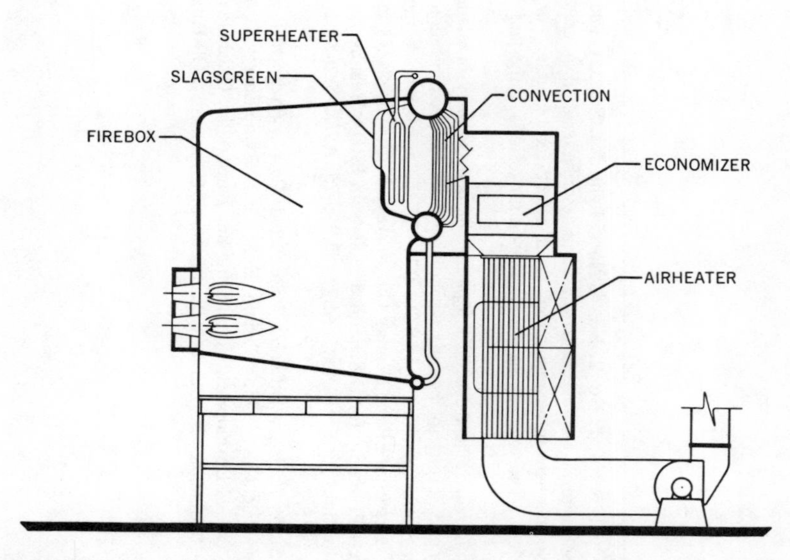

Figure 2-3. Typical oil-fired boiler. The auxiliary equipment (water treatment, oil storage, etc.) is not shown. (Source: *Fundamentals of Boiler Efficiency*, Exxon Company U.S.A., 1976.)

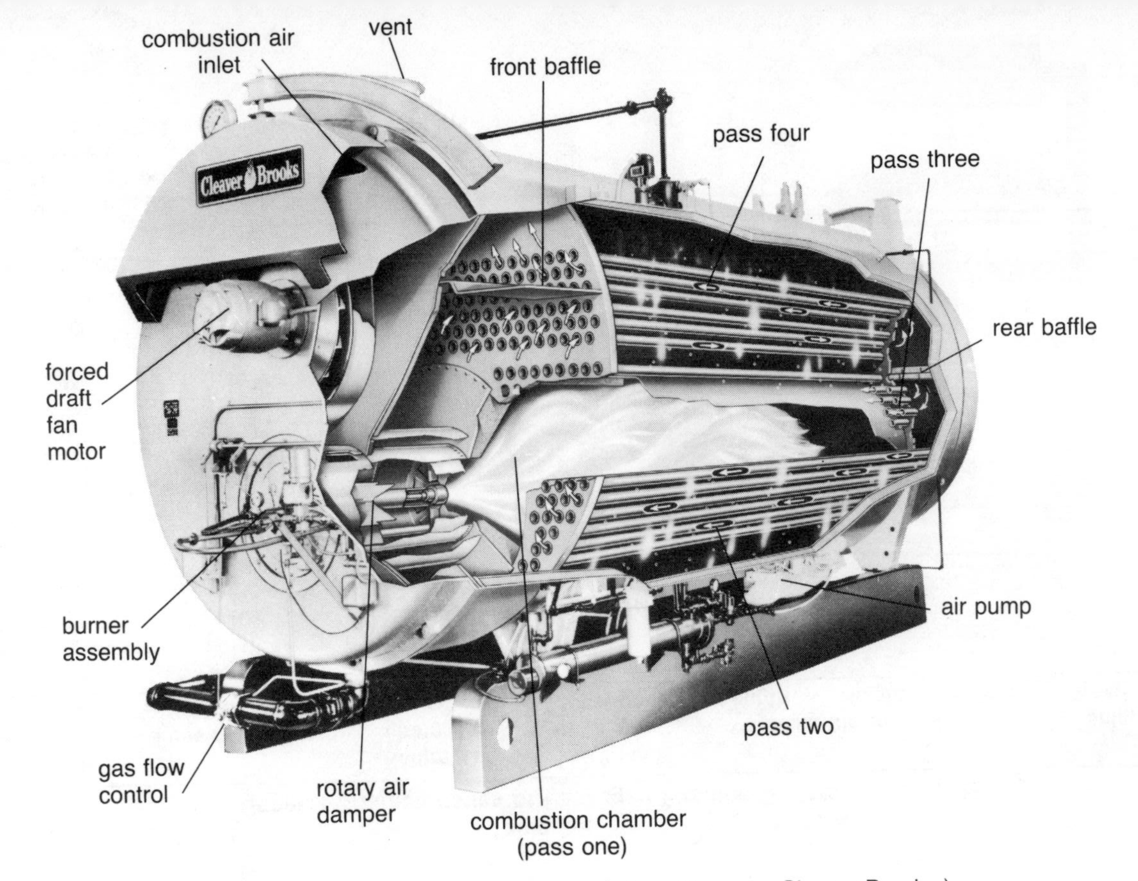

Figure 2-4. Four-pass, packaged, firetube boiler. (Courtesy Cleaver-Brooks.)

Table 2-2
General Characteristics of Packaged Firetube Boilers

Boiler Size, hp	Steam Output, lbs/hr	Fireside Heating Surface, sq ft	Natural Gas Firing Rate, cfh	No. 2 Fuel Oil Firing Rate, gph	Boiler Dimensions, ft			Shipping Weight, lbs
					Length	Height	Width	
20	690	100	835	6	10	5	4	3,300
30	1,035	150	1,255	9	10	5	4½	4,700
40	1,380	200	1,675	12	11	5	5	5,250
50	1,725	250	2,100	15	11½	5½	5	6,050
60	2,070	300	2,510	18	14	6	5	6,750
80	2,760	400	3,350	24	13	6	5½	7,600
100	3,450	500	4,185	30	14	6½	6	9,800
125	4,312	625	5,230	37	15	7	6	11,200
150	5,175	750	6,280	45	16	7	6	12,000
200	6,900	1,000	8,375	60	17½	7½	6½	16,000
250	8,625	1,250	10,500	75	16	7½	7	22,100
300	10,350	1,500	12,600	90	18	7½	7	25,500
350	12,075	1,750	14,600	105	19	8	7½	27,200
400	13,800	2,000	16,800	120	20	9	8	35,700
500	17,250	2,500	21,000	150	23	9	8	37,400
600	20,700	3,000	25,200	180	23	9½	8½	42,000
700	24,150	3,500	29,400	210	25	10	9	50,000

Note: Energy units in rating boilers:
1 hp = 34.5 lbs of steam per hour = 33,472 Btu/hr
1 therm = 100,000 Btu

Source: Holzhauer, Ron, ''Packaged Boilers,'' *Plant Engineering*, December 11, 1980, p. 72.

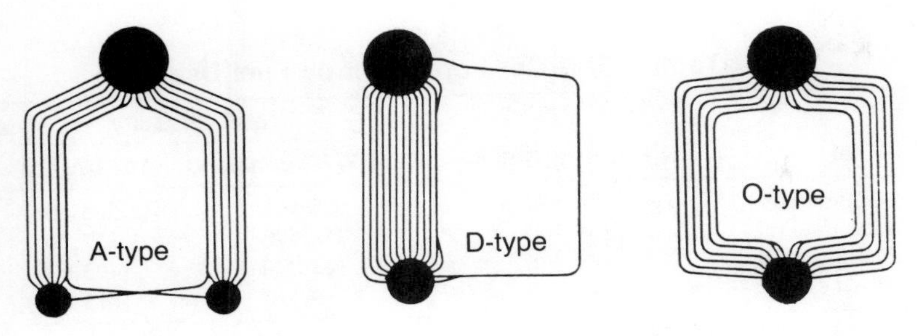

Figure 2-5. Classification of watertube boilers.

Table 2-3
Derivation of Fuel Consumption from Horsepower Rating

Fuel Type	Consumption
No. 2 fuel oil	0.3 gal/hr × hp
No. 6 fuel oil	0.28 gal/hr × hp
Gas @ 500 Btu/ft^3	84 ft^3/hr × hp
Gas @ 800 Btu/ft^3	53 ft^3/hr × hp
Gas @ 1,000 Btu/ft^3	42 ft^3/hr × hp

The following data compare the efficiencies of firetube and watertube boilers:[5]

	Firetube (1000 hp)	*Watertube (956 hp)*
Natural gas	81.2%	78.5%
No. 2 fuel oil	84.7%	81.0%

Electrical boilers. These are available in different sizes also. They are less efficient than fuel-fired boilers; however, they have no stack losses and no burner cleaning or adjustment. The typical power-generating plant can convert only about one-third of the fuel energy into electricity that will be delivered to the user.

The fuel consumption for steam boilers can be estimated from its hp rating, as shown in Table 2-3. In these estimates, the boiler runs at 80% fuel-to-steam efficiency. Tables 2-4 and 2-5 show the number of boilers classified by fuel use, heat-transfer distribution, and capacity.

Operating Procedures

Boiler manufacturers publish operating manuals on their boilers. For instance, an operating manual for packaged boilers of 125–350 hp using light oil, heavy oil, gas, or a combination can be ordered from Cleaver-Brooks Company in three different languages: English, Spanish, and French.[6]

Table 2-4
Total U.S. Boiler Population by Fuel Use

Fuel	Number of Boilers	Total Capacity	
		MW Thermal	10^6 Btu/hr
Natural gas	954,350	588,590	(2,008,800)
Residual oil	389,104	358,570	(1,223,800)
Distillate oil	244,206	127,040	(433,600)
Coal	214,400	239,110	(815,830)

Source: EPA-600/7-79-178a, August 1979, "Population and Characteristics of Industrial/Commercial Boilers in the U.S.," Interagency Energy/Environment R&D Program Report, National Technical Information Service, Springfield, Virginia 22161.

Table 2-5
U.S. Boiler Distribution by Heat-Transfer Distribution and Capacity

Heat Transfer Configuration	Number of Boilers	% of Total Capacity
Watertube	50,500	57
Firetube	275,000	23
Cast iron*	1,500,000	20

*Cast iron boilers are used in domestic or small commercial operations to produce either low-pressure steam or hot water.
Source: EPA-600/7-79-178a, August 1979, "Population and Characteristics of Industrial/Commercial Boilers in the U.S.," Interagency Energy/Environment R&D Program Report, National Technical Information Service, Springfield, Virginia 22161.

Appendix B summarizes the usual boiler room daily, weekly, monthly, and yearly maintenance requirements. Additional information on boiler designs, their operation and codes, can be obtained from the literature.[7,8,9,10,11,12,13]

Combustion

Before discussing specific methods to increase combustion efficiency, the basic principles of combustion must be considered.

Chemical reaction. Combustion, or burning of a fuel, is a chemical reaction. Combustion is the rapid combination of oxygen with a fuel, resulting in the release of heat.

The most common fuels, such as fuel oil, natural gas, and coal, consist of carbon and hydrogen (with small amounts of sulphur existing in many fuels). Trace amounts of other elements, such as vanadium and sodium in

Table 2-6
Effect of Excess Air on Combustion

% CO_2 in Flue Gas	% Excess Combustion Air	Total Combustion Air, ft^3/gal of Fuel	% Fuel Wasted
15.3	0	1484*	0
14.2	10**	1628	0**
12.5	25	1850	1.5
10.5	50	2220	4.0
8.0	95	2886	10.0
6–7	150	3700	18–20

*Approximate stoichiometric value.
**Usually the minimum needed to prevent smoke formation and to stay within 100–400 ppm CO. This is the "base" line for practical boiler operations.

fuel No. 6, are also present. For practical purposes, it is the oxidation of three elements—carbon, hydrogen, and sulfur—that is involved in combustion. Primarily, three chemical reactions are involved:

carbon (C) + oxygen (O_2) → carbon dioxide (CO_2) + heat

hydrogen (H_2) + oxygen (O_2) → water vapor (H_2O) + heat

sulfur (S) + oxygen (O_2) → sulfur dioxide (SO_2) + heat

Putting it differently:

$$100 \text{ lb air } (79 \text{ lb } N_2 + 21 \text{ lb } O_2) + 7 \text{ lb oil } (6 \text{ lb C} + 1 \text{ lb } H_2) = 19 \text{ lb } CO_2 + 79 \text{ lb } N_2 + 9 \text{ lb } H_2O$$

Likewise:

$$\text{air } (79\% \ N_2 + 21\% \ O_2) + 1 \text{ lb fuel oil (containing 15\% hydrogen by weight)} = 84.7\% \ N_2 + 15.3\% \ CO_2 + \text{water vapor}$$

The water vapor is not considered in carbon-dioxide analysis.

Theoretically, about 1,484 cubic feet of air is required to supply the oxygen necessary to burn one gallon of residual oil. If excess air is used, there will be more flue gas, and the CO_2 will be diluted, as shown in Table 2-6. The combustion obtained by reacting the exact proportions of fuel and oxygen (air) to effect complete conversion to carbon dioxide, water vapor, and sulfur dioxide (if sulfur is present) is called perfect combustion, or stoichiometric combustion (see Table 2-7).

Table 2-7
Approximate Stoichiometric Air Fuel Ratios for Quick Estimates

Unit	Fuel	Air Required		Gross Heating Value
		ft^3 or lb	ft^3/1000 Btu*	
1 ft^3	Natural gas	10 ft^3	9.4	1000 Btu/ft^3
1 ft^3	Propane (gas)	23.5 ft^3	9.5	2500 Btu/ft^3
1 gal	No. 2 fuel oil**	1345 ft^3	9.8	—
1 lb	No. 2 fuel oil	192 lbs	9.8	19,500 Btu/lb
1 gal	No. 6 fuel oil	1484 ft^3	9.9	—
1 lb	No. 6 fuel oil	181.8 lbs	9.9	18,300 Btu/lb

*Each cubic foot of air produces about 100 Btu with almost any fuel.
**Most fuel oils from light to heavy contain almost the same percentages of carbon (84%) and hydrogen (13%–14%), plus 1%–2% of other chemical species such as sulfur.
Source: TID-27600, "Maintenance and Adjustment Manual for Natural Gas and No. 2 Fuel Oil Burners," Bureau of Natural Gas, Federal Power Commission, and U.S. Department of Energy, 1977.

Table 2-8
SO_2 Produced Per 1% Sulfur in Fuel

Fuel Oil	kg/m^3 Fuel	lb/U.S. gal Fuel	ppm in Flue Gas 15% Excess Air
No. 1	16.05	0.134	575
No. 2	17.01	0.142	581
No. 4	17.97	0.150	591
No. 5	18.69	0.156	596
No. 6	19.29	0.161	602

Note: 1000 ppm = 0.1%
Example: If a No. 6 fuel has a 0.5% S it would have:
$0.5 \times 602 = 301$ ppm SO_2
$0.5 \times 19.29 = 9.65$ kg SO_2/m^3 of fuel
$0.5 \times 0.161 = 0.08$ lb SO_2/gal of fuel
Source: Fundamentals of Boiler Efficiency, Exxon Company U.S.A., 1976.

Excess air does not help fuel combustion, since there is no more fuel to burn; it merely cools down the hot gases, and some of the air further reacts with SO_2 to form sulfur dioxide (SO_3). Table 2-8 gives an estimate of SO_2 produced for 1% sulfur on the fuel. Only a small percentage, 1%–3% of SO_2, is transformed into SO_3.

When less air is used than the stoichiometric air needed, other combustion products can form. The most significant of these products under these conditions is carbon monoxide (CO).

Nitrogen oxides (NO_X) are formed in very small amounts with the high temperatures present in the boiler. Depending on the boiler and the operating condition, NO_X vary between 50–600 ppm for fuel oils and natural gas. Coal boilers will vary between 200–1500 ppm NO_X.[14]

Before combustion can start, three elements are needed: fuel, oxygen, and heat. Below a certain temperature, fuels will not burn.

Air. O_2 for combustion usually comes from air, which contains approximately 21% O_2 and 79% nitrogen (N_2). The nitrogen is of minor importance in the production of heat, because only a very small percentage takes part in the chemical reactions of combustion. It does affect boiler efficiency, however, because some of the heat released by the combustion reaction must go to heat the nitrogen to the same temperature as the combustion products.

Air mixed with the fuel at or in the burner is known as primary air, which is about 10% of the stoichiometric (or theoretical) air required for complete combustion. Air that diffuses into the flame from the atmosphere is called secondary air.

Heat. The amount of heat obtained from burning a fuel depends on the fuel and its composition. Tables 2-9 and 2-10 list the approximate compositions and heating values of some commercial fuels. In most cases, both gross and net heating values are given. The difference between these is related to the heat of vaporization of water. For example, when water is condensed:

$$2\ H_2 + O_2 \rightarrow 2\ H_2O + 61\ 002\ \text{Btu/lb}$$

and when water escapes as vapor:

$$2\ H_2 + O_2 \rightarrow 2\ H_2O + 52\ 002\ \text{Btu/lb}$$

More usable heat is released when water is condensed than when it escapes as a vapor. This is because the vapor contains an amount of heat equal to the heat of varporization, which is the heat necessary to transform water from a liquid to a gas.

When water is removed as a vapor, the heat of vaporization is lost. If water is allowed to cool to a liquid before it leaves the stack, the heat of vaporization is released. It can then be recovered and utilized to increase boiler efficiency.

In general, fuels are specified and sold by gross heating value, but most commercial applications realize only the net heating value because the combustion gases are vented at a temperature above which condensation would occur. For quick heat estimates, use any of the values shown in Tables 2-7, 2-9, and 2-10.

Table 2-9
Comparative Analysis of Typical Fuels

Fuel	Analysis, % by Weight				Heating Value, Gross	Btu/lb, Net
	C	H	S	Ash		
Natural gas	75	25	—	—	23,850	21,492
Propane	82	18	—	—	21,564	19,854
Butane	83	17	—	—	21,240	19,620
Naphtha	85	15	0.03	—	20,700	19,152
Kerosene	86	13.7	0.07	—	19,800	18,558
Heating oil	86	13.2	0.3	—	19,584	18,378
No. 4 fuel oil	87	12.5	0.7	0.02	19,224	18,090
No. 6 fuel oil	87	12.0	1.0	0.03	19,098	17,982
No. 6 fuel oil	86	11.5	2.5	0.08	18,640	17,784
Coal, bituminous	80	5.5	1.5	5.0	14,094	13,590
Coke	85	0.5	1.0	12.0	12,600	12,510
Charcoal	—	—	—	—	14,400	*
Wood	—	—	—	—	8640	*
Bagasse	—	—	—	—	2700	*

Note: Natural gas, propane, and butane have higher hydrogen contents per unit of weight than liquid or solid fuels, which results in higher heating values. But when equal volumes of gas and oil are burned, the fuel oil gives up more heat because fuel oil contains more hydrocarbons per unit volume (higher specific gravity) than the gas. Also, notice that the heating value of petroleum products is about 19,000/lb.
*Water content varies widely.
Source: Fundamentals of Boiler Efficiency, Exxon Company U.S.A., 1976.

Table 2-10
Typical Heating Values of Various Fuels

Fuel Oil	Heat Content
No. 1	137,400 Btu/gallon
No. 2	139,600
No. 3	141,800
No. 4	145,100
No. 5	148,800
No. 6	152,400

Natural gas	950–1,150 Btu/ft^3
Purchased steam	1012 Btu/lb (based on 10-psig supply; condensate returned at 180°F)
Coal:	
Bituminous	11,500–14,000 Btu/lb
Sub-bituminous	9500–11,500 Btu/lb
Lignite, brown coal	6300–8300 Btu/lb

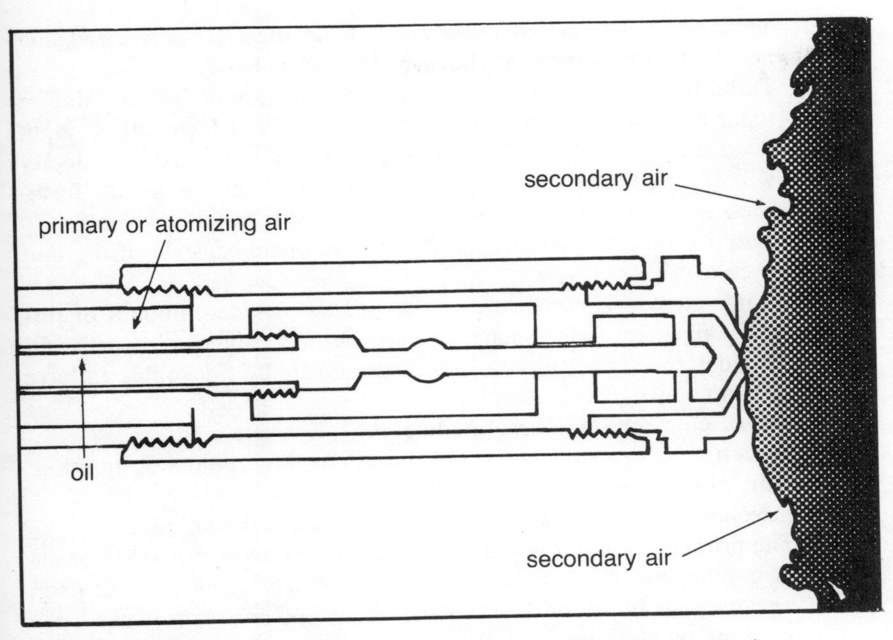

Figure 2-6. Schematic of a boiler nozzle. (Courtesy Cleaver-Brooks.)

Ignition temperature. Before the fuel can burn in the presence of sufficient air, the temperature of the air-fuel mixture must be high enough to start the combustion. This is called the ignition temperature.

Flame temperature. This is the highest temperature produced in combustion. Flame temperatures of most fuels are in the range of 3360°F (1850°C) to 3800°F (2100°C).

Combustion control. Combustion is controlled by regulating the flows of air and fuel. Several variables influence combustion:

1. Burner or nozzle design. This is the heart of a boiler. Oil must be atomized and vaporized in order to burn. Fuel oil will not burn as a liquid; it must first be converted into a gas.

   The function of the burner, shown schematically in Figure 2-6, is to mix sufficient air (primary air) with the fuel to atomize and vaporize it and provide a continuous ignition of the mixture. Gas is very easy to burn, since it is already vaporized. Coal must be pulverized and burned in an air suspension. Burners are defined by:

a. Turndown ratio, which is the ratio of the maximum fuel-air mixture input rates at which the burner can operate satisfactorily. If input flows are too large, the flame will go out, since the mixture velocity will exceed flame velocity. If the input flows are too small, flashback will occur when flame velocity exceeds mixture velocity.

b. Stability. In other words, the flame is maintained, even if the unit is cold.

c. Flame shape. The pressure of the input flows and amount of primary air affect flame shape. In most burners, higher pressures will broaden the flame; more air will shorten it. It is desirable to have a short, bushy flame.

2. Types of atomization. These include:

a. Steam. Use dry steam to avoid pulsations that could stop ignition.

b. Air: primary air.

c. Mechanical: less efficient than steam or air atomization.

The primary air for low-pressure atomizers is about 5%–15% of the total combustion air. For high-pressure atomizers, say 30–150 psig, the primary air is 3% of the combustion air. When steam is used, the rule of thumb is to use 1–2 lbs of steam per gallon of light oil, and 2–3 lbs of steam per gallon of heavy oil.[15]

For oils, the design viscosity must be used for proper atomization.

3. Firebox. Poor combustion is caused by insufficient air or poor atomization, producing smoke or soot. Turbulent air is required to ensure proper mixing.

References

1. "Specifications for Fuels Oils," ASTM, D-396-75, ASTM Standards for Petroleum Products (Part 17), American Society for Testing and Materials, 1975.
2. "Guidelines for Burner Adjustments of Commercial Oil-Fired Boilers," Report EPA-600/2-76-088, March 1976.
3. Quillman, B., "Present and Future of Coal Utilization at DuPont," *Chemical Engineering Progress,* p. 43, March 1980.
4. Bender, R. J., "Steam Generation," *Power,* McGraw-Hill, Inc.
5. Holzhauer, Ron, "Packaged Boilers," *Plant Engineering,* p. 72, December 11, 1980.
6. Cleaver-Brooks, Division of Aqua-Chem, Inc.
7. *ASME Boiler and Pressure Vessel Codes,* Sections I through VI, IX, and XI, American Society of Mechanical Engineers, New York.

8. Elonka, S. M.: *Standard Plant Operator's Manual,* McGraw-Hill Book Company, New York, 1975.
9. Elonka, S. M., and A. L. Kohan: *Standard Boiler Operators' Questions and Answers,* McGraw-Hill Book Company, New York, 1969.
10. *National Board Inspection Code,* National Board of Boiler and Pressure Vessel Inspectors, Columbus, Ohio.
11. *National Fire Protection Codes,* National Fire Protection Association, Boston. *Power Piping Code,* ANSI B31.1, American National Standards Institute, New York, 1977.
12. *State, County, and City Synopsis of Boiler and Pressure Vessel Laws on Design, Installation, and Reinspection Requirements,* Uniform Boiler and Pressure Vessel Laws Society, Hartford, Conn.
13. Spring, H. M., and A. L. Kohan, "Boiler Operator's Guide," McGraw-Hill Book Co., 1981.
14. TID-27600, Maintenance and Adjustment Manual for Natural Gas and No. 2 Fuel Oil Burners, Bureau of Natural Gas, Federal Power Commission and U.S. Department of Energy, 1977.
15. Seabold, J. G., "How to Conserve Fuel in Furnaces and Flares," *Hydrocarbon Processing,* November 1981.

3

How to Increase Boiler Efficiency

This chapter deals exclusively with boilers, both firetube and watertube. The basic methods used to increase boiler efficiency can be applied to any kind of fossil fuel and other energy-related equipment such as furnaces and dryers. The latter two are covered in Chapters 5 and 6, respectively.

Proven Methods

After testing and calibrating several hundred boilers and many glass and steel furnaces throughout the world, certain methods were found to improve their efficiencies. These methods supported those results found independently by government-funded studies in the U.S.[1,2,3,4] These proven methods can be divided into two separate groups:

1. Efficiency improvements without expenses or new capital investment.
2. Efficiency improvements with expenses and/or additional equipment.

Some variables inherent in these methods are excess air, carbon dioxide content in the flue gases, and fuel treatment. In each method presented, keep in mind that there is a synergistic effect between variables. For instance, if the fuel rate is halfed, the steam rate and the stack temperature will also go down. Also, as the amount of air supplied to the boiler changes, the following parameters will also change: the gas opacity (smoke), temperature, unburned hydrocarbons, O_2, CO_2, CO, NO_X, SO_2, and SO_3.

Determination of Boiler Efficiencies

Boiler efficiencies can be determined using two analytical techniques.

Direct technique. The efficiency of the boiler is tested measuring the input and output flowrates of fuel and water, including temperatures, pressures, and fuel composition. Calibrated instruments are used. From the heating values of the fuel and steam, and from the data thus obtained, the boiler efficiency is calculated. Dyer and Maples[5] give an example of how to perform these calculations. Boiler efficiencies are calculated according to the American Boiler Manufacturers Association *Engineering Manual*.[6] Typical fuel analyses are shown in Tables 2-7, 2-9, and 2-10.

Indirect technique. This technique measures the combustion efficiency indirectly. Heat losses such as radiation and convection are not included. This test is performed by obtaining an exhaust-gas analysis (O_2, CO_2, and CO or smoke) and the exhaust-gas temperature. This analysis can be obtained using an Orsat or the Bacharach Combustion Tester[7] (see Figures 11-1 and 7-1, respectively). The combustion efficiency is calculated from the $\%CO_2$ and exhaust-gas temperature at the stack. From Tables 3-1 and 3-2, the combustion efficiency is calculated. Specially designed slide rules are available to perform the calculations quickly.[7] *Caution:* For gas boilers, the CO content must also be determined, as discussed later in this chapter.

The boiler must be tested at several load conditions over the range used. The errors involved in the analysis are discussed in Chapter 10.

The step-by-step analytical techniques to follow are discussed in Chapter 7. Appendix A includes a data sheet called "Boiler Efficiency Test Form," which is used in the evaluation of a boiler.

Procedure

Each of the proven methods to increase boiler efficiencies discussed in this chapter is described as follows:

1. *Method number and objective*. In this heading, the objective or basic technique used in this method is defined, such as "Reduce Excess Air" or "Reduce Boiler Pressure." The implication is that increased boiler efficiency will be obtained if either excess air or boiler pressure is reduced.
2. *Potential savings*. The potential savings are given as guidelines of what is possible to achieve using the method described. Savings are given as percentage points. Thus, "5%" means that 5% of the fuel consumed can be saved using this method.

p. 25 (text continued on page 28)

Table 3-1
Combustion Efficiency for Fuel Oils

STACK LOSS - % - NO. 2 OIL

%	DIFFERENCE BETWEEN FLUE GAS AND ROOM TEMPERATURES IN DEGREES FAHRENHEIT																														
CO_2	200	220	240	260	280	300	320	340	360	380	400	420	440	460	480	500	520	540	560	580	600	620	640	660	680	700	750	800	850	900	950	1000
3.0	24.1	25.8	27.7	29.3	31.3	33.9	34.8	36.4	38.2	40.0	42.9	44.8	45.5	47.0	49.0	50.8	52.4	54.3	56.0	57.9	59.6	61.5	63.5	65.0	66.8	68.8						
3.5	21.7	23.1	24.8	26.2	27.8	29.2	31.7	32.5	33.9	35.3	36.9	38.5	40.0	41.7	43.1	44.8	46.1	47.8	49.4	50.9	52.2	53.9	55.7	57.0	58.3	60.0	63.8	67.8				
4.0	19.9	21.2	22.5	24.9	25.2	26.5	27.9	29.2	31.7	32.0	33.3	35.8	36.0	37.3	38.7	40.0	41.4	42.9	44.1	45.5	46.9	48.1	49.8	50.9	52.1	53.8	57.0	60.2	63.9	67.1		
4.5	18.4	19.7	20.8	22.0	23.2	24.4	25.6	26.9	28.0	29.3	30.4	31.8	32.9	34.2	35.6	36.7	37.8	39.0	40.1	41.2	42.5	43.8	45.0	46.3	47.4	48.8	51.8	54.6	57.8	60.9	63.9	66.9
5.0	17.2	18.5	19.5	20.7	21.7	22.7	23.8	24.9	26.0	27.1	28.2	29.4	30.3	31.5	32.7	33.8	34.9	35.9	36.8	38.0	39.2	40.1	41.7	42.4	43.7	44.7	47.4	50.1	52.9	55.8	58.3	61.2
5.5	16.3	17.4	18.4	19.4	20.4	21.3	22.3	23.3	24.3	25.4	26.3	27.3	28.4	29.4	30.6	31.4	32.4	33.6	34.5	35.3	36.4	37.4	38.4	39.6	40.3	41.7	44.0	46.5	49.0	51.8	54.1	56.5
6.0	15.6	16.5	17.4	18.3	19.3	20.4	21.2	22.0	23.0	23.9	24.9	25.8	26.8	27.7	28.6	29.5	30.4	31.4	32.3	33.1	34.2	35.0	36.0	36.9	37.9	38.9	41.0	43.5	45.8	48.0	50.3	52.8
6.5	14.9	15.7	16.7	17.5	18.4	19.3	20.1	20.9	21.8	22.7	23.6	24.5	25.3	26.1	27.0	27.8	28.8	29.6	30.6	31.3	32.3	33.0	34.1	34.8	35.7	36.5	38.7	40.8	42.9	45.1	47.5	49.7
7.0	14.4	15.3	16.0	16.8	17.8	18.4	19.3	20.1	20.9	21.7	22.4	23.2	24.1	24.9	25.7	26.5	27.3	28.1	28.9	29.8	30.5	31.4	32.3	33.0	33.8	34.6	36.5	38.6	40.5	42.7	44.7	46.6
7.5	13.9	14.6	15.4	16.2	16.9	17.7	18.5	19.2	20.1	20.7	21.3	22.2	23.0	23.8	24.5	25.2	26.0	26.8	27.5	28.2	29.0	29.8	30.6	31.3	32.2	32.9	34.8	36.5	38.5	40.3	42.3	44.2
8.0	13.5	14.3	14.9	15.7	16.3	17.1	17.7	18.5	19.3	20.0	20.7	21.4	22.1	22.8	23.5	24.2	25.0	25.7	26.3	27.0	27.8	28.5	29.2	30.0	30.8	31.5	33.2	35.0	36.8	38.5	40.2	42.1
8.5	13.2	13.8	14.5	15.2	15.8	16.5	17.3	17.8	18.6	19.3	20.0	20.6	21.3	21.9	22.6	23.3	23.9	24.6	25.3	25.9	26.7	27.3	28.0	28.8	29.4	30.1	31.8	33.5	35.2	36.9	38.7	40.2
9.0	12.8	13.4	14.1	14.7	15.4	16.0	16.7	17.3	17.9	18.6	19.3	20.0	20.6	21.2	21.8	22.4	23.1	23.8	24.4	25.0	25.7	26.3	27.0	27.7	28.3	28.9	30.5	32.1	33.8	35.3	37.0	38.5
9.5	12.5	13.2	13.7	14.3	14.9	15.7	16.3	16.8	17.4	18.1	18.6	19.3	19.9	20.5	21.1	21.7	22.4	22.9	23.5	24.1	24.8	25.4	26.0	26.7	27.2	27.9	29.4	31.0	32.5	34.0	35.5	37.2
10	12.3	12.8	13.4	14.0	14.6	15.2	15.7	16.3	16.9	17.5	18.1	18.7	19.3	20.0	20.5	21.0	21.6	22.2	22.8	23.4	24.0	24.6	25.1	25.8	26.3	27.0	28.3	29.9	31.4	32.9	34.4	35.7
11	11.8	12.4	12.8	13.4	13.9	14.5	15.0	15.5	16.2	16.7	17.2	17.8	18.3	18.7	19.4	20.0	20.5	20.9	21.5	22.0	22.6	23.1	23.7	24.2	24.8	25.3	26.7	28.0	29.4	31.8	32.1	33.5
12	11.4	11.8	12.5	12.9	13.4	13.9	14.4	14.9	15.4	15.9	16.4	16.9	17.4	17.9	18.4	18.9	19.5	20.0	20.5	20.9	21.4	22.9	22.4	22.9	23.5	24.0	25.2	26.5	27.8	29.0	30.2	31.7
13	11.2	11.6	12.1	12.5	12.9	13.4	13.9	14.3	14.7	15.3	15.8	16.3	16.7	17.2	17.7	18.1	18.6	19.1	19.6	20.1	20.5	21.1	21.3	21.8	22.3	22.8	24.0	25.2	26.3	27.5	28.8	30.0
14		11.3	11.8	12.2	12.6	13.0	13.4	13.8	14.3	14.8	15.3	15.6	16.2	16.5	16.9	17.4	17.8	18.3	18.7	19.2	19.7	20.2	20.6	21.0	21.4	21.8	22.9	24.1	25.2	26.2	27.4	28.6
15			11.4	11.7	12.4	12.6	13.1	13.5	13.8	14.3	14.8	15.3	15.6	15.9	16.4	16.7	17.3	17.7	18.1	18.4	18.9	19.4	19.8	20.3	20.6	21.0	22.0	23.1	24.2	25.2	26.2	27.3

STACK LOSS - % - NO. 6 OIL

%	DIFFERENCE BETWEEN FLUE GAS AND ROOM TEMPERATURES IN DEGREES FAHRENHEIT																														
CO_2	200	220	240	260	280	300	320	340	360	380	400	420	440	460	480	500	520	540	560	580	600	620	640	660	680	700	750	800	850	900	950	1000
3.0	24.5	26.5	28.5	30.2	32.2	34.5	36.5	38.2	40.4	42.2	44.4	46.4	48.2	50.0	52.3	54.3	56.3	58.2	60.3	62.0	64.1	66.2	68.1	70.1								
3.5	21.8	23.4	25.2	26.8	28.6	30.4	32.1	33.8	35.5	37.4	39.0	40.6	42.2	44.0	45.6	47.5	49.2	51.0	52.8	54.0	56.0	57.8	59.9	61.1	63.0	64.9	69.0					
4.0	19.8	21.2	22.8	24.2	25.7	27.3	28.8	30.2	31.6	32.5	34.8	36.3	37.8	39.4	40.8	42.2	43.8	45.1	46.9	48.2	49.8	51.2	52.9	54.2	56.0	57.8	61.1	65.0	68.9			
4.5	18.2	19.4	20.8	22.2	23.5	24.8	26.2	27.4	28.8	30.4	31.5	33.0	34.2	35.4	37.0	38.1	39.4	41.0	42.2	43.5	45.0	46.3	47.9	49.0	50.1	51.9	55.0	58.2	61.8	65.1	68.5	
5.0	16.8	18.0	19.3	20.4	21.7	22.8	23.2	25.3	26.6	27.8	29.0	30.3	31.4	32.6	33.8	35.3	36.2	37.5	38.8	39.8	41.0	42.3	43.8	44.9	46.1	47.5	50.1	53.6	56.3	59.8	62.3	65.8
5.5	15.8	16.8	18.0	19.2	20.3	21.3	22.5	23.5	24.6	25.8	26.9	28.0	29.2	30.2	31.4	32.5	33.5	34.7	35.8	37.0	37.9	39.2	40.1	41.3	42.3	43.8	46.1	49.1	52.0	54.7	57.8	60.1
6.0	14.8	15.8	16.9	18.0	19.0	20.0	21.1	22.0	23.1	24.2	25.2	26.3	27.3	28.2	29.3	30.3	31.3	32.3	33.5	34.3	35.3	36.5	37.5	38.3	39.7	40.5	43.0	45.8	48.2	50.9	53.5	56.0
6.5	14.3	15.2	16.1	17.1	18.0	18.9	19.9	20.8	21.8	22.8	23.7	24.6	25.5	26.5	27.5	28.5	29.4	30.4	31.4	32.3	33.4	34.3	35.1	36.1	37.1	38.0	40.2	42.8	45.1	47.6	49.9	52.1
7.0	13.5	14.4	15.3	16.2	17.1	17.9	18.8	19.7	20.6	21.5	22.4	23.3	24.2	25.0	25.8	26.8	27.7	28.6	29.0	30.2	31.2	32.2	33.0	33.9	34.9	35.8	37.9	40.1	42.1	44.4	46.8	49.0
7.5	13.0	13.8	14.6	15.5	16.3	17.3	18.0	18.8	19.7	20.5	21.4	22.2	22.9	23.7	24.6	25.4	26.3	27.2	27.9	28.8	29.6	30.5	31.2	32.1	33.0	34.9	35.9	37.9	40.0	42.0	44.1	46.1
8.0	12.5	13.3	14.1	14.8	15.7	16.4	17.3	18.0	18.8	19.6	20.4	21.2	21.9	22.7	23.5	24.2	25.0	25.8	26.6	27.4	28.2	29.0	29.9	30.6	31.5	32.1	34.1	36.0	38.0	40.0	41.9	43.9
8.5	12.2	12.8	13.6	14.4	15.1	15.7	16.6	17.3	18.0	18.7	19.6	20.3	21.0	21.6	22.5	23.3	23.9	24.7	25.5	26.2	26.8	27.6	28.2	29.1	29.9	30.8	32.6	34.2	36.2	38.0	39.9	41.8

(table continued on next page)

% CO₂	200	220	240	260	280	300	320	340	360	380	400	420	440	460	480	500	520	540	560	580	600	620	640	660	680	700	750	800	850	900	950	1000
	DIFFERENCE BETWEEN FLUE GAS AND ROOM TEMPERATURES IN DEGREES FAHRENHEIT																															
9.0	11.7	12.4	13.2	13.8	14.6	15.3	15.9	16.6	17.4	18.1	18.8	19.5	20.2	20.8	21.6	22.3	22.9	23.7	24.4	25.0	25.7	26.5	27.1	27.9	28.7	29.4	31.1	32.9	34.6	36.3	38.0	39.9
9.5	11.4	12.1	12.7	13.4	14.1	14.7	15.4	16.0	16.7	17.5	18.1	18.7	19.4	20.0	20.7	21.4	22.1	22.8	23.5	24.0	24.7	25.4	26.1	26.8	27.5	28.1	29.8	31.2	33.2	34.9	36.4	38.1
10	11.2	11.7	12.3	13.0	13.7	14.4	14.8	15.5	16.2	16.8	17.5	18.2	18.7	19.4	20.0	20.6	21.3	21.9	22.6	23.2	23.8	24.5	25.1	25.8	26.4	27.0	28.7	30.1	31.8	33.5	35.0	36.7
11	10.6	11.3	11.8	12.4	12.9	13.5	14.2	14.7	15.3	15.8	16.5	17.0	17.6	18.2	18.8	19.4	20.0	20.6	21.2	21.7	22.3	22.9	23.5	24.1	24.8	25.2	26.8	28.1	29.8	31.2	32.5	34.1
12	10.2	10.7	11.3	11.7	12.3	12.8	13.4	13.8	14.5	15.1	15.6	16.2	16.7	17.2	17.8	18.3	18.8	19.4	19.9	20.4	21.0	21.6	22.1	22.7	23.1	23.8	25.0	26.4	27.9	29.1	30.5	31.9
13		10.3	10.8	11.3	11.8	12.3	12.8	13.3	13.8	14.4	14.8	15.4	15.8	16.3	16.8	17.3	17.9	18.4	18.9	19.3	19.8	20.4	20.9	21.4	21.9	22.4	23.8	24.9	26.2	27.5	28.9	30.0
14		9.8	10.4	10.8	11.4	11.8	12.3	12.8	13.3	13.7	14.3	14.7	15.2	15.6	16.2	16.6	17.1	17.5	18.0	18.5	18.8	19.4	19.9	20.4	20.9	21.2	22.5	23.7	24.9	26.1	27.2	28.5
15			10.2	10.6	11.0	11.4	11.8	12.4	12.7	13.3	13.7	14.2	14.6	15.0	15.4	15.8	16.4	16.8	17.3	17.7	18.2	18.6	19.0	19.5	19.9	20.3	21.5	22.6	23.8	24.9	25.9	27.1
16				10.3	10.7	11.1	11.5	11.8	12.3	12.8	13.3	13.7	14.0	14.4	14.8	15.3	15.7	16.2	16.6	16.9	17.4	17.9	18.2	18.8	19.1	19.5	20.6	21.6	22.7	23.8	24.8	25.9

Note: No. 2 fuel oil—140,000 Btu/gal; No. 6—150,000 Btu/gal.
Source: Cleaver-Brooks.

Table 3-2
Combustion Efficiency for Natural Gas (1000 Btu/ft³)

STACK LOSS - % - NATURAL GAS

% CO₂	200	220	240	260	280	300	320	340	360	380	400	420	440	460	480	500	520	540	560	580	600	620	640	660	680	700	750	800	850	900	950	1000
	DIFFERENCE BETWEEN FLUE GAS AND ROOM TEMPERATURES IN DEGREES FAHRENHEIT																															
3.0	23.1	24.4	25.9	27.2	28.6	30.0	31.3	32.8	34.1	35.8	36.9	38.2	39.8	41.0	42.2	43.8	45.0	46.3	47.8	49.0	50.0											
3.5	21.2	22.5	23.8	24.9	26.1	27.2	28.4	29.6	30.9	32.0	33.2	34.4	35.8	36.8	37.9	39.2	40.3	41.6	42.8	43.8	45.0	46.2	47.7	48.3	49.8							
4.0	19.9	20.9	22.0	23.1	24.1	25.1	26.2	27.2	28.3	29.4	30.4	31.8	32.5	33.8	34.8	35.8	36.8	37.8	38.8	39.9	40.9	42.1	43.0	44.1	45.2	46.2	48.8					
4.5	18.9	19.9	20.9	21.8	22.7	23.6	24.5	25.5	26.4	27.3	28.3	29.2	30.2	31.2	32.2	33.0	34.0	34.9	35.9	36.8	37.8	38.6	39.8	40.4	41.5	42.6	44.8	47.2	49.8			
5.0	18.0	18.9	19.8	20.6	21.4	22.2	23.1	24.0	24.9	25.8	26.8	27.5	28.3	29.1	30.1	30.9	31.8	32.5	33.6	34.3	35.7	36.2	36.9	37.8	38.8	39.7	41.8	43.8	46.0	48.2		
5.5	17.4	18.1	18.9	19.8	20.5	21.2	22.1	22.9	23.8	24.5	25.2	26.2	26.9	27.8	28.5	29.2	30.0	30.8	31.8	32.3	33.2	34.1	34.9	35.8	36.3	37.3	39.2	41.0	43.0	45.3	47.2	49.0
6.0	16.8	17.4	18.2	18.9	19.6	20.4	21.1	21.8	22.7	23.3	24.1	24.9	25.5	26.2	27.0	27.8	28.4	29.2	30.0	30.8	31.5	32.2	32.9	33.8	34.3	35.2	36.8	38.8	40.4	42.5	44.3	46.2
6.5	16.3	16.9	17.6	18.4	19.0	19.8	20.4	21.1	21.8	22.4	23.2	23.8	24.5	25.2	25.9	26.5	27.2	27.9	28.7	29.2	30.0	30.9	31.4	32.1	32.8	33.5	34.6	36.8	38.4	40.3	42.0	43.8
7.0	15.8	16.5	17.1	17.8	18.4	19.1	19.8	20.4	21.0	21.8	22.3	22.9	23.6	24.2	24.9	25.5	26.2	26.8	27.4	28.0	28.8	29.4	30.0	30.8	31.2	32.0	33.8	35.3	36.8	38.3	40.0	41.8
7.5	15.5	16.1	16.7	17.2	17.9	18.5	19.1	19.8	20.3	20.9	21.5	22.2	22.8	23.3	24.0	24.6	25.2	25.8	26.4	26.9	27.7	28.2	28.8	29.4	30.1	30.8	32.2	33.8	35.2	36.8	38.3	39.9
8.0	15.2	15.7	16.3	16.9	17.4	18.0	18.6	19.2	19.8	20.3	20.9	21.5	22.1	22.8	23.2	23.8	24.4	25.0	25.5	26.0	26.7	27.2	27.8	28.4	29.0	29.5	31.0	32.4	33.8	35.4	36.8	38.2
8.5	14.9	15.4	15.9	16.5	17.1	17.6	18.2	18.7	19.3	19.8	20.4	20.9	21.4	22.0	22.5	23.1	23.7	24.2	24.8	25.3	25.8	26.4	26.9	27.4	28.1	28.6	29.9	31.3	32.8	34.2	35.4	36.8
9.0	14.6	15.2	15.7	16.2	16.6	17.2	17.8	18.3	18.8	19.3	19.9	20.4	20.9	21.4	21.9	22.5	23.0	23.5	24.1	24.5	25.2	25.8	26.2	26.7	27.2	27.8	29.0	30.3	31.8	33.0	34.3	35.7
9.5	14.4	14.9	15.4	15.9	16.4	16.9	17.4	17.9	18.4	18.9	19.5	19.9	20.5	20.9	21.4	21.9	22.4	22.9	23.4	23.8	24.4	24.9	25.4	25.9	26.4	26.9	28.2	29.4	30.8	32.0	33.3	34.5
10	14.2	14.6	15.2	15.6	16.1	16.6	17.1	17.5	18.1	18.5	19.0	19.5	20.0	20.4	20.8	21.4	21.8	22.4	22.8	23.3	23.8	24.2	24.8	25.2	25.8	26.2	27.4	28.6	29.8	31.2	32.2	33.4
11		14.4	14.7	15.2	15.6	16.1	16.5	16.9	17.4	17.8	18.4	18.8	19.3	19.6	20.2	20.5	20.9	21.4	21.9	22.3	22.8	23.2	23.7	24.2	24.6	25.0	26.2	27.2	28.3	29.5	30.8	31.8
12			14.4	14.8	15.2	15.6	16.1	16.5	16.9	17.3	17.8	18.2	18.6	19.0	19.4	19.8	20.2	20.6	21.1	21.4	21.9	22.3	22.8	23.2	23.6	24.0	25.1	26.1	27.2	28.3	29.2	30.3

Source: Cleaver-Brooks.

3. *Problem*. The problem is defined in general terms so that the reader can relate it to his own operation.
4. *Impact*. Two examples are given: the potential yearly savings on a 100-hp boiler and the savings on a 100,000-lbs/hr steam boiler. If an expense or capital investment is required to use this method, it is mentioned here.
5. *Solutions*. Proposed techniques to implement this method are described in brief.
6. *Discussion*. A full discussion on the techniques suggested is included.

Increase Efficiency with *No* Expenses or Capital Investment

Method 1: Reduce Excess Air

Potential savings: 5%–10%.

Problem. Unneeded excess air is used, probably to dilute the flue gas so that smoke is not seen.

Impact: 100-hp boiler—$30,700 yearly savings; 100,000-lbs/hr steam boiler—$888,000 yearly savings. Table 3-3 gives an estimate of savings when the efficiency of a boiler is increased 10%.

Solutions. Reduce excess air to the minimum 10%–15% required.[8] For specific situations, adjust air as recommended in Table 3-4. See also Chapter 11, "Instrumentation and Boiler Controls."

Discussion. Excess air has a very strong influence on the boiler efficiency. Thus:

% CO_2	*% Excess Air*	*% Efficiency*	*Difference in Efficiency*	*Actual Decrease in Efficiency**
14.0	10	84.8	—	—
7.5	100	78.5	6.3%	7.5%
6.0	140	75.0	9.8%	11.6%

Stack temperature was kept constant at 450°F; air at 50°F. Since it is not uncommon to find boilers running at 100%–200% excess air (in order to

---

*See Table 7-3; in other words, 100% excess air lowered the original 84.8% boiler efficiency by 7.5%, or down to 78.5% efficiency.

Table 3-3
Annual Savings by Improving Boiler Efficiency by 10%

Steam Production		Fuel Price, $/gal			
hp	**lb/hr**	**0.75***	**1.00**	**1.25**	**1.50**
100	3,450	$ 23,000	$ 30,700	$ 38,380	$ 47,970
200	6,900	46,000	61,330	76,700	95,830
400	13,800	92,000	122,664	153,330	191,660
800	27,600	184,000	245,270	306,590	383,200
—	60,000	400,000	533,300	666,300	832,820
—	100,000	666,700	888,000	1,110,000	1,388,000
—	200,000	1,333,332	1,777,330	2,221,300	2,776,600
—	400,000	2,666,700	3,555,000	4,444,000	5,556,000
—	800,000	5,333,300	7,110,000	8,888,000	11,109,400

*By pricing natural gas at $5.00 per million Btu or per thousand cubic feet, similar savings are obtained for these boilers.

Table 3-4
Excess Air Recommended

General Guidelines[5]

Fuel	*% Excess air*
Natural gas	10
Fuel oil No. 2	12
Fuel oil No. 6	15

Excess Oxygen Recommended vs. Controls[8]

Control	*% Excess Oxygen* *Gas Boiler*	*Oil Boiler*
Portable analyzer, weekly check	5.0	6.0
Continuous oxygen recorder	4.5	5.0
Automatic damper control	3.0	4.0
CO Improved Controller	1.5	2–2.5

Typical Minimum and Excess Oxygen Levels[2]

Fuel Type	*Minimum*	+	*"Cushion" Excess Recommended*	=	*Total O_2*
Natural gas	0.5–3.0%		0.5–2.0%		1.0– 5.0%
Fuel oils	2.0–4.0		0.5–2.0		2.5– 6.0
Pulverized coal	3.0–6.0		0.5–2.0		3.5– 8.0
Coal stoker	4.0–8.0		0.5–2.0		4.5–10.0

When in doubt use 10%–15% excess air (unless the operating manual of your boiler states otherwise) keep smoke down in the stack), 5%–10% fuel savings are possible by controlling the excess air alone.[2] Rule of thumb:

100% excess air reduces boiler efficiency by 5%.

Boilers are usually run at high excessive air because manufacturers recommend operating boilers at 10%–20% excess air (thus operators see "no harm" in adding excess air); and because two of the boiler man's main responsibilities are to keep the boiler running and to decrease the appearance of smoke in the stack by diluting it with excess air. However, there is a scientific and very practical reason to operate boilers at 10%–15% excess air: if they were operated at 0% excess air, any change in atmospheric conditions (humidity, temperature, or pressure) could easily change the air flow from excess air into deficient air and drastically reduce boiler efficiency with the appearance of smoke, i.e. unburned fuel. With 0% excess air and stack temperature at 300°F, the boiler efficiencies under different conditions are estimated as follows:

Atmospheric Pressure	*% Boiler Efficiency*
15.0 psia	85.7%
14.7	85.8
14.5	84.0
14.0	80.6

If the temperature changed (keeping 14.7 psia pressure and 0% excess air):[5]

Ambient Temperature	*% Boiler Efficiency*
75°F	85.7%
100°F	82.2

Changes in air humidity would have similar effects: The more humidity, the lower the efficiency at 0% excess air. In addition, other variations, such as fuel rate, could easily make an effect on the air consumed and cause deficient air at some point.

It is clear that the excess air offers a cushion to allow variations in air conditions without affecting boiler efficiency significantly. Therefore, the recommendations shown in Table 3-4 for excess air must be followed to control boiler efficiency.

CO. The CO in the flue gas (measured in ppm of CO) stays fairly constant at high excess air. However, as excess air is reduced below the optimum level, the CO content rises sharply. Smoke or unburned hydrocarbons appear in the stack. This is shown in Figure 3-1 for gas, oil, and coal used as fuels.

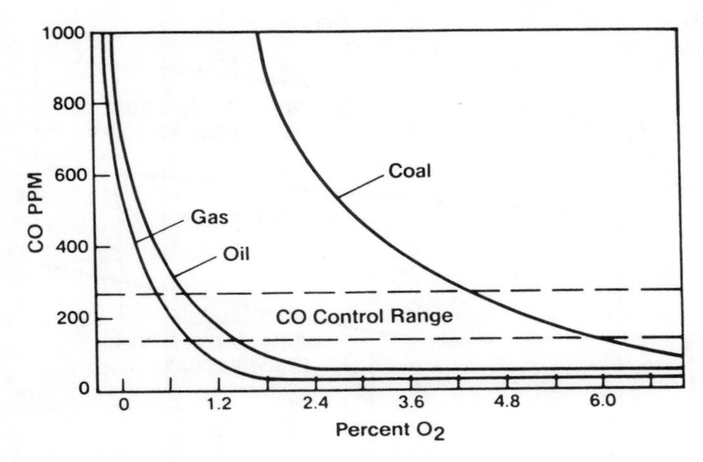

Figure 3-1. Excess-air controls: CO vs. O_2.[2]

Maximum combustion efficiency is obtained at the correct CO content.[2] From Figure 3-1, the following rules of thumb are derived:

> Control your CO at 100–400 ppm. In calibrating your boiler, once you determine the minimum amount of air required, add a little excess air to serve as a cushion (see Table 3-4).

Infrared sensors can control CO within a 150–250 ppm range, which is near the optimum combustion efficiency using a minimum amount of excess air.

Always determine CO content in gas-fired boilers. In oil-fired boilers, this is not necessary because smoke appears as soon as the excess air is around zero.

Calibration of a boiler always involves obtaining a curve of excess air vs. smoke for oil-fired boilers and excess air vs. CO for gas-fired boilers. Then, the minimum and "cushion" excess air are determined from these curves.

Caution: When air falls below the stoichiometric value, unburned hydrocarbons can lead to explosion,[9] boiler efficiencies can be reduced drastically, and soot will accumulate on the fireside.

See also Chapter 11, "Instrumentation and Boiler Controls."

Method 2: Decrease Flue-Gas Temperature

Potential savings: 1% for every 40°F decrease. The potential savings are 3%, since many stacks run 120°F too high.[2]

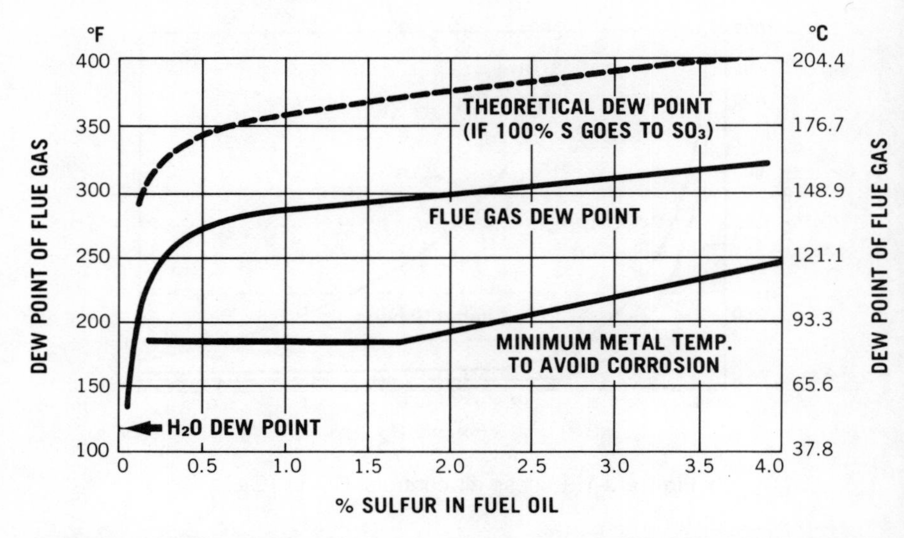

Figure 3-2. Dewpoint vs. sulfur in fuel oil. (Source: *Fundamentals of Boiler Efficiency*, Exxon Company U.S.A., 1976.)

Problem: Either excess air or fouled water and/or fireside tubes.

Impact: 100-hp boiler—$3,100 yearly savings; 100,000-lbs/hr steam boiler—$89,000 yearly savings.

Solutions. Tackle the possible problems one by one (at this point the cause is unknown):

1. If there is excess fuel, the fuel rate can be decreased to decrease the temperature. *Caution:* The required steam production must be maintained. If steam production suffers, this is not a good solution to the problem.
2. If the tubes are fouled up, steam production will suffer, and the only solution is to shut down and clean them up or blow soot on the fireside if a sootblower is available.
3. An economizer may be economical (see Method 15).

Discussion.

Stack temperature. A good stack temperature to minimize sulfuric acid corrosion is 300°F–350°F; the higher temperature is used for fuels with sulfur content higher than 2.0%. Figure 3-2 shows the dewpoint of sulfuric acid at different temperatures. A temperature below 250°F would quickly invite acid corrosion. Many boilers have stack temperatures of 600°F or higher. Determine whether excess fuel is being used. This is often the simplest solution.

Fouled heat-transfer surfaces. Scale and/or rust deposits on either side of the tubes may be responsible. You may have to blow the soot and/or descale the boiler, or treat the fuel properly (see Chapter 8).

Economizers. If you determine that an insufficient heat-transfer surface exists, i.e. by maintaining the required steam production you cannot lower the stack temperature below 400°F, then an economizer should be considered. An economizer both lowers the stack temperature and raises the temperature of the feed water to the boiler (see Method 15). Economizers are expensive, and an economic analysis is recommended before installing one (see Chapter 13). Costs run around $2,000 per MBtu/hr duty.

Don't forget the rule of thumb:

> For every 40°F decrease in stack temperature, you gain 1% efficiency.

Method 3: Reduce Boiler Pressure

Potential savings: 1% for every 70-psig reduction.[1]

Problem. The boiler is being operated at a pressure higher than necessary.

Impact: 100-hp boiler—$3,100 yearly savings; 100,000-lbs/hr boiler—$89,000 yearly savings.

Solution. Slowly reduce boiler pressure to a point where the amount of steam produced is sufficient to fulfill plant requirements.

Discussion. Some of us are not fully aware that reducing the operating pressure could have an impact on fuel savings. Even if you lower the pressure a mere 35 psig, you still gain 0.5% in boiler efficiency. Why not do it? However, consider the following savings and problems:

Steam (fuel) savings. Lowering the operating pressure results in the following savings:

1. Lower flue gas temperature due to improved heat transfer.
2. Lower heat losses from boiler and piping.
3. Smaller steam leaks due to lower steam pressure.

Problems and cost of reducing steam pressure.

1. Boiler circulation may be upset.
2. Relief valves may have to be changed.
3. Mud blowdown may have to be timed so that steam demand is at its lowest point.

Alternative boiler operations. If more than one boiler is available, consider running one at low pressure for low-pressure requirements; run the other boiler at high pressure. Rule of thumb:

> Reducing boiler pressure by 70 psig yields a 1% efficiency gain

Method 4: Increase Fuel No. 6 Temperature

Potential savings: 5%.

Problem. Atomize the fuel at the right viscosity. Too high or too low a viscosity will yield poor atomization and poor efficiency.

Impact: 100-hp boiler—$15,400 yearly savings; 100,000-lbs/hr boiler—$444,000 yearly savings.

Solution. Preheat the fuel at 212°F–230°F or more so that the fuel viscosity will be 100–300 SUS.

Discussion.

Preheater. Boilers designed to burn fuel No. 6 have an oil preheater to control the fuel viscosity between 100–300 SUS. The temperature required is usually 212°F–230°F, as shown in Figure 3-3.

Fuel atomization. The boiler nozzles are designed to facilitate the blending of air and fuel so that combustion efficiency will be optimized. If the fuel is too viscous or too thin, the air-fuel blend will suffer, resulting in lower combustion efficiency.

Instruments. Preheaters come equipped with a dial thermometer. The viscosity can be measured with a portable viscometer at different temperatures. See Chapter 11 for a list of relatively inexpensive analytical instruments.

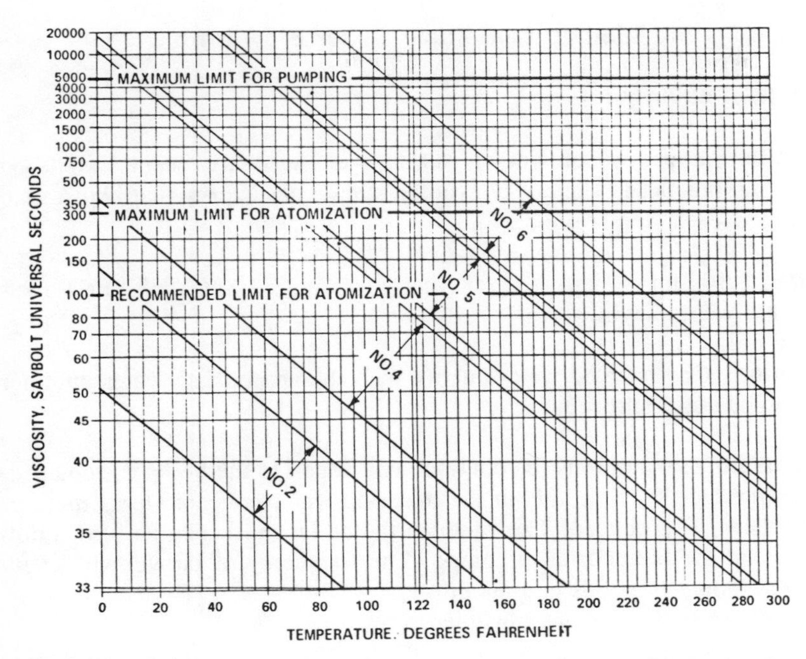

Figure 3-3. Fuel-oil viscosity vs. temperature. (Source: "Guidelines for Burner Adjustments of Commercial Oil-Fired Boilers," EPA-600/2-76-088, March 1976.)

Method 5: Optimize Fuel Atomization Pressure

Potential savings: 1%.

Problem. The fuel atomization pressure is lower or higher than that specified by the nozzle or burner designer.

Impact: 100-hp boiler—$3,100 yearly savings; 100,000-lbs/hr boiler—$89,000 yearly savings.

Solution. Adjust fuel pressure according to the nozzle operating instructions.

Discussion. Fuel atomization is a function of fuel pressure, fuel viscosity, primary air or steam (whichever is used), and nozzle design. Operating the burner at higher or lower pressures will decrease combustion efficiency.

Method 6: Reduce Boiler Blowdown

Potential savings: 1%.

Problem: Excessive blowdown due to poor water treatment and/or poor operating practices. The hot blowdown stream has energy that is lost unless it is recovered (see Method 16).

Impact: 100-hp boiler—$3,100 yearly savings; 100,000-lbs/hr boiler—$89,000 yearly savings.

Solution. Control feed-water quality with the appropriate water treatment; review operating procedures.

Discussion. Water must be continuously added to the boiler to replace the steam produced. The solids in the make-up water concentrate as more and more steam is produced. The water in the boiler must then be blown down to prevent the boiler from plugging. The frequency of blowdown depends on the amount of solids and alkalinity.

There are two types of blowdown:

1. The mud blowdown designed to remove heavy sludge at the bottom of the boiler (it is blown for a few seconds at specified intervals).
2. Continuous or skimming blowdown designed to remove solids dissolved in the water.

The potential energy recovery from reducing the percent blowdown is shown in Table 3-5.[1,5,10] For instance, in reducing the percent blowdown from 10% to 5%, there is a 1.7% gain in boiler efficiency for a 200-psig boiler. The feed-water quality must be reviewed to check the appropriate water treatment (see Chapter 9). Regular checks of total dissolved solids, alkalinity, etc. in the blowdown and feed water will determine how much blowdown is really necessary.

Table 3-5
Percent Blowdown Vs. Percent Efficiency Lost

	Boiler Pressure, psig			
	200	400	600	800
% Blowdown	% Efficiency Lost			
10	3.3%	4.0%	4.5%	5.1%
5	1.7	2.0	2.2	2.5
2	0.7	0.8	0.9	1.0

Method 7: Optimize Single-Boiler Firing

Potential savings: 5%–10%.

Problem. A boiler may come on for a few minutes and then be off for several minutes, resulting in large energy losses due to the removal of useful heat when the boiler is off; or a boiler may "hunt," i.e. the firing rate is continually adjusting, resulting in much more excess air.

Impact: 100-hp boiler—$30,700 yearly savings; 100,000-lbs/hr boiler—$888,000 yearly savings.

Solutions. For an on-off boiler, fire the boiler at an intermediate rate or buy a smaller boiler; for boilers that "hunt," adjust the firing rate so that larger steam-pressure fluctuations are allowed.

Discussion. In either case, the boiler controls must be used carefully to operate at optimum efficiency.

Method 8: Optimize Multiple-Boiler Operations

Potential savings: 2%–5%.

Problem. A plant having two or more boilers operates them without distributing the load according to the efficiency of each boiler.

Impact: 100-hp boiler—$15,400 yearly savings; 100,000-lbs/hr boiler—$444,000 yearly savings.

Solutions. Obtain the efficiency of each boiler vs. load; adjust each boiler to operate at peak efficiency.

Discussion. Boiler efficiencies vary with boiler design, load, age, and many other factors. Peak operating efficiency occurs at a load less than 100% because of interaction among stack temperature, excess air flow, and surface radiation losses. Therefore, the most efficient boilers should be used most (or all) of the time. This can be accomplished by never operating *all* boilers at reduced load and performance; scheduling boilers for specific needs according to their efficiencies and loads; and considering purchasing smaller boilers. Boiler selection can be done by obtaining the efficiency of each boiler vs. load; analyzing the steam peak demands; coordinating boiler and plant operations; and adjusting each boiler to operate at peak efficiency.

Table 3-6
Steam Losses Through Piping Leaks

Steam Pressure: 100 psig

Hole size, in.	*Steam wasted, lb/month*	*Total cost, $/year*
1/2	835,000	100,000
7/16	636,600	76,400
3/8	470,000	56,400
5/16	325,000	39,000
1/4	210,000	25,200
3/16	116,600	14,000
1/8	53,300	6,400

Basis: $10 per 1000 lb of steam; 8000 hr per year

Steam at Different Pressures

Steam-system pressure, psig	*Loss rate (lb/hr) for hole size indicated*				
	1/16 in.	*1/8 in.*	*1/4 in.*	*1/2 in.*	*1 in.*
100	18.00	73.00	290.00	1,161.00	4,645.00
200	34.00	137.00	543.00	2,171.00	—
400	66.00	262.00	1,048.00	—	—
850	137.00	546.00	—	—	—

Source: Yarway Corporation.

Method 9: Stop Steam Leaks

Potential savings: 5%–10%.

Problem: Piping leaks.

Impact: For a quarter-inch steam leak at 100 psig—$25,200 per year. For a quarter-inch steam leak at 400 psig—$83,800 per year (see Table 3-6).

Solution. Plug leaks as soon as they appear.

Discussion. A very small leak can turn out big losses. If instead of a quarter-inch leak there were two or more leaks, steam losses would be staggering.

Method 10: Stop Steam-Trap Leaks

Potential savings: 5%–10%.

Problem: Steam trap malfunctioning and leaking steam.

Impact: 100-psig steam-trap leak through half-inch pipe—$100,000 per year; at 200 psig the loss is $173,000 per year. It has been reported that in some instances, 50% of the steam produced has literally gone down the drain.[3]

Solution. Check steam traps routinely and repair them.

Discussion. The proper installation and care of steam traps by their manufacturers must be followed.[12]

There are three types of conventional traps: mechanical, thermostatic, and thermodynamic. Each type operates under different principles, respectively: difference between density of steam and condensation; temperature difference; and difference of energy.

Steam-trap testing must be part of a regular maintenance program. Testing can be done using one or more of the following methods:

1. *Visual*. Keep in mind that live steam is invisible at the source and leaks at high velocity. Do not confuse it with flash steam, which is visible at the source.
2. *Temperature difference*. As long as the difference across the trap is known, temperature-sensitive tape can be used to detect the temperature difference between the steam (upstream) and condensate (downstream).
3. *Sonic method*. Listen with sonic detectors or stethoscope to detect trap leaks. The user of this method must become familiar with the sounds of the three different conventional steam traps.

Increase Efficiency with Some Expenses and/or Capital Investment

Method 11: Reduce Deposits in Burner

Potential savings: 1%–5%.

Problem. Organic and inorganic deposits build up in the burner, reducing atomization efficiency and therefore reducing combustion efficiency.

Impact: 100-hp boiler—$15,000 yearly savings; 100,000-lbs/hr boiler—$444,000 yearly savings. To obtain net savings, the cost of the fuel treatment must be subtracted from the gross savings in fuel.

Solution. Use a fuel additive with detergent dispersants to keep burners clean.

Discussion. Fuel additives should have four benefits to be cost-effective:

1. Keep burners clean.
2. Reduce burner maintenance.
3. Improve atomization and boiler efficiency.
4. Cost less than the gross savings.

The fuel-treatment cost with good additives is equivalent to 1%–3% boiler efficiency; in other words, the fuel additive must increase the boiler efficiency by at least 1% to break even. Chapter 8 discusses the types of additives available. The services of an additive company representative are usually required.

Method 12: Reduce Scale and Soot Deposits on Fireside

Potential savings: 2%–9%.

Problem. Soot and/or vanadium-based deposits decrease the heat-transfer rate. If all conditions are constant, this decrease is noticeable when the flue gas temperature increases with time.

Impact: 100-hp boiler—$27,600 yearly savings; 100,000-lbs/hr boiler—$800,000 yearly savings. The cost of fuel treatment and equipment needed must be deducted from the gross savings.

Solutions. Treat fuel with additives to minimize either soot deposits or vanadium-based scale. Use soot blowers if available.

Discussion. Soot deposits reduce combustion efficiencies (see Chapter 7). Figure 3-4 shows that a one-eighth-inch soot deposit could decrease boiler efficiency by 9%. The presence of vanadium and sodium in fuel oil No. 6 is the rule rather than the exception; these deposits can reduce boiler efficiencies by 3%. Figure 3-5 gives the reduction in efficiencies for different types of deposits.[3,10,11]

Specialized additives to minimize soot deposits are on the market. Soot blowers are used to remove both fly ash and soot deposits from oil- and coal-fired boilers. The installation of a simple system costs $20,000–$60,000, depending on the size of the boiler. Naturally, particulate emissions may temporarily increase while operating a soot blower.

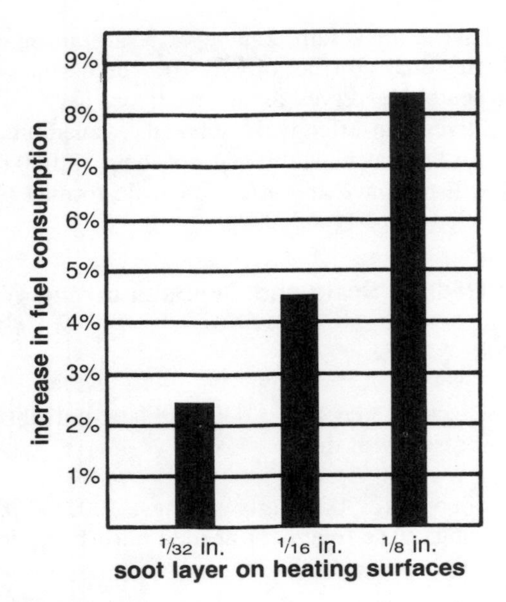

Figure 3-4. Effect of soot on fuel consumption—fireside. (Source: United Technologies—Bacharach.)

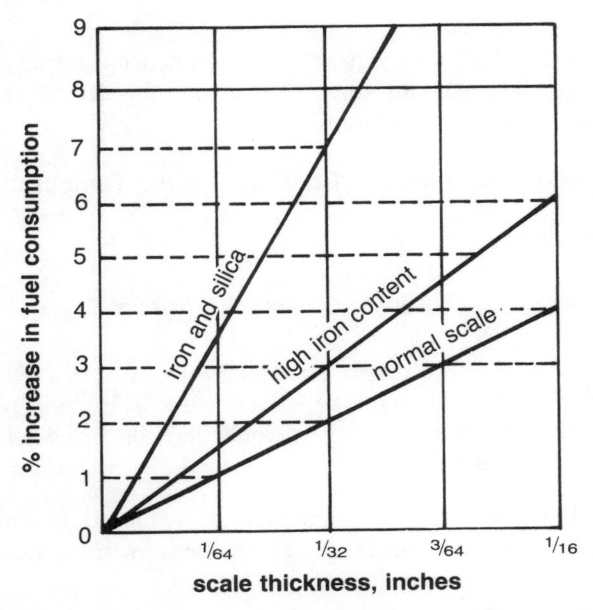

Figure 3-5. Effect of scale thickness on fuel consumption. (Source: National Bureau of Standards in Cooperation with the Federal Energy Administration, Conservation, and Environment, Supplement 1, December 1975.)

Certain crudes have a vanadium content of 350 ppm or even more that accumulates in the residual oil (fuel oil No. 6). The corrosive scale formed by the vanadium pentoxide serves as an insulator, decreasing boiler efficiencies. Fuel additives can effectively solve the vanadium problem. The equipment needed for the special additive costs about $3,000 (in 1983). Note that soot blowers will not have any effect on scale formed by the presence of vanadium.

Method 13: Reduce Scale and Deposits on the Waterside

Potential savings: 2%–4%.

Problem. Inorganic scale and deposits decrease heat-transfer rate; i.e. flue-gas temperature increases with time.

Impact: 100-hp boiler—$12,300 yearly savings; 100,000-lbs/hr boiler—$355,200 yearly savings. The treatment needed must be deducted from the gross savings.

Solution. Treat feed water properly using the guidelines indicated in Chapter 9.

Discussion. The boiler may have to be cleaned up before an improved water treatment provides any improvement. Poor water treatment may be almost as bad as no treatment at all.

Method 14: Increase Combustion Air Temperature

Potential savings: 1%.

Problem. Preheat combustion air. Every 40°F rise yields a 1% gain in efficiency.

Impact: 100-hp boiler—$3,100 yearly savings; 100,000-lbs/hr boiler—$89,000 yearly savings. The cost of the equipment needed must be deducted from the gross savings.

Solution. There are several possible solutions: relocate air intake duct so that a maximum air duct temperature is obtained, or install an air preheater if possible and economically feasible.

Discussion. Relocating the air intake duct may easily raise the air temperature by 40°F or more. Material and labor expenses run under $1,000.

If an air preheater is feasible, economics will decide whether it is advisable. It costs about $2,000 per MBtu/hr duty (1983 dollars). Rule of thumb:

40°F rise in air temperature increases boiler efficiency 1%.

Method 15: Increase Feedwater Temperature

Potential savings: 3%.

Problem. Raise water temperature.

Impact: 100-hp boiler—$10,700 yearly savings; 100,000-lbs/hr boiler—$296,000 yearly savings.

Solution. The equipment needed is an economizer that uses the heat energy from the flue gas, if economically feasible. Also, the water can be preheated using the waste-heat energy from the blowdown (Method 16).

Discussion. Economizers are generally preferred over air preheaters on small units. A combination air preheater and economizer is used on larger units operating above 400 psig. An economizer accomplishes two things: it raises the feed-water temperature and lowers the temperature of the flue gas. Economizers cost $2,000 per MBtu/hr. Rule of thumb:

10°F rise in feed-water temperature raises efficiency 1%.

Method 16: Recover Heat Energy from Blowdown

Potential savings: 1%.

Problem. Preheat water by recovering energy from blowdown.

Impact: 100-hp boiler—$3,000 yearly savings; 100,000-lbs/hr boiler—$89,000 yearly savings. The cost of additional equipment must be deducted from the gross savings.

Solutions. Add flash tank to system. The blowdown is flashed by lowering the pressure in the flash tank; the steam produced is then vented into the feed water to the boiler. Some 50% of the heat in the blowdown is recovered. Send blowdown at 220°F to waste.[3,10]

Also, a flash tank may be added, followed by a heat exchanger to extract one-third more energy from the blowdown before going to waste.[3,10]

Discussion. The flash-tank system is very simple and costs about $1,000 (1983 dollars). The combination flash tank and heat exchanger costs $10,000. An economic analysis will reveal the most convenient alternative.

Method 17: Energy Recovery from Excessive Steam Pressure

Potential savings: Variable.

Problem. Steam is required at a lower pressure than that provided by the boiler.

Impact: Variable.

Solution. Use a throttling or back-pressure turbine.

Discussion. The idea is to utilize the energy from steam rather than to decrease its pressure through a throttling valve.[3,9]

Method 18: Reduce Heat Losses in Boiler, Steam, and Valves

Potential savings: 5%–8%.

Problem. Heat is lost by radiation and convection through the walls of uninsulated or poorly insulated boiler surfaces and piping.

Impact: 100-hp boiler—$24,000 yearly savings; 100,000-lbs/hr boiler—$712,000 yearly savings. The cost of the insulation must be deducted from the gross savings.

Solution. Use a surface thermometer (see Chapter 11) and determine where heat losses are present; then insulate.

Discussion. There are two methods to minimize heat losses by insulation:

1. The "quick and dirty" method. Add a layer or two of insulating material and keep feeling the surface until it is cool enough that you can comfortably put your hand on it (the boiler, pipe, or valve).
2. The engineering method. Take as many surface-temperature readings as needed to calculate energy losses. Then determine the economic insulation thickness, plotting the data as shown in Figure 3-6. Example calculations are given in Chapter 13.

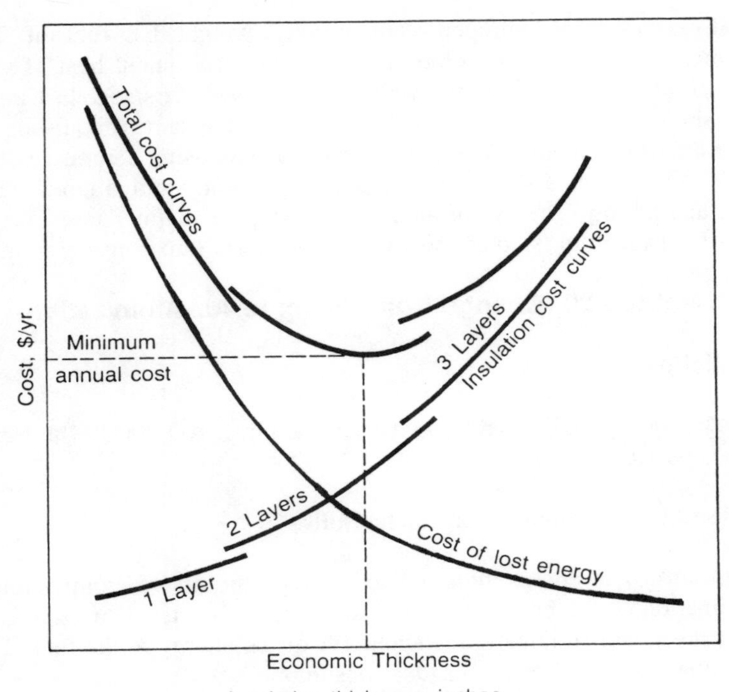

Figure 3-6. Determining economic insulation thickness.

Method 19: Use Fuel Oil Rather Than Natural Gas

Potential savings: 2%.

Problem. If the price of natural gas is the same or higher than that of fuel oil, which is more economical?

Impact: 100-hp boiler—$6,000 yearly savings; 100,000-lbs/hr boiler—$190,000 yearly savings. New equipment will be needed for the changeover.

Solution. Use fuel oil.

Discussion. Even if both fuels cost the same per MMBtu, fuel oil gives about a 2% higher efficiency than natural gas. The reason for this is that natural gas has more hydrogen atoms per unit weight than fuel oil. Therefore, more water vapor is formed from the gas. The latent heat of vaporization of water is lost when the water vapor leaves the stack. In Chapter 2 it was shown that two similar boilers producing the same amount of steam were 2.5%–3.0% more efficient using fuel oil than using natural gas. Burning fuel oil requires a significant capital investment: storage tanks, pumps, filters, and piping. The use of additives and higher maintenance make fuel oil costly. Feasibility studies must take these costs into consideration.

Method 20: Change from Steam to Air Atomization

Potential savings: 1%.

Impact: 100-hp boiler—$3,100 yearly savings; 100,000-lbs/hr boiler—$89,000 yearly savings.

Solution. Use air atomization for fuel oils.

Discussion. The energy required to produce the air of atomization is a small fraction of the energy required to produce the steam of atomization. In fact, the steam atomization is about 1% of the energy in the fuel. Therefore, using air will save 1% in fuel consumption.

References

1. PB-262 577. *Industrial Boiler Users' Manual*—Volume II. KVB, Inc. FEA Contract C-04-50085-00, Report No. FEA/D-77/026, January 1977.
2. PB-264 543. *Guidelines for Industrial Boiler Performance Improvement (Boiler Adjustment Procedures to Minimize Air Pollution and to Achieve Efficient Use of Fuel)* by M. W. McElroy and D. E. Shore, KVB, Inc. EPA Contract 68-02-1074, Report No. EPA-600/8-77-003a, January 1977.
3. PB-265 713. *Measuring and Improving the Efficiency of Boilers—A Manual for Determining Energy Conservation in Steam Generating Power Plants,* by Engineering Extension Service, Auburn University, Alabama. FEA Contract FEA-CO-04-50100-00, Report No. FEA/D-77/132, November 1976.
4. TID-27600. *Maintenance and Adjustment Manual for Natural Gas and No. 2 Fuel Oil Burners.* Bureau of Natural Gas, Federal Power Commission, and U.S. Department of Energy, 1977.
5. Dyer, David F., and Glennon Maples, ''Boiler Efficiency Improvement,'' Boiler Efficiency Institute, Auburn, Alabama, 1981.

6. *Packaged Firetube Boiler Engineering Manual,* American Boiler Manufacturers Association, 1971.
7. Bacharach Instrument Company.
8. Brannan, Carl, *The Process Engineer's Pocket Handbook,* Gulf Publishing Co., 1976.
9. Ostroot, G., "Case History: A Two-Burner Boiler Explosion," *Hydrocarbon Processing,* p. 85, December 1976.
10. Yaverbaum, L. H., "Energy Saving by Increasing Boiler Efficiency," Noyes Data Corporation, Park Ridge, New Jersey, 1979.
11. Phelan, J. V., and L. P. Gelosa, "How To Control Boiler Iron Deposits," *Chemical Engineering,* p. 174, March 1975.
12. *Steam Trap Application Data,* Yarway Corporation, Blue Bell, Pennsylvania, 19422.

4

Increasing the Efficiency of Ship Boilers

Basically, the *combustion* efficiency of a ship boiler can be increased using the same principles discussed previously for stationary boilers.

Increasing *boiler* efficiency is something else. A ship boiler has many more variables than a stationary boiler, such as the directions of the wind and tide, condition of the hulls (smooth or rough), propeller designs, and other such conditions peculiar to ships. To make things more complicated, ships have both boilers and diesel engines. Fuel savings also involves coordination of boilers and diesel-engine operations.

An article in the *Marine Engineering/Log* journal[1] using the Maritime Energy Efficiency Measures Matrix prepared by the Argonne National Laboratory for the United States Department of Energy and the Maritime Shipping Industry, determined that there were at least 63 ways of increasing fuel efficiency on ships. Some of the suggested methods include the following:

1. *Use an electric in-port feed pump.* Fuel savings of up to 15% are claimed. Less steam is wasted because one boiler can be shut down.
2. *Employ vacuum pumps instead of air ejection.* At reduced power, fuel savings of 2% are possible.
3. *Improve fluid-flow controls.* Fuel savings of 2%–10% is possible.
4. *Improve fuel atomization.* Controlling fuel viscosity with the proper temperature adjustment improves combustion efficiency significantly. Preheating Bunker *C* fuel to temperatures as high as 280°F have been recommended.[2]

Figure 4-1. Ships can improve their fuel-burning efficiency.

5. *Use fuel additives*. Efficiency improvements ranging from 4% to 8% have been reported by the Ferrous Corporation.[3] The Ferrous computer software program evaluates the input data that come directly from the engine room. After applying several correcting factors, the net fuel efficiency improvement due to the additive effect is determined.
6. *Install dual-economizer steam-air heaters*. Save 1.5%–3.5% fuel.
7. *Operate only one boiler whenever possible*. Savings up to 3% have been reported.
8. *Improve CO and CO_2 controls*. Minimize excess air. Use variable-speed forced-draft fan-motor controllers.

These are only a few examples of how to save on fuel. The methods used include techniques requiring no investment and techniques requiring some or substantial capital investment. An economic analysis is called for in the latter instance.

Figure 4-1 shows ships hauling Bunker fuel.

References

1. "63 Ways To Burn Fuel More Efficiently," *Marine Engineering/Log,* p. 46, January 1982.
2. "Hot Bunker *C* a Must for Diesels," *Marine Engineering/Log,* p. 112, February 1982.
3. Ferrous Corporation, P.O. Box 1764, Bellevue, Washington, 98009.

5

Increasing the Efficiency of Process Furnaces

The 1981–1982 *Directory of Industrial Heat Processing and Combustion Equipment* lists the following types of equipment:[1]

- Atmosphere equipment
- Furnaces
  - Batch
  - Car
  - Continuous
  - Electric
  - Melting
  - Multiple-type
  - Rotary hearth
  - Rotary retort
  - Salt bath
  - Shaker hearth
  - Vacuum
  - Walking beam
  - Wire
- Incinerators
- Induction heating equipment
- Kilns

Figure 5-1. A commercial furnace.

Ovens
- Batch
- Foundry
- Infrared
- Multiple-type

Because of the high cost of fuel, the directory also includes equipment and controls to minimize fuel consumption, such as:

Combustion equipment
- Burners, elements, and accessories
- Controls, instruments, and meters
- Valves

Heat transfer/Recovery
- Gas coolers
- Heat exchangers
- Recuperators and regenerators
- Startup heaters
- Waste-heat boilers

Because of the numerous designs available and their various uses it is hard to pick a typical furnace design. Figure 5-1 shows a commercial steel furnace. As shown in Table 5-1, process-furnace temperatures and average efficiencies vary widely.[2]

Table 5-1
Typical Operating Temperature of Process Furnaces

Type of Industry	Operation	Approximate Temperature °F	Approximate Temperature °C	Average Efficiency
Aluminum	Melting	1380	740	59
Brass	Melting	2100	1150	42
Cement	Calcining	2800	1535	20
Copper	Melting	2200	1205	38
Glass	Melting	2400	1315	33
Iron	Smelting (blast furnace)	3000	1645	15
	Melting (cupola)	2600/2800	1425/1535	20
Lead	Smelting	2200	1205	38
	Melting	720	380	73
Lime	Burning	2100	1150	41
Steel (carbon steel)	Forging	1900/2400	1035/1315	33
Steel (stainless)	Forging	1650/2300	900/1200	35
Zinc	Melting	740/930	395/500	70

Source: Combustion Technology Manual, Industrial Heating Equipment Association, 1980.

Table 5-2
Flue-Gas Temperature of Process Furnaces Versus Efficiencies

Flue-Gas Temperature °F	Flue-Gas Temperature °C	Efficiency %
3000	1645	15
2800	1535	20
2400	1315	33
2200	1205	38
2100	1150	41
1800	980	48
1400	760	59
970	515	69
720	380	73

Source: Combustion Technology Manual, Industrial Heating Equipment Association, 1980.

From the data in Table 5-1, it follows that the lowest efficiencies are obtained in process furnaces with the highest flue-gas temperatures, as shown in Table 5-2. Unless the heat from the flue gases is partially recovered, the efficiencies will remain low. The best possible efficiency is obtained when the flue-gas exit temperature is approximately 100°F higher than the maximum process temperature using the stoichiometric air-fuel ratio.[2]

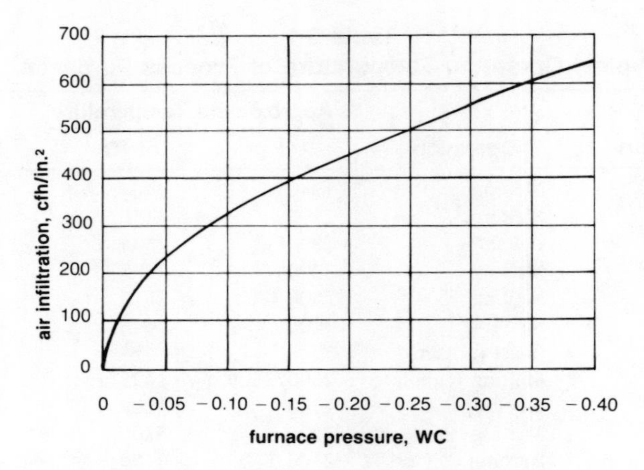

Figure 5-2. Air infiltration into a process furnace vs. furnace pressure. (Courtesy Industrial Heating Equipment Association.)

The following rules of thumb can be used to estimate heat losses in a furnace:

1. Reduce excess air. It is common to find process furnaces running with 100% or more excess air.
2. Reduce infiltration air. This air alone may account for most of the excess air found in the furnace. Figure 5-2 can be used to estimate the amount of air infiltration as a function of the negative pressure present in the system.
3. Balance burners to balance furnace heat.
4. Decrease flue-gas temperature. Is that high excess margin of safety really needed?
5. Increase fuel-oil temperature—improve atomization and combustion efficiency.
6. Reduce deposits in burners—use fuel additives.
7. Increase combustion-air temperature.
8. Reduce heat losses on furnace—insulate.
9. Reduce heat losses on stack—insulate.
10. Recover heat from product.
11. Recover heat from flue gases.
12. Eliminate radiation losses through openings.
13. Preheat load.
14. Operate at high loads. Reduced production decreases furnace efficiency.

The general procedures used for boilers can be used for process furnaces.

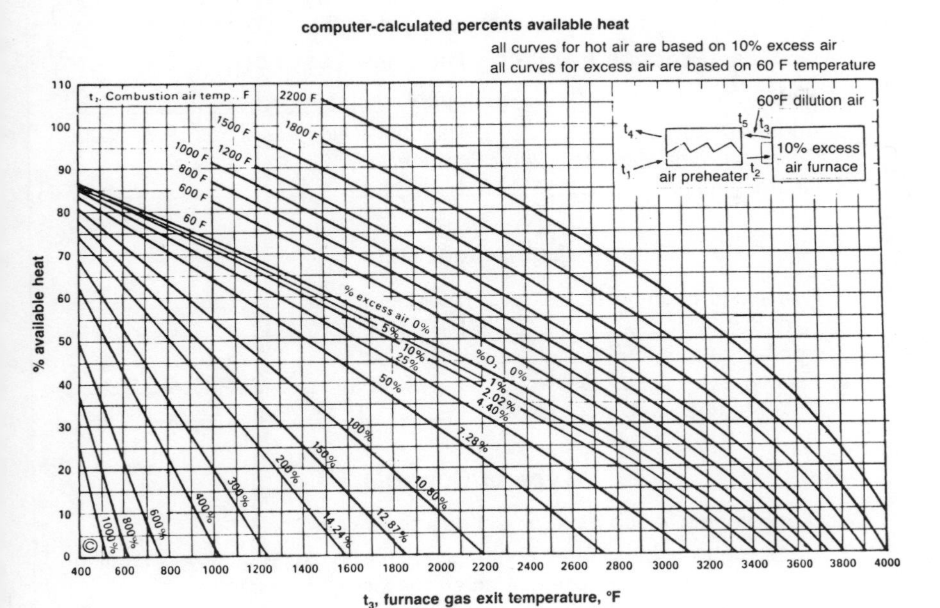

Figure 5-3. Process-furnace efficiencies vs. flue-gas temperature. (Courtesy Industrial Heating Equipment Association.)

The Industrial Heating Equipment Association[1] has published tables to estimate the fuel savings for natural gas, coke-oven gas, and residual fuel (fuel oil No. 6) when the air is preheated in the range of 600°F–1500°F. Fuel-savings estimates of heat-loss recovery in the flue gas can also be made using Figure 5-3. For example, if the exit temperature of the flue gas is 1400°F, the efficiency of the furnace is 64%. If the gases are used to generate steam at 100 psig in a waste-heat-recovery boiler, the gases would leave at 500°F rather than at 1400°F, and furnace efficiency would go up to 84%. Therefore, the fuel savings would be:

$$\frac{84 - 64}{84} \times 100 = 23.8\%$$

Heat-Recovery Equipment

The higher the waste-gas temperature, the more significant heat recovery becomes. For metallic recuperators, waste-gas temperatures above 1400°F are economically practical. Preheating the combustion air to approximately

Process Furnace Efficiency Test Form

1. Company: __________ Type of industry __________ Date: __________
2. Address: ____________________ Telephone No. ______________
3. Contact (*Plant Manager, Plant Engineer*) ______________
4. Operator: ______________________________

Equipment Data

5. Process furnace no. __________ Designer __________ Year __________
6. Heat duty: ______________ MMBTU/hr; ______________ Kcal/hr
7. Number of fuel feed nozzles: ______________
8. Fuel type: _______ ; Gal/hr: _______ $MMft^3/hr$; _______ ; Cost _______
9. Outlet temperature of product: __________ F; Product: __________ lb/hr

Type of Controls

Check the control which most nearly describes your unit:

10. _____ Highly sophisticated control system with oxygen analysis feeding back to individually characterized fuel and air control.
11. _____ Complex control system which measures and controls both fuel and air according to a set of characteristics for various loads.
12. _____ Control from outlet temperature of product with fuel and air operated in parallel open loop and individually characterizable.
13. _____ Control from product outlet temperature with modulating fuel control and manually adjusted air flow.
14. _____ Other controls (CO, CO and O_2, etc.).

Test Data

Comments:

15. Test no. ______________________________
16. Fuel pressure return ______________________________
17. Fuel pressure to burner ______________________________
18. Fuel preheater temp. ______________________________
19. Fuel viscosity @ preheater temp. ______________________________
20. Air temperature ______________________________
21. Flue gas temp. ______________________________
    A. Before heat recovery ______________________________
    B. After heat recovery *(Stack)* ______________________________
22. % CO_2 ______________________________
23. % O_2 ______________________________
24. % Excess air ______________________________

Figure 5-4. Process-furnace efficiency test form.

Test Data

Comments:

25. CO ppm ________________
26. Smoke color no. ________________
27. SO_2 ________________
28. Draft ________________
29. % Combustion eff. ________________

Operations

30. Average furnace load in percent of capacity of the unit: ________
31. Days per year furnace operates: ________________
32. Type of heat recovery equipment: ________________
33. Attach sketch with temp. profile of process furnace.
34. Burners: cleaning frequency: ____ per week; new installed: ____ per year.
35. Frequency fuel filters cleaned or changed: ________________
36. When was the last time tanks were cleaned?: ________________
    Pictures taken?: ________________

Fuel Data

37. Attach recent fuel analysis.
38. Heating value:
    A. natural gas: ________________ Btu/ft^3
    B. fuel oil: ________________ Btu/gal

For fuel oil analysis:

39. API** ________________
40. SP. GR.** ________________
41. Heating value* ________________
42. Water** and sediment % ________
43. Viscosity** ________________
44. Pour point ________________
45. Sulphur %** ________________
46. Vanadium, ppm ________________
47. Nickel, ppm ________________
49. Ash % ________________
50. Fuel treatment (*if any*) ________
    ____
    ____
51. Storage tank capacity: ________
52. Capacity day tank: ________
53. Monthly fuel consumption: ____
    ____

* Estimate if not available.
** Obtain on the spot if fairly recent analysis is not available.

Test data witnessed by: ________________ Date: ________
(signature)

Tests conducted by: ________________ Date: ________
(signature)

Figure 5-4 continued.

50% of the flue-gas temperature is advisable.[1,2] Thus, it is generally economical to preheat the air to 750°F if the flue gas is at 1500°F.

The total cost of a heat-recovery system is four to five times the cost of the recuperator itself, since the cost of engineering, installation, etc. must be included.

It has been recommended that the wall temperature of a recuperator be maintained at about 1200°F for continuous operation when fuel oil No. 6 is used to protect the recuperator against vanadium and sulfur corrosion.[2]

Data Sheet Form

The data sheet form in Figure 5-4 can be used to carry out an energy audit on a furnace.

References

1. *The Directory of Industrial Heat Processing and Combustion Equipment, U.S. Manufacturers, 1981–1982,* The Energy Edition, Information Clearing House, Inc.
2. *Combustion Technology Manual,* Industrial Heating Equipment Association, Arlington, Virginia, 22209, 1980.

6

Increasing Dryer Efficiency

The basic principles covered in Chapters 3, 4, and 5 apply also to the design and operation of dryers, and will increase their efficiencies and save on fuel costs.

Consider the following ways to increase dryer operating efficiency:

1. Reduce vent-gas temperature.
2. Preheat air to dryer.
3. Reduce infiltration air.
4. Pretreat in-product to reduce water content. The lower the water content, the less fuel required to vaporize water.
5. Raise humidity in dryer. If a higher humidity is acceptable, this is a good way to increase efficiency.
6. Reduce radiation losses.
7. Stop steam leaks.
8. Stop steam-trap leaks.
9. Recover condensate.

Some of these methods require an engineering or scientific background. Chapter 13, on engineering calculations, presents typical problems and how to go about solving them using basic engineering formulas. However, notice that many of the methods suggested here require no real engineering calculations and no capital investment. They do, however, require practical experience and common sense.

7

Test Methods and Boiler Operation Diagnosis

This chapter includes certain tests that must be performed to correct boiler problems and increase boiler efficiency. Each test is divided into four parts:

1. How to perform the test, and the instruments recommended.
2. The problems present in the boiler.
3. Causes for the problems.
4. Solutions.

Gas samples at the stack must be taken very close to the boiler, say six inches (15 cm). Failure to do so may result in false temperature readings, especially if the stack is poorly insulated. The hole where the stack thermometer is located can be used for gas sampling, or a 0.25-inch hole can be drilled in the stack. The results of individual tests must be used in conjunction with all of the other tests, since one usually affects one or more of the other variables.

Carbon Dioxide (CO_2), Oxygen (O_2), and Carbon Monoxide (CO) Tests

Flue-gas analysis is used to determine the air-to-fuel ratio and the completeness of combustion. The gas components usually measured are CO_2, O_2, and CO. The CO and CO_2 contents can be used as indicators of the completeness, or efficiency, of combustion. The O_2 content indicates the

Table 7-1
Effect of Excess Air on CO_2 in Combustion Products

		% Excess Air								
Fuel		**0**	**10**	**20**	**40**	**60**	**80**	**100**	**150**	**200**
Natural gas	$\%CO_2$	12.0	10.7	9.8	8.3	7.2	6.3	5.7	4.5	3.7
Propane	$\%CO_2$	14.0	12.6	11.5	9.8	8.5	7.5	6.7	5.3	4.4
Butane	$\%CO_2$	14.3	12.9	11.7	10.0	8.6	7.6	6.8	5.4	4.5
Distillate oil	$\%CO_2$	15.2	13.8	12.6	10.7	9.3	8.2	7.4	5.9	4.9
Residual oil	$\%CO_2$	15.6	14.1	12.9	11.0	9.6	8.5	7.6	6.1	5.0
Bituminous coal	$\%CO_2$	18.4	16.7	15.3	13.0	11.4	10.1	9.0	7.2	6.0
Anthracite coal	$\%CO_2$	19.8	18.0	16.5	14.1	12.4	11.0	10.0	7.9	6.9

Note: At 0% excess air, the % CO_2 content in the flue gas is the ultimate or maximum possible for that fuel. This % CO_2 content is also known as ''stoichiometric mixture.''
Caution: Gas samples from a furnace processing materials containing CO_2 (Na_2 CO_3, for instance) would yield higher % CO_2 than those shown under 0% excess air because of the decomposition of the carbonates under elevated temperatures.
Source: Fundamentals of Boiler Efficiency, Exxon Company U.S.A., 1976.

amount of excess combustion air. Under ideal conditions, the flue-gas analysis shows no CO or O_2 and has a maximum value for CO_2, termed ultimate CO_2. For greatest efficiency, the fuel-air mixture should be adjusted until the maximum CO_2 level is obtained. Table 7-1 lists the ultimate CO_2 readings for varying amounts of excess air in the combustion of common fuels. A low CO_2 or high O_2 reading may indicate the presence of too much excess air.

These tests can be carried out with an Orsat, with electronic instruments, or with the portable Bacharach or Dwyer Combustion Testers.

CO_2 Measurement

Of the many available instruments on the market, both the portable Bacharach and Dwyer Combustion Testers have proven to be reliable, need practically no maintenance, and are relatively inexpensive. In addition, their operation is so simple that anyone can learn to operate them in just a few minutes (see Figure 7-1). On the other hand, an Orsat requires a skilled operator. An electronic instrument may cost several thousand dollars and requires frequent checking and maintenance. The Bacharach Combustion Tester is used as follows:

1. Insert sampling tube into a hole in the stack.
2. Place the rubber cap end on top (on the plunger valve) of the CO_2 indicator and hold the plunger valve in a depressed position, i.e. open.

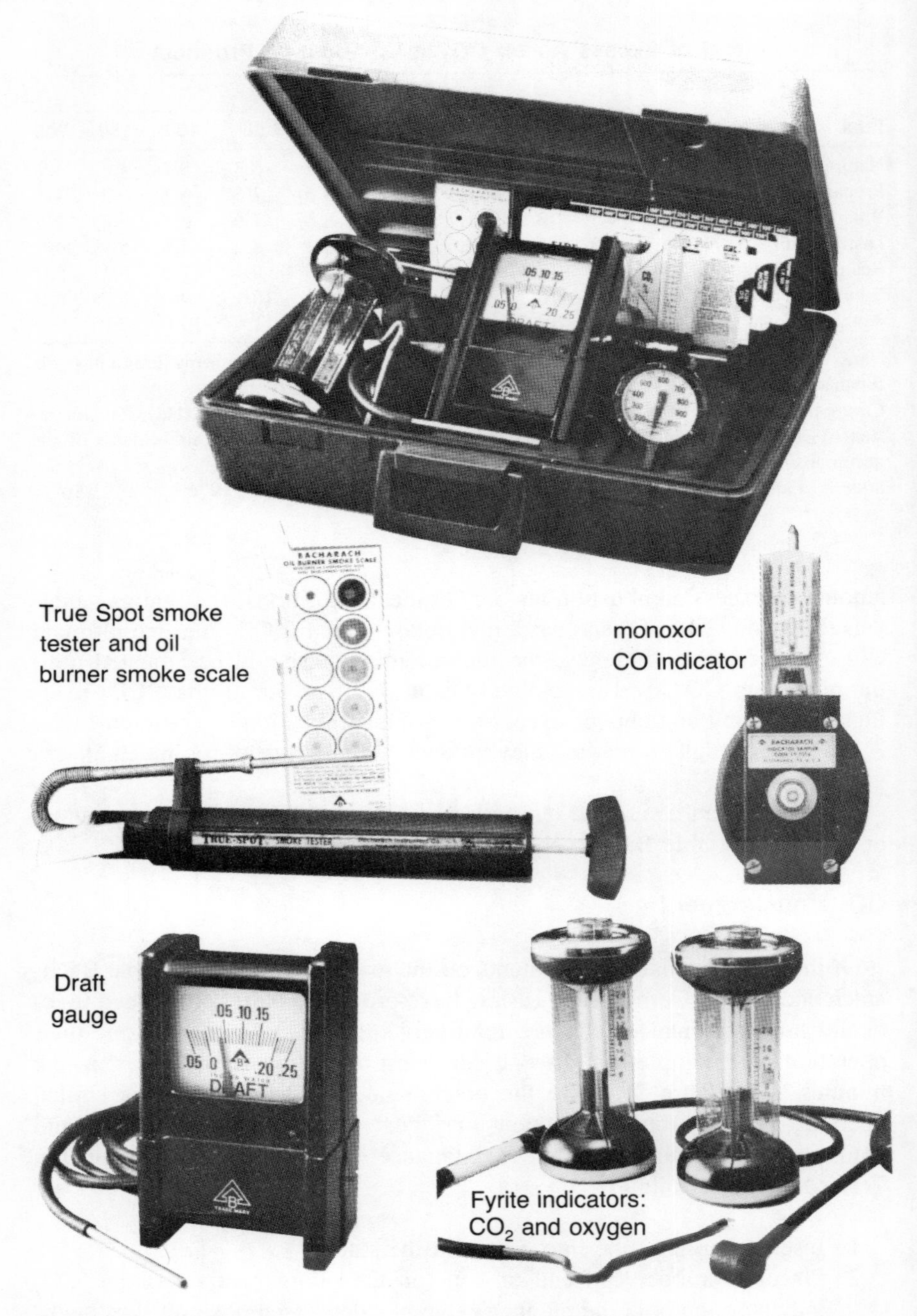

Figure 7-1. Bacharach combustion tester. (Courtesy United Technologies—Bacharach.)

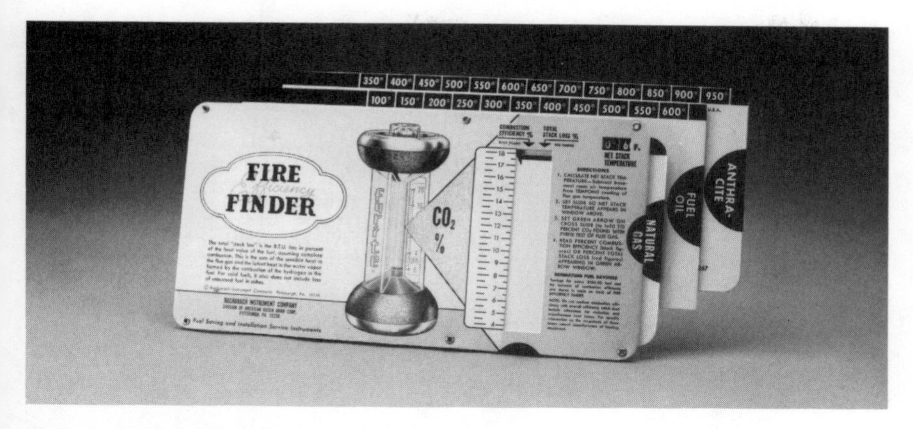

Figure 7-2. Bacharach slide rule to calculate combustion efficiency. Basis: (1) fuel oils containing 0.11–0.19 H_2/C ratios; (2) natural gas with 1060–1190 Btu/ft^3; (3) bituminous coal with 68% C, 5% H_2, and 13,500 Btu/lb dry coal; and (4) anthracite coal with 13,000 Btu/lb. (Source: United Technologies—Bacharach.)

3. Release the plunger valve after the eighteenth squeeze.
4. Turn the CO_2 indicator twice to allow the proper gas mixing.
5. Read % CO_2 directly on the scale calibrated for that purpose.

Table 7-1 shows the maximum (ultimate or stoichiometric) %CO_2 in the stack gas for different fuels. At 0% excess air, the data show the maximum CO_2 obtainable for a particular fuel. The % CO_2 falls off as more excess air is present. From the CO_2 reading, we can obtain the following information.

Combustion efficiency. The %CO_2 and the flue-gas temperature are used to estimate the combustion efficiency. This can be done quickly using specially designed slide rules (the Bacharach slide rule is shown in Figure 7-2), or it can be estimated using specially prepared tables similar to Tables 3-1 and 3-2 and those published by Dyer and Maples.[1]

Boiler efficiency. As discussed in Chapter 1, the boiler efficiency can be estimated as follows:

$$\text{Boiler efficiency (B.E.)} = \text{combustion efficiency (C.E.)} - \text{heat losses (from the boiler)}$$

Depending on the load, type of boiler, and insulation quality, radiation and convection heat losses may run as high as 7%. To these losses, add 1% heat

Table 7-2
Radiation and Convection Heat Losses as a Function of Load for Cleaver-Brooks Model CB Boilers

	Boiler	
	100–350 hp	400–800 hp
	Operating Pressure	
% of Load	125 psig	125 psig
100	1.4%	1.1%
50	2.6	2.2
25	5.1	4.4

loss in the boiler blowdown. Therefore, to estimate boiler efficiency, use the following guidelines:

B.E. = C.E. − 3% for a boiler running at full load.

B.E. = C.E. − 5% for a boiler running at half load.

B.E. = C.E. − 8% for a boiler running at low load.

For instance, Cleaver-Brooks determined radiation and convection losses for their Model CB Boilers[2] as shown in Table 7-2. Therefore, if we estimated 80% C.E. running the first boiler at 25% firing load, the B.E. would be:

80 − 5.1 − 1.0 = 73.9%.

Of course, boilers should be run at peak efficiency and not at such low loads.

Efficiency losses. These losses, or fuel wasted, can be estimated as follows:

100 − C.E. (these are minimum losses)

100 − B.E. (these are total energy losses around boiler)

Potential savings. Since most boilers can be brought up to 80%–85% C.E. with the techniques shown in this book (unless there is something mechanically wrong with the burner or boiler itself), the potential savings

Table 7-3
Percent Fuel Savings by Increasing Combustion Efficiency

From an Original Efficiency of	To an Increased Combustion Efficiency of								
	55%	60%	65%	70%	75%	80%	85%	90%	95%
50%	$9.10	16.70	23.10	28.60	33.30	37.50	41.20	44.40	47.40
55%		8.30	15.40	21.50	26.70	31.20	35.30	38.90	42.10
60%			7.70	14.30	20.00	25.00	29.40	33.30	37.80
65%				7.10	13.30	18.80	23.50	27.80	31.60
70%					6.70	12.50	17.60	22.20	26.30
75%						6.30	11.80	16.70	21.10
80%							5.90	11.10	15.80
85%								5.60	10.50
90%									5.30

Note: Assume constant radiation and other unaccounted-for losses. The percent savings in fuel costs is greater than the nominal percent increase in fuel-to-steam efficiency. Thus, a 10% increase in efficiency, say from 70% to 80%, drops fuel costs by 12.50%. The manual calculation is:

$$\frac{\text{increase in efficiency}}{\text{efficiency increase}} \times 100, \quad \text{or} \quad \frac{10}{80} \times 100, = 12.50\%$$

can be estimated either using Table 7-3 or manually as follows (as an example, assume C.E. is 70% and the potential is 85% C.E.):

$$\frac{85 - 70}{85} = 17.6\%$$

In other words, 17.6% fuel is now being wasted in this example. If all heat losses remain approximately constant, i.e. if nothing is done to recover the heat energy being wasted, the *improvement* in C.E. is approximately equal to B.E.

Excess air. There are two ways to estimate excess air: Using Figure 7-3, if CO_2 is 8.0%, for instance, the excess air is 80%. Or, the O_2 can be determined and the excess air estimated from it (see Table 7-4). This is especially important when fine-tuning a boiler.

O_2 in flue gas. There are two ways to estimate excess O_2: From Figure 7-3, for 8.0% CO_2, the O_2 content is 10%. Or, as mentioned previously, the O_2 content can be determined with a manual or electronic analyzer.

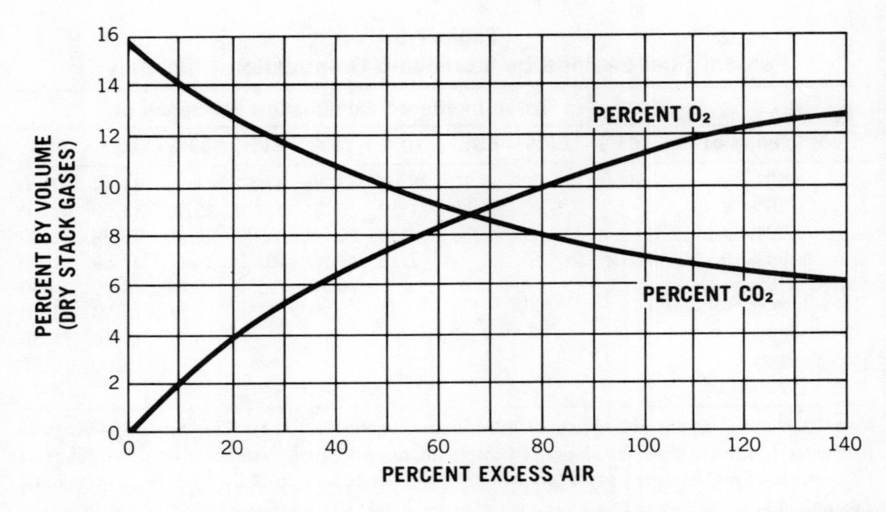

Figure 7-3. Percent O_2 and CO_2 vs. excess air. (Source: *Fundamentals of Boiler Efficiency*, Exxon Company U.S.A., 1976.)

Table 7-4
Percent Oxygen in Flue Gas

	% Excess Air							
Fuel	**0**	**1**	**5**	**10**	**20**	**50**	**100**	**200**
Natural gas	0	0.25	1.18	2.23	4.04	7.83	11.4	14.7
Propane	0	0.23	1.08	2.06	3.76	7.38	10.9	14.3
No. 2 fuel oil	0	0.22	1.06	2.02	3.69	7.29	10.8	14.2
No. 6 fuel oil	0	0.22	1.06	2.01	3.67	7.26	10.8	14.2

Source: Fundamentals of Boiler Efficiency, Exxon Company U.S.A., 1976.

O_2 Measurement

An Orsat, the Bacharach Combustion Tester, or an electronic analyzer can be used to measure the oxygen content. The Bacharach is used here similarly to the way it was used in the CO_2 determination.

CO Measurement

The portable, easy-to-handle instruments from suppliers such as the Bacharach Instrument Co. or Mine Safety Appliances can be used to measure CO.

CO must be determined when natural gas is used as a fuel. The CO determination helps in fine-tuning a boiler, whether it burns gas or No. 2 or No. 6 fuel oil (also known as Bunker *C*, Bunker fuel, or residual oil).

Table 7-5
Boiler Rating Vs. Flue-Gas Analysis

Rating	Gas	No. 2 Fuel Oil	No. 6 Fuel Oil
Excellent	10.0% CO_2	12.8% CO_2	13.8% CO_2
Good	9.0	11.5	13.0
Fair	8.5	10.0	12.5
Poor	< 8.0	< 9.0	<12.0

Note: Assume that CO content is 100–400 ppm maximum.
Source: Cleaver-Brooks.

Problems

The highest possible %CO_2 reading is desired as long as the smoke rating remains low. Low CO_2 readings mean low combustion efficiency (see Tables 7-1 and 7-5). CO readings above 400 ppm also mean low combustion efficiency. In gas boilers, if air is deficient, the danger of an explosion in the stack exists when a fuel-rich flue gas at a high temperature encounters a spark in the presence of an air pocket (air leaking in, for instance). Excess air should be maintained at a minimum, such as 10%–15%, to ensure complete combustion (see Table 3-4).

Causes

Causes for low %CO_2 and high %O_2 are:

1. High draft, causing poor combustion.
2. Excess combustion air.
3. Air leakage.
4. Poor oil atomization.
5. A worn, plugged, or incorrect nozzle.
6. Poor boiler design.
7. Erratic draft-regulator operation.
8. Incorrectly set oil pressure.

Solutions

1. Decrease excess air while maintaining high CO_2 and acceptable smoke.
2. Decrease CO (especially for a gas boiler) by balancing air flow until CO content is 100–400 ppm.
3. Check mechanical parts for proper operation.
4. Check boiler operating instructions for proper oil pressure to ensure optimum oil atomization.
5. Use a fuel additive to maintain nozzle cleanliness.

Smoke Test

Smoke in the flue gas, as measured by the Bacharach Smoke Tester, for example, is an indication of how efficiently fuel is being burned (see Figure 7-1). With this method, a total of 110.39 cu in. of smoke-laden flue products are drawn through a 0.049 sq in. area (0.250 in. or 6.4 mm diameter) of a standard-guide filter paper. The color of the resultant smoke stain on the filter paper is matched to the closest color spot on the standard graduated smoke scale, and the results are interpreted according to a 1 through 9 color scale (the low numbers mean little or no smoke). Using this scale, the following smoke colors are acceptable for fuel oils:

fuel oil No. 2: 3 or less

fuel oil No. 6: 4 or less

Problems

Smoke is primarily unburned carbon particles. These particles coat the walls of the boiler and/or tubes and soon act as insulation, reducing heat-transfer efficiency and increasing fuel consumption. In other words, soot reduces boiler efficiency. Fig. 3-4 shows the effect of soot on fuel consumption.

Causes

Excessive smoke is evidence of incomplete fuel combustion. The causes for smoke include one or more of the following:

1. Improper air delivery.
2. Insufficient draft.
3. Improper fuel viscosity.
4. Oil-pump malfunction.
5. Defective or incorrect fuel nozzle.
6. Improper fuel-to-air ratio.
7. Excessive air leaks.
8. Improper preheat.
9. Improper fuel atomization.
10. Too-short burner "on" periods.
11. Worn, clogged, or incorrect nozzle.
12. Improperly adjusted oil pressure to nozzle.
13. Leaking cut-off valve, allowing after-drip of fuel oil.

Caution: While the absence of smoke may show clean combustion, it does not necessarily mean efficient combustion. CO_2 must also be checked. For one thing, large amounts of excess air may give the impression of a clean combustion when in fact it could be very poor.

Solutions

The possible solutions to a large amount of smoke in the flue gas include:

1. Balancing the CO_2 (with more or less air, depending on the case) until the smoke number is 4 or less.
2. Controlling fuel-oil temperature so that the recommended viscosity will be reached.
3. Treating fuel oil properly to keep the burner clean and improve atomization.
4. Determining any mechanical problems through further testing.

Stack Temperature

To estimate the combustion efficiency, at least two parameters must be known: stack temperature and $\%CO_2$.

The net stack temperature is obtained by subtracting the ambient temperature from the flue-gas temperature.

Problems

A high net stack temperature indicates that heat is being wasted to the atmosphere. Good practice dictates that stack temperature be kept as low as possible without causing cold-end corrosion. Cold-end corrosion is generally caused by the formation of sulfuric acid (H_2SO_4) when SO_3 from the combustion of fuel comes in contact with water vapor. Sulfur in fuel usually burns to SO_2. In the presence of large amounts of excess air, some of this will be converted to SO_3, usually 1%–3%, which combines with water vapor to form sulfuric acid. If the temperature of the air heater, air ducts, and stack falls below the acid dewpoint, the sulfuric acid will condense on the metal surfaces (see Figure 3-2). Also, steam production could be lower than what the design calls for.

Causes

Temperatures higher than 400°F (204°C), or higher than those required by the boiler manufacturer, are caused by:

1. Excessive draft.
2. Dirty, carbon-covered heating surfaces.
3. Poor design of heat exchanger surfaces and lack of sufficient baffling.

4. Undersized furnace.
5. Incorrect or defective combustion chamber.
6. Overfiring of boiler or furnace.
7. Improper adjustment of draft regulator.

Solutions

1. Repair the broken or corroded baffle that could be bypassing the hot gases to the stack.
2. Adjust draft.
3. Decrease soot deposits on heating surfaces by improving fuel combustion efficiency.

Draft

Draft determines the rate at which combustion gas passes through the boiler or furnace. Readings are in terms of 2–3 mm (0.1 inch) water column below atmospheric pressure.

For this test, use a standard draft gauge and adjust it to zero. Check the boiler operating manual for the recommended draft. Test the draft both at the front of the boiler and at the stack.

Problems

Excessive draft can increase the stack temperature and reduce the percentage of CO_2 in the flue gas.

Inadequate draft can result in insufficient combustion air and smoky operation. At perfect draft, the flame bushes out to nearly fill the firebox, giving maximum time for clean burn-out.

Insufficient draft can cause pressure in the combustion chamber, leading to escape of smoke and odor in the boiler room.

Causes

1. Too much or too little air, as the case may be.
2. Poor draft regulation.
3. Stack plugged up with soot or foreign objects.
4. High draft loss through the boiler (means it is partially plugged with deposits).

Solutions

1. Adjust air and/or damper.
2. Inspect stack.
3. Clean the boiler if it is plugged up.

Sulfur Dioxide, SO_2

The higher the sulfur content in the fuel oil, the higher the SO_2 content in the flue gas. This, in turn, may produce more SO_3 and cold-end corrosion.

The same portable instrument used for the CO determination can also be used for SO_2 (see Figure 7-1), but a new scale and a new tube must be used.

Table 2-8 can be used to estimate the sulfur content in the fuel. Of course, oil samples can be sent out to a local laboratory for analysis, or the oil vendor may be able to give you an analysis of the oil delivered to your plant.

Problems

Problems with sulfuric acid corrosion in the stack and local health regulations can be encountered.

Causes

1. Sulfur in fuel.
2. Excess air oxidizing SO_2 into SO_3.
3. SO_3 plus water will condense as H_2SO_4 if the temperature is low enough.

Solutions

1. Keep stack temperatures above sulfuric acid dewpoint, i.e. above 300°F (149°C) (see Figure 3-3).
2. Use only enough excess air to minimize the formation of SO_3.
3. Treat fuel with the proper additive.

Dewpoint Temperature of Sulfuric Acid (DPT)

There are several instruments on the market to determine the DPT. This test is only recommended if high sulfur fuels are being used; if temperatures in the system might be low enough to reach the DPT; and if fuel additives are being evaluated that will react with sulfur oxides before any harm has been done to the equipment.

Problem

Cold-end corrosion is a problem here.

Causes

Cold-end corrosion is caused by flue gas temperatures low enough to reach the DPT of sulfuric acid.

Solution

1. Use temperatures higher than the DPT, say 350°F (177°C).
2. Use the proper fuel additives.
3. Use small amounts of excess air to minimize formation of SO_3.

Fuel-Oil Viscosity

All oils are not exactly the same. To assure the highest combustion efficiency, the oil pressure, viscosity, and preheater temperature must be checked periodically.

As shown in Figure 3-3, an oil viscosity of 100–300 SUS is desirable. To reach the proper viscosity, the preheater temperature must be maintained between 180°F (82°C) and 230°F (110°C) for No. 6 fuel oil. No preheater is needed for No. 2 fuel oil (see Table 2-1).

Viscosities at different temperatures and their effect on the combustion efficiency, i.e. atomization, can be studied using portable viscometers. A certain viscosity will be reached where the combustion efficiency is highest.

Problems

Poor combustion efficiency and high smoke in flue gas are potential problems. If fuel oil is too thick, there may be problems even pumping it.

Causes

Viscosities outside the limits recommended in Figure 3-3 will cause poor oil atomization.

Solution

Maintain preheater temperatures between 180°F (82°C) and 230°F (110°C). If oil viscosity is unknown, preheat fuel oil to 250°F (121°C) and reduce temperature in 10°F steps until the best combustion is obtained.

References

1. Dyer, David F., and Glennon Maples, ''Boiler Efficiency Improvement,'' Boiler Efficiency Institute, Auburn, Alabama, 1981.
2. *Boiler Efficiency Facts,* Cleaver-Brooks, 1981.
3. *Power Handbook,* McGraw-Hill, Inc.

4. Slide rules to estimate combustion efficiency, Bacharach Instrument Company, Dwyer Instruments, Inc., among others.
5. Stoa, T. A., "Calculating Boiler Efficiency and Economics," *Chemical Engineering*, p. 77, July 16, 1979.

8

Fuels and Fuel Treatment

Fuels used in boilers and furnaces include:

1. Gas fuels—natural gas.
2. Liquid fuels—diesel no. 2 and Bunker *C*.
3. Solid fuels—coal, wood chips, garbage, and baggasse.
4. Nuclear. This fuel is not covered here because this book is only concerned with those fuels that burn air (oxygen).

Economics determines the fuel to use in a particular case. Local regulations, which translate into economics, also weigh upon the choice of fuel burned in a particular boiler installation.

Fuel Analysis

The fuel characteristics are expressed as either *ultimate* or *proximate* analyses. The ultimate analysis lists all of the various chemical constituents, as shown in Tables 8-1 and 8-2. The proximate analysis is a common form for coal analysis. It lists fuel constituents by weight percent of:

1. Moisture.
2. Volatile matter—mostly hydrocarbons.
3. Fixed carbon.
4. Ash. Sulfur weight is included in the ash content.

Typical Residual Fuel Oil Analysis

	High Sulfur	Mid-Continent	West Coast	West Coast†	Gulf Coast	Venezuela	Venezuela†
Gravity, °API	17.3	9.3	8.3	6.8	12.0	14.1	15.5
Viscosity SUS @ 38°C (100°F)	3800	1900*	1600*	1100*	1700*	1900*	1200*
Carbon, wt %	86.23	—	—	—	—	—	—
Hydrogen, wt %	11.23	—	—	—	—	—	—
Sulfur, wt %	2.14	1.45	1.3	1.5	2.1	2.37	1.5
Nitrogen, wt %	0.45	—	—	—	—	—	—
Conradson carbon residue, wt %	12.46	—	—	12.6	—	9.6	—
Heptane insoluble, wt %	10.37	—	—	—	—	—	—
Water and sediment, wt %	—	0.04	0.3	—	—	0.10	—
Flash point, °F	—	315	—	265	155	206	265
Pour point, °F	—	30–40	—	5	35	20	—
Ash, wt %	0.08	0.22	0.09	0.06	0.02	0.05	—
Vanadium, ppm	350	117	—	79	—	—	236
Sodium, ppm	15	18	—	—	—	—	—
Lead, ppm	—	<1.0	—	—	—	—	—
Manganese, ppm	—	<1.0	—	—	—	—	—
Magnesium, ppm	—	6.2	—	—	—	—	—
Calcium, ppm	—	34.4	—	78	—	—	—
Nickel, ppm	50	26.9	—	—	—	—	86
Iron, ppm	—	—	—	—	—	—	—

*Viscosity obtained at 122°F using Saybolt Seconds Furol in the original publication. Translated into SUS in this table.

†Shows variations in published data.

Sources: Goldstein, H. L., and C. W. Siegmund, "Particulate Emissions from Residual Fuel Fired Boilers: Influence of Combustion Modification," Paper No. 76-WA/APC-3, *Transactions of the ASME, Journal of Engineering for Power,* December 1976; and Barrett, R. E., *et al,* "Residual Fuel Oil—Water Emulsions," Report PB 189 076, Batelle Memorial Institute, Columbus, Ohio, 43201, January 12, 1970.

Table 8-2
Spectrochemical Analysis of a Residual Oil

Spectrochemical	Result expressed in ppm by wt
Iron	001.
Lead	000.
Copper	000.
Chromium	000.
Aluminum	004.
Nickel	036.
Silver	000.
Tin	000.
Silicon	000.
Boron	000.
Sodium	010.
Phosphorus	000.
Zinc	000.
Calcium	000.
Barium	000.
Magnesium	000.
Titanium	000.
Vanadium	081.

Lab #05432
Unit: Bunker #9001
Sample date: 5/14/81

Viscosity @ 210°F (ASTM D445): 144.2 SUS
Bottom sediment and water (ASTM D96): 0.1% by volume
Specific gravity (ASTM D1298): 1.0505 @ 60°F
Sulfur (ASTM D1552): 1.62% by wt

For fuel oils, several ASTM tests are used, such as those for viscosity and pour point. A summary of the tests is included in Appendix C. Fuel analysis can be expressed on either a *mole* or *weight* basis. Thus:

$$\frac{\text{moles of chemical } A}{\text{total moles of mixture}} = \text{mole fraction of } A$$

and:

$$\frac{\text{weight of chemical } A}{\text{total weight of mixture}} = \text{weight fraction of } A$$

In more technical terms, weight is referred to as *mass*.

A mole of any substance is defined as a weight in pounds (or kilograms) equal to the molecular weight. Thus, for pure carbon with a molecular weight of 12, one mole has:

$$\frac{12 \text{ lbs}}{12 \text{ molecular weight}} = 1 \text{ mole}$$

i.e. 12 lbs of pure carbon are equivalent to one mole of carbon. Likewise, one mole of pure carbon is equivalent to 12 pounds of carbon. A mole fraction is thus:

$$\frac{6 \text{ lbs of carbon}}{12 \text{ molecular weight}} = 0.5 \text{ moles of carbon}$$

For all gases, the volume of a mole is the same at a given temperature and pressure. At 60°F and 14.7 psia, the volume of a ''perfect'' gas is 379 cubic feet.

The *heating value* is reported separately as Btu/lb (or calories per kilogram), or can be estimated from the ultimate analysis and the heating values, shown in Table 8-3.

Table 8-3
Heating Value of Combustibles

			For Gases*		Heating value, Btu	
Element or Compound	**Formula**	**Molecular Weight**	**Weight lb/ft³**	**Volume ft³/lb**	**per lb**	**per ft³***
Carbon	C	12	—	—	14,093	—
Sulfur	S	32	—	—	3,983	—
Hydrogen	H_2	2	0.00531	188	61,100	325
Carbon monoxide	CO	28	0.07385	13.53	4,347	322
Oxygen	O_2	32	0.08440	11.85	—	—
Nitrogen	N_2	28	0.07421	13.48	—	—
Air	—	29†	0.07636	13.10	—	—
Carbon dioxide	CO_2	44	0.1167	8.58	—	—
Water	H_2O	18	—	—	—	—

*At 60°F and 14.7 psia.
†Equivalent molecular weight.
Source: Power Handbook, McGraw-Hill, Inc.

Combustion Reactions

By definition, combustion is the combination of oxygen and a combustible material, resulting in rapid evolution of heat. For a fuel to burn, the following conditions must be present:

1. The fuel must be gasified.
2. The oxygen-fuel mixture must be within flammable range, i.e. not too lean or too rich.
3. The temperature of the air-fuel mixture must be *above* the ignition temperature.

When pure carbon is burned completely with pure oxygen, the following reaction occurs:

$$2\ C + 2\ O_2 \rightarrow 2\ CO_2$$

24 parts + 64 parts → 88 parts (molecular weights)

1 lb + 2.67 lbs → 3.67 lbs + 14,093 Btu (see Table 8-3)

If there is not enough air, the following reaction occurs:

$$2\ C + O_2 \rightarrow 2\ CO$$

24 parts + 32 parts → 56 parts

1 lb + 1.33 lbs → 2.33 lbs + 3950 Btu

With more oxygen, the CO can be burned to CO_2 as follows:

$$2\ CO + O_2 \rightarrow 2\ CO_2$$

56 parts + 32 parts → 88 parts

2.33 lbs + 1.33 lbs → 3.66 lbs + 10,143 Btu (see Table 8-3)

In other words, 3950 Btu + 10,143 Btu = 14,093 Btu when the combustion is complete, as shown in the first reaction.

Hydrogen burns as follows:

$$2\ H_2 + O_2 \rightarrow 2\ H_2O$$

4 parts + 32 parts → 36 parts

1 lb + 8 lbs → 9 lbs + 61,100 Btu

Sulfur burns as follows:

$S + O_2 \rightarrow SO_2$

32 parts + 32 parts → 64 parts

1 lb + 1 lb → 2 lbs + 3983 Btu

These basic combustion reactions are summarized in Table 8-4.

From the preceding, it can be seen that the heating value of a fuel can be estimated from its ultimate analysis and from the values in Table 8-3:

Calculation Examples

Example 8-1. Based on ultimate analysis of a coal found in Table 8-5, estimate the combustion of this coal.

1. *Moisture content.* Assume that all oxygen in the fuel is present as moisture:

$$2\,H_2 + O_2 \rightarrow 2\,H_2O$$

4 lbs + 32 lbs → 36 lbs

Therefore, 1 lb H_2 requires 8 lbs O_2 to yield 9 lbs H_2O:

1 lb + 8 lbs → 9 lbs

Therefore:

$$\left(\frac{0.0880}{8}\right)(1) = 0.011 \text{ lb hydrogen}$$

(i.e. this is the hydrogen tied up as water).

Table 8-4
Air and Products for Perfect Combustion

	For 1 mol of Fuel					For 1 cu ft of Fuel					For 1 lb of Fuel					
	Air		Other Products (than N_2)			Air		Other Products (than N_2)			Air		Other Products (than N_2)			
Fuel	O_2	N_2	CO_2	H_2O	SO_2	O_2	N_2	CO_2	H_2O	SO_2	O_2	N_2	CO_2	H_2O	SO_2	
	1.0	3.76	1.0	—	—	—	—	—	—	—	.0833	.313	.0833	—	—	mols
C	379	1425	379	—	—	—	—	—	—	—	31.6	118.8	31.6	—	—	cu ft
	32.0	105	44.0	—	—	—	—	—	—	—	2.67	8.78	3.67	—	—	pounds
	0.5	1.88	—	1.0	—	.00132	.00496	—	.00264	—	.250	.940	—	0.5	—	mols
H_2*	189.5	712	—	—	—	0.5	1.88	—	—	—	94.8	356	—	—	—	cu ft
	16.0	25.6	—	18	—	.0422	.139	—	.0475	—	8.0	26.3	—	9.0	—	pounds
	1.0	3.76	—	—	1.0	—	—	—	—	—	.0312	.1176	—	—	.0312	mols
S	379	1425	—	—	379	—	—	—	—	—	11.84	44.6	—	—	11.84	cu ft
	32.0	105	—	—	64	—	—	—	—	—	1.0	3.29	—	—	2.0	pounds
	0.5	1.88	1.0	—	—	.00132	.00496	.00264	—	—	.179	.0672	.0357	—	—	mols
CO	189.5	712	379	—	—	0.5	1.88	1.0	—	—	6.77	25.4	13.53	—	—	cu ft
	16.0	52.6	44.0	—	—	.0422	.139	.116	—	—	.571	1.88	1.57	—	—	pounds

*Using 2 as the molecular weight of hydrogen rather than 2.02.
Source: Power Handbook, McGraw-Hill, Inc.

Table 8-5
Coal Analysis for Example 1

Ultimate Analysis	% by weight
Carbon	72.41
Oxygen	8.80
Hydrogen	6.58
Sulfur	1.31
Nitrogen	1.18
Ash	9.72
Total	100.00

Total moisture: 0.0110 + 0.0880 = 0.099 lb, or 9.90%

2. *Free hydrogen.* In other words, this is hydrogen not tied up as water

$$0.0658 - 0.0110 = 0.0548, \text{ or } 5.48\%$$

3. *Heating value.* This is also very simple. Simply multiply the individual constituents that burn by their heating value. Thus:

$$\begin{aligned} C &: 0.7241 \times 14{,}093 = 10{,}204 \text{ Btu/lb} \\ H &: 0.0548 \times 61{,}100 = 3{,}348 \\ S &: 0.0131 \times 3{,}983 = \underline{\quad 52 \quad} \\ &\text{heating value:} \quad 13{,}604 \text{ Btu/lb} \end{aligned}$$

Of course, the heating value can be obtained by actual measurement of the flue gas using a calorimeter.

4. *Theoretical air required.* Estimate the amount of oxygen needed to burn each individual constituent (from Table 8-4):

$$\begin{aligned} C &: 0.7241 \times 2.67 = 1.9330 \text{ lbs of } O_2 \\ H &: 0.0658 \times 8 = 0.5264 \\ S &: 0.0131 \times 1 = \underline{0.0131} \\ &\text{theoretical } O_2 = 2.4725 \text{ lbs per lb of fuel} \end{aligned}$$

Since air has 21% oxygen by volume, which is equivalent to 23% by weight:

$$\text{theoretical air required} = \frac{2.4725}{0.23} = 10.75 \text{ lbs of air per lb of fuel}$$

From the theoretical air (stoichiometric air), the excess air can be estimated. For instance, if 20% excess air is recommended:

$$1.20 \times 10.75 = 12.9 \text{ lbs of total air}$$

and

$$1.20 \times 2.4725 = 2.697 \text{ lbs of total } O_2$$

5. *Combustion products*. Since all calculations have been based on 1 lb of coal, the total weight of the combustion products is:

$$1 \text{ lb coal} + 12.9 \text{ lbs air} = 13.9 \text{ lbs per lb of fuel}$$

To be more exact, the ash must be deducted:

$$1.39 - 0.1 = 13.8 \text{ lbs combustion products}$$

Assuming moisture brought in by air is negligible, the total water appearing in the flue gas will be:

$$0.0658 \times 9 = 0.5922 \text{ lb per lb of fuel}$$

Therefore, the dry-gas composition is:

$$13.8 - 0.5922 = 13.21 \text{ lbs, made up of (see Table 8-4):}$$

	lbs		%
CO_2:	0.7241×3.67	= 2.657	20.3
SO_2:	0.0131×2	= 0.026	0.2
O_2:	0.2×2.4725	= 0.494	3.8
N_2:	$0.011 + (12.9 - 2.967)$	= 9.922	75.7
total		13.099	100.0

In this example, it was assumed that CO was negligible, perhaps in the range of 100–400 ppm.

There are nomograms and specially constructed slide rules available to facilitate these calculations.

Example 8-2. For natural gas, the calculations would be similar. The combustion heat for components other than those shown for coal would have to be taken into account—mainly methane (CH_4) and other hydrocarbons present in various amounts.

Example 8-3. Many times it is impractical or impossible to obtain the ultimate analysis of the fuel, flue gas, and ash. This is especially true for smaller plants and in cases where the fuel supply, such as Fuel No. 6 (Bunker *C*), changes with each delivery. In such cases, estimates can be made within 2% using a simple flue-gas analysis and the known heating value of the fuel. Specially prepared charts and scales have been devised to perform those estimates (i.e. air and excess air required, water produced per lb of fuel, etc.). Typical properties of the various fuels can be found in Tables 2-1, 2-5, and 2-6.

Simplified Methods to Estimate Liquid Fuel Properties

Heating Value

First obtain the specific gravity of the fuel. Then use Figure 8-1 to estimate the heating value.

Type of Fuel (Fuel No.)

From the specific gravity obtained previously (Figure 8-1), get the API value. The fuel no. can be estimated from Table 8-6.

Viscosity

With the Shell cup viscometer shown in Figure 8-2, viscosity can be obtained in less than five minutes using the following method: Submerge the cup in the fluid to be measured until it is full; then hold it above the fluid surface until it is empty. Record the time required for the cup to empty, using a stop watch. The viscosity can then be read from a chart. The viscosity value will tell at what temperature the fuel oil preheater must be set to yield optimum atomization.

Sulfur Content of Fuel

Sulfur content can be estimated, when necessary, from an analysis of the SO_2 content in the flue gas, using a portable instrument such as the Bacharach apparatus shown in Figure 7-1. From Table 2-2, the sulfur content in the fuel can be determined.

Effect of Water on Fuel Viscosity

Water is present in fuels because of condensation, leaks, bacterial action, etc. The presence of emulsified water increases the viscosity of the fuel.

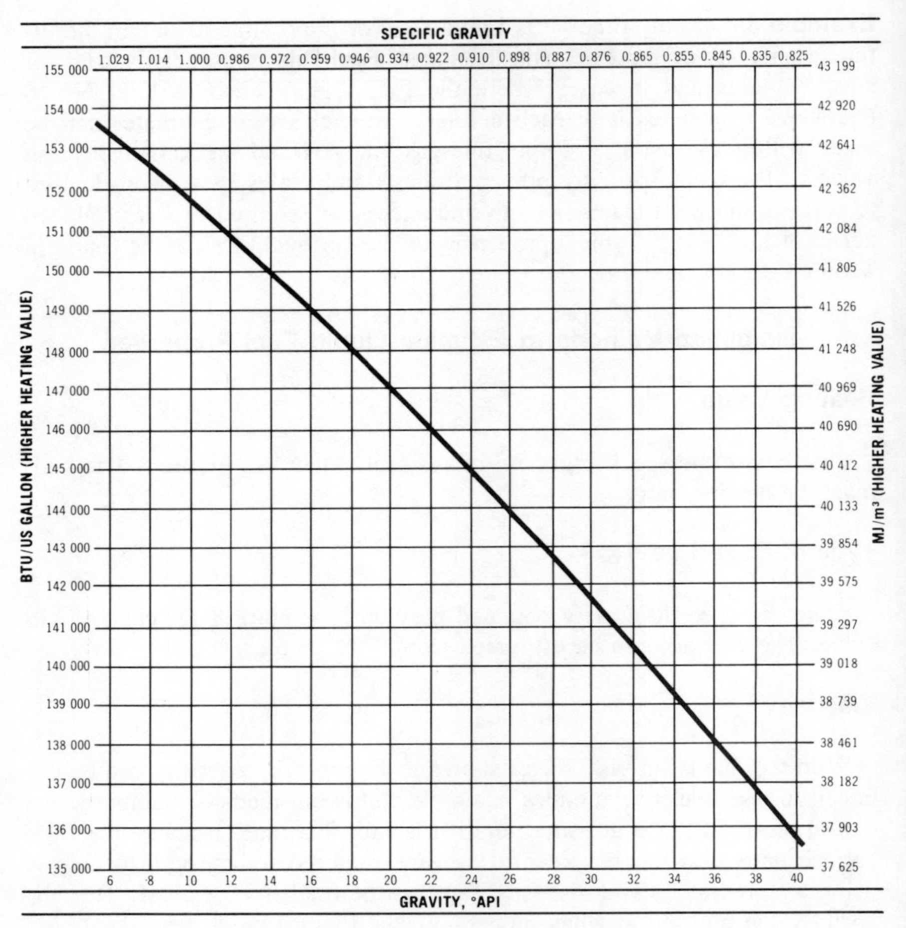

Figure 8-1. Specific gravity of fuel oil vs. heat content. (Source: *Fundamentals of Boiler Efficiency*, Exxon Company U.S.A., 1976.)

Table 8-7 lists some published data giving fuel viscosities at different temperatures and water contents.[1,2] Using these data, the possible effect of emulsified water on the fuel atomization and combustion efficiency can be estimated. For instance, 10% water emulsified in the fuel will increase the viscosity by 50% when compared to the same fuel with little or no water.

Certain fluids, such as water emulsified in fuel, are known as non-Newtonian fluids because at higher shear rates their apparent viscosity decreases.[3] This can rapidly be observed using a Brookfield viscometer at different rpm's.[4]

Table 8-6
API Gravity Range of Fuel Oils

Type	API Gravity Range
#2	29° to 39°
#4	24° to 28°
#5	16° to 22°
#6	6° to 15°

Note: API gravity $= \dfrac{141.5}{\text{specific gravity @ 16°C}} - 131.5$

Source: Dyer, D. F., and G. Maples, "Boiler Efficiency Improvement," Boiler Efficiency Institute, P.O. Box 2255, Auburn, Alabama, 36830, 1981.

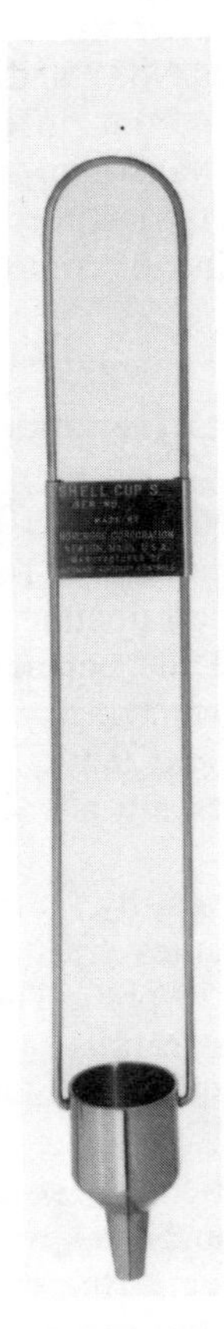

Figure 8-2. Fuel-oil viscosity cup, developed by Shell Chemical Company and manufactured and sold by Norcross Corporation. (Courtesy Norcross Corporation.)

Table 8-7
Fuel Viscosity As a Function of % Water and Temperature[1,2]

	Viscosity, Centipoises	
% Water	@60°C	@120°C
40	950	75
20	400	42
10	300	32
0	200	22

Typical Fuels

The commercial standards for fuel oils are shown in Appendix C. Table 8-8 summarizes the typical characteristics of the different fuels. Table 8-9 presents a typical analysis of fuels No. 2 and No. 6, and Table 8-10 presents an actual analysis of fuel No. 6 from different crudes around the world. Notice the wide variation in the vanadium and sulfur contents.[5]

Table 8-11 shows the range of constituents in residual fuel oil and ash. Table 8-12 gives the melting points of components typically found in residual oil ash.

Fuel Treatment

Fuels and oils that come on the market are already, at least partially, treated with additives by the oil company. This is often ignored by the user. Perhaps the best known fuel and oil treatments are the tetraethyl lead added to gasolines (leaded) to increase the octane number, the specially treated winter fuels with lower pour-point temperatures, and the motor oils with detergents and multi-viscosity agents that contain viscosity index improvers (polymers). Many other additives are added to the fuels and oils from a refinery.

For instance, under fuel additives, by far the largest volume of protective products added by the oil companies are the antioxidants (such as phenylenediamines) in the range of 3–30 ppm. They are added to stabilize color and minimize gum formation that results from the oxidation of the olefins and diolefins that are produced during catalytic cracking.[6]

Metal deactivators (such as disalicylidene amines) are also added in the range of 9–15 ppm to neutralize copper so that gums will not form. Corrosion inhibitors (alkyl-amine phosphates and dodecenyl succinic acid, for instance) are added by the refiner in the range of 15–24 ppm to prevent pipeline and tank leaks and the formation of rust which can plug filters and accelerate engine wear.

Many other additives are also added in lesser or greater amounts: dyes, biocides, conductivity improvers (antistatic materials), dispersants (alkyl succinimides), detergents (amine-type), pour-point depressants (polymeric materials), combustion catalysts (organo-metallic compounds), and cetane and octane improvers (mixed primary nitrates and an organo-manganese compound, respectively).[6]

Since all fuel oils are not exactly the same, fuels must be treated individually to solve specific problems in specific boiler applications.

Today, many fuel users are increasing the reserve capacity of their tank farm as insurance against a sudden cut in fuel supplies and extreme fluctuations in prices. Because of the high prices for petroleum products, sophisticated catalytic cracking methods are used to draw as much distillate as possible from each barrel of crude oil. This often produces a less-stable product and a lower-quality No. 6 fuel oil. Fuel quality depends on refinery methods, crude source, handling, and storage (temperature and exposure to air). Therefore, the use of chemical additives has found wider applications as fuel-oil prices go up and fuel-oil quality goes down.

Additives are used to:

1. Improve combustion to improve efficiency and lower fuel costs.
2. Lower the flue-gas temperatures below sulfuric-acid dewpoint by neutralizing the acid. Lower temperatures mean higher boiler efficiencies.
3. Reduce stack emissions (to avoid violating the law and being sued by neighbors).
4. Reduce boiler fouling, thereby improving boiler efficiency.
5. Reduce corrosion, thereby reducing maintenance and forced boiler shutdowns.
6. Reduce sludge, gum, and microorganism build-up to maintain high boiler efficiencies.

A list of commonly used additives has been published.[7] Unfortunately, those lists are of limited use to the person responsible for boiler operations because:

1. The chemical formulations are best understood by chemists. For instance, few boiler engineers would like to struggle to figure out how alkyl esters could help them solve corrosion problems.
2. To be effective, additives must be formulated to give maximum protection to the user. This may involve more than one chemical mixed in a carrier, or the addition of a small, particle-size inorganic additive. This can not be done by a company not equipped for it. Specialists and additive suppliers can do it; it is their only function and business.

(text continued on page 91)

Table 8-8
National Bureau of Standards Specifications for Fuel Oils[a]

Grade of fuel oil[b]		Flash point, °F	Pour point, °F	Water and sediment, %	Carbon residue on 10% residuum, %	Ash, %	Distillation temperatures, °F			Viscosity								Gravity, °API
										Saybolt				Kinematic centistokes at				
										Universal at 100°F		Furol at 122°F		100°F		122°F		
Number	Description						10% point	90% point	End point									
		min	max	max	max	max	max	max	max	max	min	max	min	max	min	max	min	min
1	Distillate oil intended for vaporizing pot-type burners and other burners requiring this grade.[c]	100 or legal	0	Trace	0.15		420		625					2.2	1.4			35
2	Distillate oil for general purpose domestic heating for use in burners not requiring No. 1.	100 or legal	[d]20	0.10	0.35		(e)	675		40				(4.3)				26
4	Oil for burner installations not equipped with preheating facilities.	130 or legal	20	.50		0.10				125	45			(26.4)	(5.8)			
5	Residual-type oil for burner installations equipped with preheating facilities.	130 or legal		1.00		.10					150	40			(32.1)	(81)		

Table 8-8 continued

Grade of fuel oil[b]		Flash point, °F	Pour point, °F	Water and sediment, %	Carbon residue on 10% residuum, %	Ash, %	Distillation temperatures, °F			Viscosity				Gravity, °API
										Saybolt		Kinematic centistokes at		
Number	Description						10% point	90% point	End point	Universal at 100°F	Furol at 122°F	100°F	122°F	
6	Oil for use in burners equipped with preheaters permitting a high-viscosity fuel.	150 or legal		[f]2.00						300	45	(638)	(92)	

[a]Recognizing the necessity for low-sulfur fuel oils used in connection with heat treatment, nonferrous metal, glass and ceramic furnaces, and other special uses, a sulfur requirement may be specified in accordance with the following table:

	Grade of fuel oil	Sulfur, max, percent
24	No. 1	0.5
25	No. 2	1.0
26	Nos. 4, 5, and 6	No limit

23

Other sulfur limits may be specified only by mutual agreement between the buyer and seller.

[b]It is the intent of these classifications that failure to meet any requirement of a given grade does not automatically place an oil in the next lower grade unless in fact it meets all requirements of the lower grade.

[c]No. 1 oil shall be tested for corrosion in accordance with standards for 3 hours at 122°F. The exposed copper strip shall show no gray or black deposit.

[d]Lower or higher pour points may be specified whenever required by conditions of storage or use. However, these specifications shall not require a pour point lower than 0°F under any conditions.

[e]The 10-percent point may be specified at 440°F. Maximum for use in other than atomizing burners.

[f]The amount of water by distillation plus the sediment by extraction shall not exceed 2.00 percent. The amount of sediment by extraction shall not exceed 0.50 percent. A deduction in quantity shall be made for all water and sediment in excess of 1.0 percent.

Table 8-9
Range of Fuel-Oil Analyses

	No. 2	No. 6
Water and sediment, Vol. %	0–0.1	0.05–2.0
Sulfur, wt. %	0.5–1.0	0.7–3.5
Ash, wt. %	0.0–0.1	0.01–0.5
Specific gravity	0.887–0.825	1.022–0.922
Pour point, °F	0 to −40	+15 to +85
Viscosity		
centistokes at 100°F	1.9–3.0	260–750
Heating value		
Btu per lb	19,170–19,750	17,410–18,990
Btu per gal	138,500	148,000

Table 8-10
Average Bunker Fuel Analysis from Different Sources

	Sulfur %	Vanadium ppm	API Gravity
USA: East Coast (and Canada)	2.36	249	13.60
USA: Gulf Coast	2.71	113	11.69
USA: West Coast	1.76	95	12.62
Central America and Caribbean	2.63	143	13.29
South America	2.68	182	14.06
Northern Europe	3.23	70	14.30
Mediterranean	3.13	110	13.05
Asia and Australia	3.22	33	16.73

Note: Individual sources may differ widely from these averages. For instance, bunker oils from Aruba may have as high as 336 ppm vanadium.
Source: Delaney, Rachelle, ''Future Bunker Fuel,'' *Marine Engineering/Log,* November 1981, p. 58.

Table 8-11
Range in Composition of Constituents in Residual fuel Oil and Ash

Constituent	Fuel (ppm)	Ash (%)
SiO_2	6.0– 86.0	0.6– 8.6
Al_2O_3	3.0– 76.0	0.3– 7.6
Fe_2O_3	0.9– 57.0	0.1– 5.7
CaO	1.0– 1.4	0.1– 1.0
MgO	1.0– 1.7	0.1– 0.2
Na_2O	5.0– 35.0	0.5– 3.5
K_2O	0.2– 1.2	0 – 0.1
V_2O_5	14.0–740.0	0.1–74.0
NiO	1.3– 25.0	0.1– 2.5

Table 8-12
Ash Fusion Temperatures of Slag-Forming Compounds

Chemical Compound	Ash Fusion Temperature, °F
Vanadium pentoxide (V_2O_5)	1274
Sodium sulfate (Na_2SO_4)	1630
Nickel sulfate ($NiSO_4$)	1545
Sodium metavanadate ($Na_2O \cdot V_2O_5$)	1165
Sodium pyrovanadate ($2Na_2O \cdot V_2O_5$)	1210
Sodium orthovanadate ($3Na_2O \cdot V_2O_5$)	1590
Nickel orthovanadate ($3NiO \cdot V_2O_5$)	1650
Sodium vanadyl vanadate ($Na_2O \cdot V_2O_4 \cdot 5V_2O_5$)	1155
Sodium iron trisulfate ($2Na_3Fe[SO_4]_3$)	1150

Source: Ferraro, R. M., "Using Fuel Additives to Increase Boiler Efficiency," *Plant Engineering,* March 17, 1977, p. 159; and Reid, W. T., *External Corrosion and Deposits,* American Elsevier Publishing Company, New York, 1971.

The experiences gained by the additive supplier are jealously guarded. In other words, you need their expertise to know how to best apply the additives. You need the counseling of the supplier who usually does not charge for this service as long as you buy his product.

Water has been mentioned as an additive to improve boiler operations.[2] One leading boiler manufacturer claims that 3%–4% water emulsified in the fuel will significantly reduce solids formation. However, addition of water means further fuel consumption due to the water heat of vaporization.

There are two main types of fuel-oil treatments: precombustion (Figures 8-3 and 8-4) and during and after combustion treatment (Figure 8-5). To better grasp the importance of proper fuel treatment, a recent history of how the quality of bunker fuel has deteriorated follows.

Quality and Availability of Bunker Fuel

Directly or indirectly, the quality and availability of bunker fuel affects all of us. For one thing, before the oil embargo, fuel was estimated to be less than 10% of the total operating costs of a ship operation; now fuel bills account for 50%–70% of the total operating costs.[5] A barrel of Bunker fuel could be bought for $2.00 in 1970; now it costs around $27.00. If your boiler or process furnace operates with Bunker fuel, you are more directly affected by the cost of the fuel and the need to extract every possible Btu from it and use it properly, i.e. increase the operating efficiency of your boiler or process furnace. Table 8-13 compares the quality of the bunker fuels between 1976 and 1980.[5]

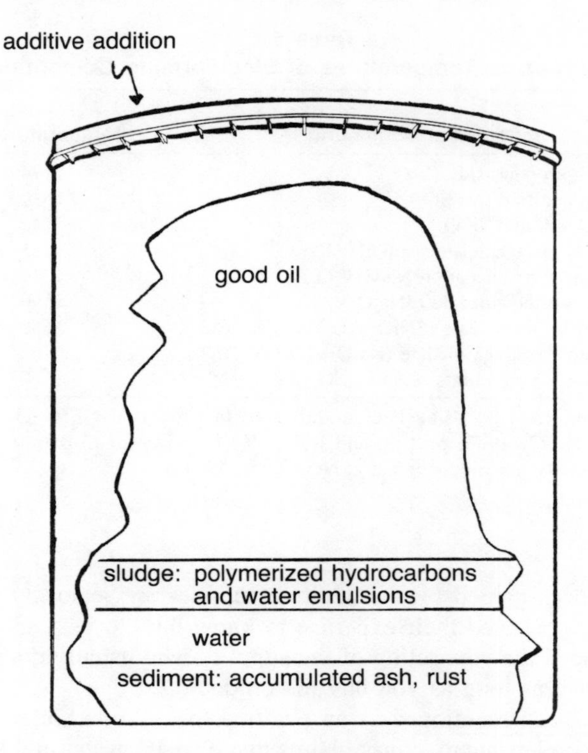

Figure 8-3. Contaminants in an oil storage tank.

The most dramatic degradation has shown up in vanadium levels, which increased from 42 to 140 ppm; ash content increased from 0.03 to 13.65; nickel content doubled to 37 ppm. All this indicates that a lower quality of fuel is now available—and the indications are it will get worse. Here is why:

Before the energy crisis, refineries produced 60% distillate and 40% residual fuel. Now the lighter-end and premium products account for 90% of production, while the residual fuel has shrunk to 10% of the barrel of crude oil because of the higher demand and more profitable diesel No. 2—which in turn causes a rise in price of the Bunker fuel, in a sort of vicious circle, until some sort of price balance is reached.

Since many suppliers do not know the origin of the fuel nor have the facilities to check the quality of the fuel, specifications mean little. And since economics are on the side of the supplier, the consumer must accept whatever he gets; all he can do is to determine the basic quality of the fuel (either in-house or outside) and treat his fuel accordingly to increase equipment efficiency.

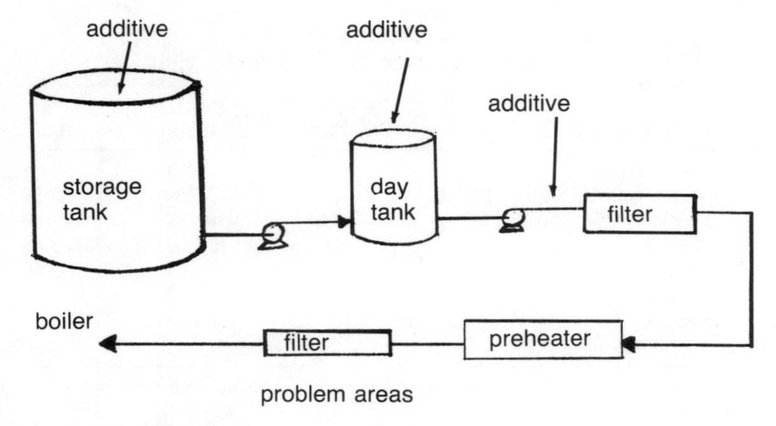

corrosion: in fuel tanks and system from the fuel and entrained salt and water

plugging: filters, preheaters, and burner from sludge, corrosion products, and sediment

Figure 8-4. Problem areas in the fuel-oil system: before combustion.

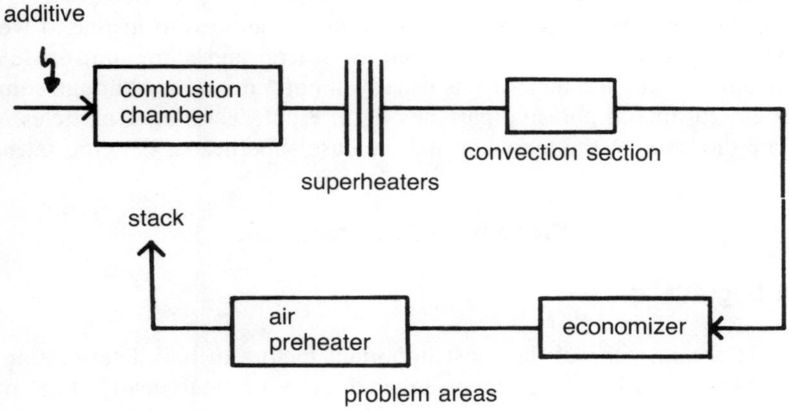

additive

combustion chamber

superheaters

convection section

stack

air preheater

economizer

problem areas

deposits: soot in combustion chamber

slag: ash in combustion chamber, superheaters, convection section; may cause plugging and fouling

corrosion: in economizer by sulfuric acid ($SO_3 + H_2O$)

emissions: acid smut (H_2SO_4 + soot) out of the stack

Figure 8-5. Problem areas in the fuel-oil system: during and after combustion.

Table 8-13
Quality of Bunker Fuel: 1976 Vs. 1980

Fuel Analysis	Industry Standard	1976	1980
API gravity	—	15.1	13.65
Viscosity (SUS @ 100°F)	—	3,264	3,982
Heating value, Btu/lb	18,500	18,780	18,408
Ash, %	—	0.03	13.65
Sulfur, %	1.20	2.97	2.60
Nickel	—	18	37
Sodium	—	15	13
Vanadium	—	42	140

Source: Delaney, Rachelle, "Future Bunker Fuel," *Marine Engineering/Log,* November 1981, p. 58.

Refiners "upgrade" bunker fuel by adding diesel oils and by blending bunkers from different sources. These techniques offer great danger: a residual fuel rich in asphalts will not blend well with a fuel rich in linear and branched paraffins. This incompatibility can lead to sludge formation and fuel-tank stratification.

Another problem facing the consumer is the possible presence of catalytic fines in the fuel. These heavy abrasive particles can lead to increased wear of the equipment used. They are a mix of silicon and aluminum oxide or aluminum silicate, and their size is usually about 5 microns, although sometimes 20–25-micron abrasive particles are present. The larger particles accelerate the wear of the atomizer and increase superheater slagging rates.[5]

Precombustion Treatment

Fuel Degradation

1. *Oxidation.* One of the most important factors in fuel deterioration is oxidation, which may occur either directly or catalytically. Fuel oxidation results in increased particulate matter suspended in the fuel, more gum, water formation, and fuel discoloration.[8,9]

   In direct oxidation, the oxygen from the air chemically combines with the hydrocarbon molecules that make up the fuel. When metals (such as iron, copper, and nickel) are dissolved or suspended in the fuel, they act as catalysts to increase oxidation.
2. *Microbial growth.* Bacteria, fungi, and other organisms grow at the water-oil interface. These microorganisms excrete acids and water, which cause sludge formation and filter plugging.

3. *Corrosion*. The corrosion of the storage tanks and fuel system, caused by the presence of acids, feeds metals to the fuel that, in turn, increase the rate of fuel oxidation. Corrosion also occurs because of the sodium chloride (Na Cl) contained in the water. Corrosion destroys the storage tanks, and its products can plug the fuel-oil filters.

Therefore, oxidation, microbial growth, and corrosion together lower the quality of fuel unless it is treated adequately before significant harm has resulted.

Figures 8-3, 8-4, and 8-5 depict the problems commonly encountered in fuel-oil systems.

Methods of Fuel Analysis

To determine the fuel quality and the proper treatment, a fuel analysis is needed. Appendix C includes the standard methods to use in fuel analysis. Some quick and inexpensive tests are also used to identify fuel quality.[8]

1. *Suspended particulate matter*. The objective is to determine how far the fuel has deteriorated. The sample is filtered and solids are weighed.
2. *Accelerated aging test*. This test determines the tendency of a fuel to oxidize. A fuel sample is filtered and heated to 300°F in a special fluid bath for 90 minutes. The sample is filtered, and the solids formed during the accelerated test weighed. A fuel treated with the right additives will show significantly less particulate matter and gums than the untreated fuel.
3. *Toluene extract of organic materials*. The oil is extracted with the toluene solvent; the oil-free extract is dyed with an indicator for microscopic inspection. The toluene separates the oil from microorganisms, asphaltenes, tars, and other organic contaminants, which are left behind and can thus be analyzed.
4. *Water in fuels*. The kit for performing this test is shown in Figure 8-6. The actual test is performed in a few minutes, with the final reading obtained after 25 minutes of reaction time.[10] This kit gives the water content of a fuel or lubricating oil in the range of 0%–10%.
5. *Pour-point test*. This kit gives a quick indication of the fuel-oil pour point, in the range of −5°C to +60°C. Generally, the higher the pour point, the higher the wax content. Once a fuel becomes solid, it may be impossible to redissolve (see Figure 8-6).
6. *Specific-gravity test*. The kit shown in Figure 8-6 is a thermostatically controlled heating unit that determines the specific gravity of fuel and lubricating oils within the range of 0.89–1.05. High specific gravities make water separation more difficult.

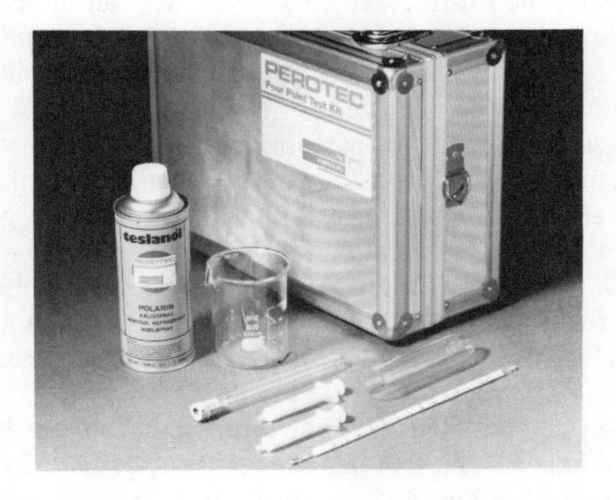

Figure 8-6. Individual test kits to measure water in fuels, pour point, specific gravity of fuels. (Courtesy The Perolin Company.)

7. *Compatibility test*. This is a quick test to assess the compatibility of fuels before blending different source fuels. The Perolin Company sells a special kit to carry out this test in minutes. Incompatibility can result in sludge formation as high as 6%.
8. *Salt test*. This is another kit from Perolin Company to test salt content up to 18,000 ppm. The presence of salt causes corrosion and combustion problems.[10]

Fuel Additives

Additives will normally contain a combination of the following compounds:

1. *Solvents*. These are used as carriers for the fuel treatment and at the same time solubilize a portion of the sludge particles.
2. *Dispersants*. These prevent particulate matter and water from agglomerating.
3. *Antioxidants*. These act to minimize the oxidation of the olefinic hydrocarbons. The role of the antioxidant is as follows:

   During the oxidation process, oxygen combines with another molecule or an atom in the fuel tank. The combination of hydrogen and oxygen yields a peroxide molecule (HOO), which in turn seeks other peroxide molecules to form water and oxygen ($4HOO \rightarrow 2H_2O + O_2$). The incomplete fuel molecule left behind is a free radical, which stabilizes itself by combining with more oxygen to form peroxy radicals (ROO). The peroxy radicals in turn combine with other fuel molecules to form hydroperoxides (ROOH), which yield additional free radicals. As long as there is oxygen present, this chain reaction will degrade the fuel, forming water and particulate matter that settle in the bottom of the tank. The corrosion inhibitor reacts with the peroxy radicals, effectively stopping the chain reaction. Obviously, the inhibitor is consumed during the process and this must be taken into account in the treatment of the fuel.
4. *Corrosion inhibitors*. Filming amines inhibit corrosion by placing a protective coating on metal surfaces that neutralizes the acids.
5. *Water emulsifiers*. These suspend the water in the fuel oil to keep the corrosive salts in the water from contacting the metal surfaces. The water that accumulates in the bottom of the tank is best drained rather than emulsified. Maintain full tanks to minimize water condensation.
6. *Microbiocides*. These destroy and prevent the growth of microorganisms.
7. *Pour-point depressants*. In colder climates, storage tanks must either be heated or pour-point depressants added. Otherwise, the fuel-oil flow

may stop completely. Depending on the supplier, additives can be added through the top of the tank while transferring the fuel from the tank car to the storage tank, or through recirculating lines.

During and After Combustion Treatment

Scale Deposits and Corrosion

As shown in Figure 8-5, problems become more acute during and after fuel combustion. Deposits can occur in high-temperature zones on waterwall tubes, the superheater, and the convection pass; and from cold-end corrosion on the economizer, air heater, fans, and stack.

In the high-temperature zone, scale is made up of low-melting point compounds of sodium, vanadium, sulfur, and oxygen (see Table 8-12). Once these compounds are deposited on the hot surfaces, they are very difficult to remove. Soot blowers have no appreciable effect on cleaning these deposits. The high-temperature corrosion reactions shown in Table 8-14 occur during combustion.[11]

Analysis

Fuel analysis must include an analysis for vanadium (V), sodium (Na), and sulfur (S) to determine the appropriate fuel treatment.

Fuel Additives[11,12,13,14,15,16]

The fuel treatment for V, Na, and S is based on the following data:

1. In the absence of magnesium (Mg), the scale formed has a relatively low melting point.
2. In the presence of Mg, compounds with significantly higher melting points are formed. These deposits are friable and easily removed.
3. When a Mg compound passes through a flame, the compound dissociates and recombines with oxygen to form magnesium oxide (MgO).
4. When MgO is also added to the flue gas, inert magnesium sulfate ($MgSO_4$) is formed in the presence of SO_3, preventing sulfuric-acid corrosion (cold-end corrosion).

To minimize operating costs, the fuel treatment must be used in conjunction with the following optimum operating conditions.

Use higher quality fuel. Avoid fuels with high V content; these are the most troublesome. Use fuels with a lower S content whenever possible.

Table 8-14
High-Temperature Corrosion Reactions[11]

Fuels Containing Sodium and Sulfur

(1) sodium oxide + sulfur trioxide → sodium sulfate

$Na_2O + SO_3 \rightarrow Na_2SO_4$

(2) sodium sulfate + ferric oxide + sulfur trioxide → sodium iron trisulfate, melting point, 1150°F

$3Na_2SO_4 + Fe_2O_3 + SO_3 \rightarrow 2Na_3Fe(SO_4)_3$

(3) iron + sodium iron trisulfate → ferrosoferric oxide + ferrous sulfide + sodium sulfate

$10\ Fe + 2Na_3Fe(SO_4)_3 \rightarrow 3Fe_3O_4 + FeS + 3Na_2SO_4$

Fuel Oil Containing Vanadium Oxides

(a) sodium oxide + sulfur trioxide → sodium sulfate, melting point, 1630°F

$Na_2O + SO_3 \rightarrow Na_2SO_4$

(b) sodium sulfate + ferric oxide + sulfur trioxide → sodium iron trisulfate, melting point, 1100–1300°F

$3Na_2SO_4 + Fe_2O_3 + 3SO_3 \rightarrow 2Na_3\ Fe(SO_4)_3$

(c) sodium iron trisulfate + iron → ferrous sulfide + ferrosoferric oxide + ferric oxide + sulfur dioxide + sodium sulfate

$4Na_3Fe(SO_4)_3 + 12Fe \rightarrow 3FeS + 3Fe_3O_4 + Fe_2O_3 + 3SO_2 + 6Na_2SO_4$

(d) sodium sulfate + vanadium pentoxide → sodium oxide-vanadium pentoxide complex + sulfur trioxide, melting point, 900–1200°F

$Na_2SO_4 + V_2O_5 \rightarrow SO_3 + Na_2O \cdot V_2O_5$ complex

(e) sodium oxide-vanadium pentoxide complex + iron → sodium oxide-vanadium tetroxide-vanadium pentoxide complex + ferrous oxide, melting point 1155°F

$Na_2O \cdot 6V_2O_5 + Fe \rightarrow Na_2O \cdot V_2O_4 \cdot 5V_2O_5 + FeO$

Since the energy crunch, the user has had little choice in fuel selection, except for price. Fuels with a lower S content draw higher prices. However, since the SO_3 content of the flue gas has little to do with the S content of the fuel, higher prices are usually not justified. Figures 8-7 and 8-8 show that even a threefold reduction in sulfur content in the fuel reduces SO_3 concentration only by 25%.

The SO_3 concentration is reduced dramatically only when oxygen concentration is reduced below 1.5%. Since all fossil fuels contain sulfur in

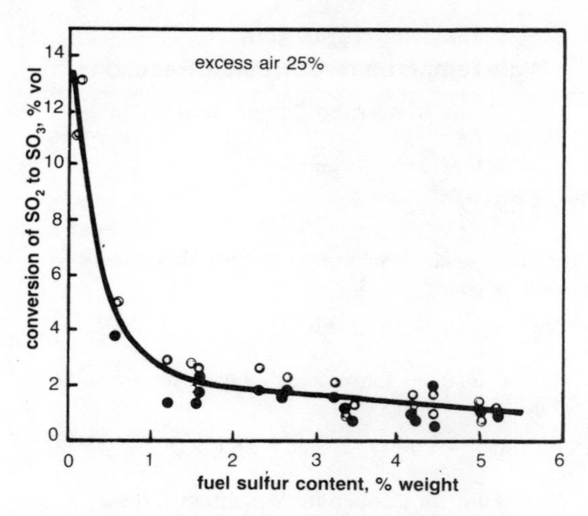

Figure 8-7. Conversion of SO_2 to SO_3 vs. sulfur content in fuel.[17,18]

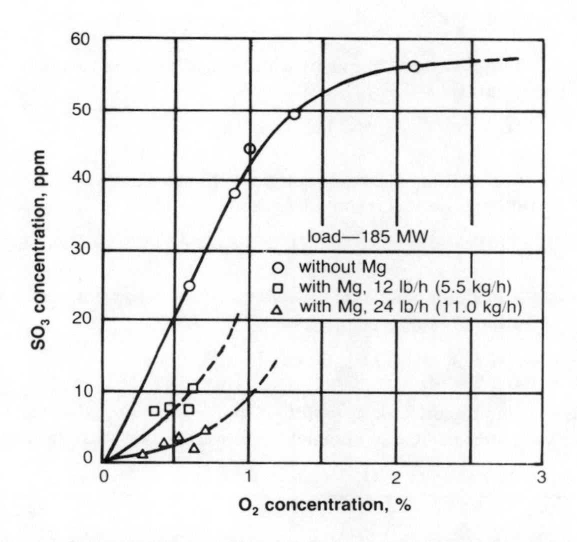

Figure 8-8. SO_3 concentration vs. O_2 concentration.[17,18]

greater or lesser amounts, the 1%–3% conversion of SO_2 into SO_3 usually yields 10–50 ppm SO_3 in flue gas. Therefore, be very careful selecting fuels with higher prices when seeking higher quality fuels.[17,18]

Modify combustion condition. Lower excess air also yields lower SO_2 and SO_3 conversions, as shown in Figure 8-8. Because many burner designs result in poor atomization, it is often difficult to operate a boiler at low excess air while maintaining acceptable smoke levels in the stack. Below 5% excess air, oxygen analyzers are necessary. These should minimize cold-end corrosion caused by sulfuric-acid condensation.

Adjust gas temperatures. To further decrease the amount of condensing sulfuric acid, use stack temperatures higher than the acid dewpoint (Figure 2-3). This can be accomplished by altering a combination of mechanical and combustion variables:

1. Insulate ducts, fan housings, etc.
2. Minimize or avoid low-load operations.
3. Use higher fuel-feed rates if necessary to maintain a stack temperature higher than the acid dewpoint.

Reduce leakage. Air leakage may significantly lower the gas temperature below the acid dewpoint. Air may leak in through the boiler casing, ducts, air preheaters, fan housings, expansion joints, etc.

Minimize precombustion problems. Any condition that requires excess air to minimize smoke in the gas will necessarily contribute to more SO_3 formation and more opportunities for cold-end corrosion. Thus, poor fuel atomization will produce smoke requiring more excess air to reach an acceptable smoke number.

The preceding items will minimize sulfuric-acid formation. However, when vanadium is also present, treating the fuels with MgO will also reduce the harmful effects of sulfur. In fact, it has been found that low excess air coupled with the MgO action can reduce 90% of the SO_3 concentration. Therefore, treating fuels with MgO is the best way to prevent the harmful effects of both vanadium and sulfur. The magnesium-based additives function in the following manner.[11–19]

1. *Inhibit catalysis.* The catalytic properties of the oxides of vanadium, iron, and manganese have been known for years. The MgO inhibits the catalysis of SO_2 to SO_3, and reacts with the iron oxide mainly in the superheater of a boiler, poisoning its catalytic properties. Manganese is sometimes one of the additive components (both as a vanadium-compound neutralizer and as a combustion catalyst), and can cause cold-end corrosion problems unless it is neutralized with MgO.
2. *Neutralize acid.* MgO will react with sulfuric acid as soon as it is formed, yielding Mg SO_4. The latter is a dry and free-flowing powder under typical operating conditions.

Table 8-15
Ash Fusion Temperatures of Nonslagging Compounds Formed by Using Fuel Additives and Low Excess Air

Chemical Compound	Ash Fusion Temperature, °F
Magnesium oxide (MgO)	5072
Aluminum oxide (Al_2O_3)	3686
Calcium oxide (CaO)	4658
Magnesium aluminate ($MgAl_2O_4$)	3875
Manganese oxide (MnO_2)	3000
Nickel oxide (NiO)	3794
Vanadium tetroxide (V_2O_4)	3572
Magnesium vanadates ($MgO \cdot V_2O_5$)	1965–2270
Sodium magnesium trisulfate ($Na_2Mg_2(SO_4]_3)$)	2059

Source: Ferraro, R. M., "Using Fuel Additives to Increase Boiler Efficiency," *Plant Engineering,* March 17, 1977, p. 159.

Introduction of Magnesium-Based Additives

The efficacy of MgO as an additive to control vanadium and sulfur deposits and corrosion depends on:

1. Point of addition.
2. Chemical form.
3. Particle size.

Boiler operating conditions and fuel quality also have an influence on these chemical reactions, as discussed previously.

Point of addition.[19] The three points of addition of MgO are:

1. *With the fuel or in the convection passes.* This is the simplest and most common technique. It has some drawbacks: burner-tip erosion is increased, and it is less efficient in neutralizing SO_3. In some units, it may not be practical to add the MgO along with the fuel, especially if the boiler is equipped with soot blowers. To keep the hot-gas passes open, soot blowers are required to remove the Mg compounds formed. This is accomplished easily, since they form soft, friable ashes (see Table 8-15).
2. *At the air preheater inlet (on the line from the economizer).* When air preheaters are available, MgO injection at this joint is preferred, to eliminate most of the SO_3 formed and to minimize the amount of ash accumulated in the boiler itself. The low gas temperatures at this point require either powdered MgO or aqueous magnesia suspension.

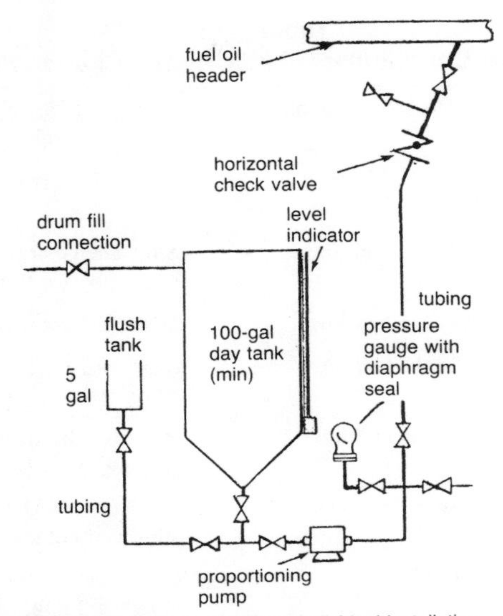

1. Tubing sizes and materials to be specified per individual installation.
2. Tubing with long radius bends to be used in lieu of pipe for all lines ½ in. or less.
3. Tubing fittings to be carbon steel compression type, Swagelok or equivalent.
4. Stop valves to be screw-fitted ball type.
5. Pressure indicator to be Bourdon tube type provided with Teflon diaphragm protective seal.
6. Check valve to be located on discharge side of proportioning pump.
7. Use tungsten carbide ball check assemblies in proportioning pumps.

Figure 8-9. Fuel-oil delivery system.

3. *At a point opposite the secondary superheater*. The injection point for the finely dispersed magnesia is selected to provide sufficient time and sufficiently high temperatures to flash off the carrier before the atomized droplet of suspended MgO contacts the boiler surfaces.[19]

Chemical form and particle size.[13,14] The most widely used additive to minimize vanadium and sulfur problems is MgO, alone or in combination with other oxides such as calcium, aluminum, manganese, or silicon.

MgO is offered in a wide range of prices and forms. This way, two additives with identical chemical content can perform differently in the treatment. The following magnesium fuel additives are available on the market:

1. *Oil soluble: magnesium sulfonate*. This is the most expensive. The typical Mg concentration is 8%–10%. It is used at 0.4 Mg/V ratios and is added directly to the fuel storage tank or is metered and added along with the fuel. Particle size is <1 μm.

Table 8-16
Fuel Conditioners: Their Primary Function

Additive Class	Problem Location	Primary Function
Dispersants	Tanks and lines	Dissolve sludge, allowing free flow of fuel and cleaning of tanks, lines, and strainers
Antioxidants	Tanks and lines	Stop sludge formation and corrosion caused by slow oxidation of fuel
Corrosion inhibitors	Tanks and lines	Stop corrosion caused by organic acids
Heavy-metal scavengers	Tanks and lines	Stop corrosion of tanks caused by heavy metal ions in fuel
Pour-point/cloud-point depressants	Tanks and lines	Prevent solidification of fuel and filter plugging at low temperatures
Emulsifiers	Tanks and lines	Hold trace water in fuel, preventing icing and collection
Emulsion breakers	Tanks	Drop water to bottom of tank for easy removal
Biocides	Tanks and filters	Kill slime-forming microorganisms which live at water-fuel interface
Slime dispersants	Tanks and filters	Disperse microorganism slime which plugs filters
Atomizers	Burners	Lower surface tension of fuel; better combustion from smaller fuel drops
Combustion catalyst	Engines, injectors, and burners	Provides better burning through chemical mixing of air and fuel

Source: Sterns, R. B., Combustion Catalysts: How They Work and How They Can Be Tested,'' *The Motor Ship,* December 1981.

2. *Water soluble: magnesium sulfate*. This costs about 20% of the cost of the oil-soluble magnesium sulfonate. Suppliers sell a 10% Mg product with <1 μm particle size. The powder is dissolved and then emulsified in the fuel. It is used at 3.0 Mg/V ratios.
3. *Oil or water dispersions: MgO or Mg $(OH)_2$*. Their costs vary widely from 8% to 80% of the cost of magnesium sulfonate. Oil or water suspensions containing 37.5% Mg and 1.7–2.0 μm particle size are offered by suppliers.

   The suspension is metered into fuel to burner or nozzle at 0.7 Mg/V ratios. The slurry approach seems to be the most effective and the least capital intensive, resulting in fewer maintenance and operating problems (see Figure 8-9).

4. *Powders: MgO or Mg $(OH)_2$*. Typical concentrations of Mg sold by suppliers are 54%–58% of 20-μm particles. The powder is fed to the boiler or economizer using a powder feeder, which is more expensive than the feeding equipment used with the other three Mg forms. The typical dosage is 3.0 Mg/V.

All of the Mg forms have good to excellent shelf life. The powder must be kept dry.

In general, the typical dosage is about 1 part MgO to 4000 parts of fuel oil, and costs in the range of 2 cents per barrel of oil. Post-flame additives are also employed to minimize acid corrosion downstream. MgO is used for this purpose, and is added to the flue gases.

Figure 8-9 shows a typical design of a fuel-oil additive-delivery system.[20,21] Table 8-16 summarizes the effect an additive has on solving fuel problems.

References

1. Dryer, F. L., G. D. Rambach, and I. Glassman, "Some Preliminary Observations on the Combustion of Heavy Fuels and Water-in-Fuel Emulsions," Aerospace and Mechanical Sciences Report No. 1271, Princeton University, April 1976.
2. Scherer, G., and L. A. Tranie, "Pollution Reduction by Combustion of Fuel-Oil Water Emulsions;" "Pollutant Formation and Destruction in Flames, and in Combustion Systems." Paper No. 83, Fourteenth International Symposium on Combustion, Penn. State University, August 20–25, 1972 (not published in the Proceedings).
3. Garcia-Borras, Thomas, "How to Interpret Data From Rotational Viscometers," *Rubber Age,* p. 84, June 1965.
4. Garcia-Borras, Thomas, "Calibrate Rotational Viscometers for Non-Newtonian Fluids," *Chemical Engineering,* p. 176, January 18, 1965.
5. Delaney, R., "Future Bunker Fuel," *Marine Engineering/Log,* p. 58, November 1981.
6. Husta, R. P., "The Use of Additives for the Efficient Utilization of Liquid Fuels," *Chemical Times and Trends,* p. 40, July 1980.
7. Eliot, R. C., "Boiler Fuel Additives for Pollution Reduction and Energy Saving," Noyes Data Corporation, New Jersey 1978.
8. O'Connor, P. J., "Protect Your Distillate Fuel Reserves," *Power,* April 1981.
9. Distillate Fuel Stability and Cleanliness, ASTM Publication Code Number (PCN) 04-751000-12, 1981.
10. The Perolin Company, 84 Danbury Road, Wilton, Connecticut, 06897.

11. Ferraro, R. M., "Using Fuel Additives To Increase Boiler Efficiency," *Plant Engineering,* Part I, p. 159, March 17, 1977; Part II, p. 53, March 31, 1977.
12. Martin, D. J., "Use Additives to Prevent Fuel-Related Problems," 1980 Energy Systems Guidebook, McGraw-Hill, Inc., Reprint 284 by Nalco Chemical Company, Oakbrook, Illinois.
13. Reid, William T., "External Corrosion and Deposits: Boilers and Gas Turbines," American Elsevier Publishing Company, Inc., New York, 1971.
14. Radway, J. E., "Selecting and Using Fuel Additives," *Chemical Engineering,* July 14, 1980.
15. Radway, J. E., "Reduction of Coal Ash Deposits with Magnesia Treatment," *Combustion,* April 1978.
16. Radway, J. E., "How More Ash Makes Less," *Environmental Science and Technology,* p. 388, April 1978.
17. Radway, J. E., and L. M. Exley, "A Practical Review of the Cause and Control of Cold-End Corrosion and Acidic Stack Emissions in Oil-Fired Boilers," ASME Publication 75-WA/CD-8, December 1975.
18. Rendle, L. K., R. D. Wilson, and G. Whittingham, "Fire-Side Corrosion in Oil-fired Boilers," *Combustion,* August 1958.
19. Radway, J. E., and R. R. Rohrbach, "SO_3 Control in Steam Generator Emissions," ASME Publication 76-WA/APC-9, December 1976.
20. Poynton, J. P., "Metering Pumps: Types and Applications," *Hydrocarbon Processing,* p. 279, November 1981.
21. Poynton, J. P., "Design Variables in Chemical Feed Systems," *Plant Engineering,* May 15, 1980.

9

Treatment of Boiler Feedwater

Water from rivers, ponds, and wells cannot be used in a boiler without previous treatment. Unless the boiler feedwater is treated properly, the life of a boiler will be shortened very quickly. Completely pure water, ready for use in a boiler, does not exist.

Boiler water is treated in three steps:

1. *Raw-water treatment*—where most of the impurities are taken out.
2. *Preboiler treatment, external or pretreatment*—where small amounts of remaining impurities are treated.
3. *Internal treatment*—boiler waters are treated directly in the boiler itself.

The impurities in natural waters include:

1. Dissolved solids (carbonates, chlorides, calcium salts, etc.).
2. Suspended (sand, sludge, etc.).
3. Gases (oxygen, carbon dioxide, etc.).

The objectives of the water treatment are the prevention of:

1. Sludge and scale deposits (mostly calcium and magnesium salts).
2. Corrosion and pitting (dissolved gases and acids).
3. Embrittlement (high caustic presence).
4. Carryover (foaming).

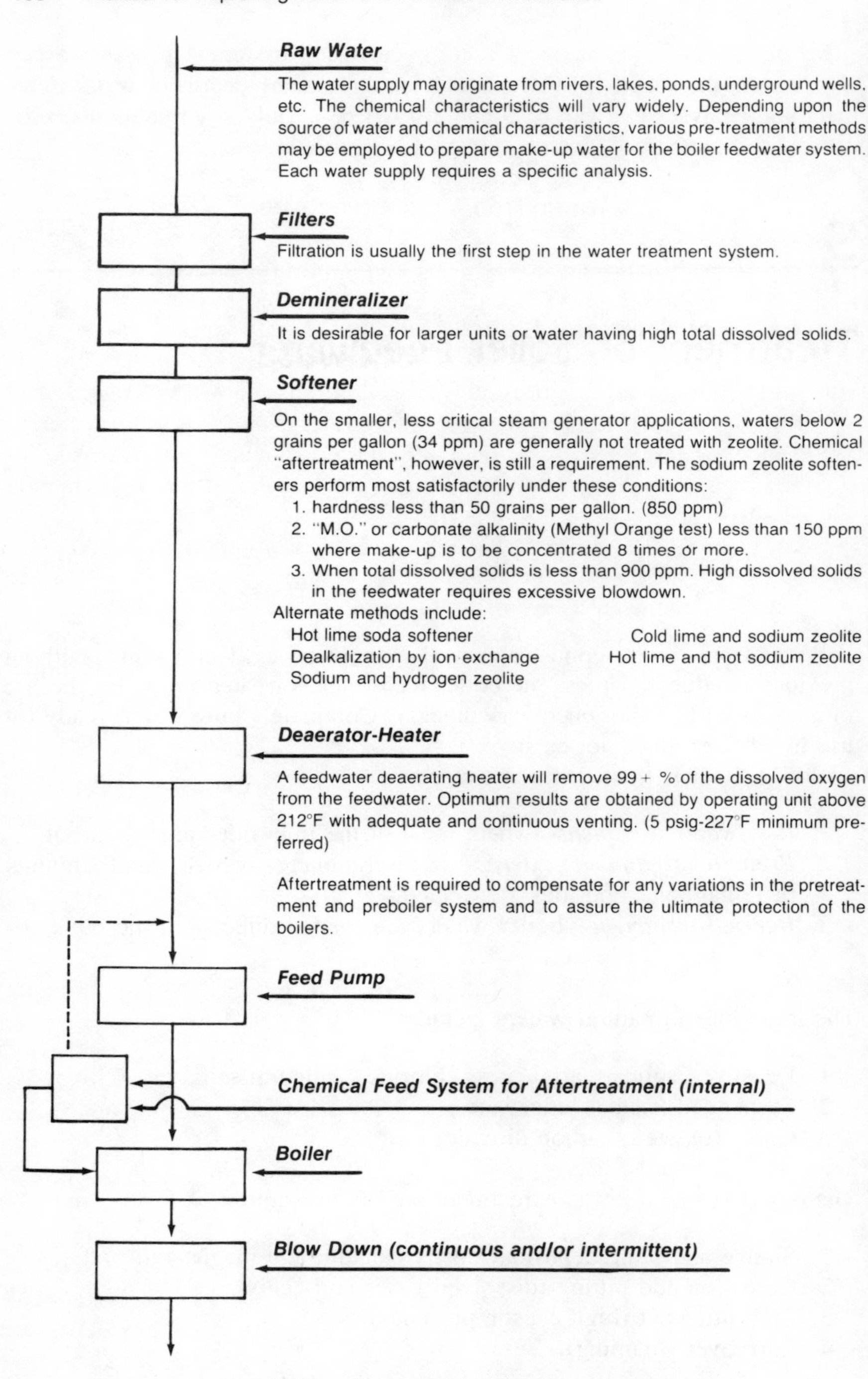

Figure 9-1. Schematic of a water-treatment system.

Figure 9-1 is a schematic of a complete water-treatment process. Since it is beyond the scope of this manual to include many details of water treatment and analyses that can be found elsewhere,[1,2] only key treatments, processes, and analytical controls will be covered here.

Water-Treatment Objectives

There are four main objectives in treating water for boiler use.

Prevention of Scale and Sludge Buildup

Hardness. Dissolved calcium and magnesium salts, known as ''hardness,'' are deposited on tube surfaces as scale when water evaporates to form steam. This happens because calcium and magnesium compounds are relatively insoluble in water and tend to precipitate out. This scale reduces heat transfer and causes overheating of the tube metal.

Hardness in natural waters varies from a few ppm to more than 500 ppm. High-pressure boilers can tolerate only 0–2 ppm.

Hardness is removed by using zeolites and/or by chemical means.

Sulfates and silica. Sulfates and silica generally precipitate on the boiler metal and do not form sludge. They are much harder to prevent from precipitating.

Iron and sodium. Dissolved or suspended iron is also found deposited on metal surfaces. Sodium compounds are usually deposited only when water is nearly evaporated, since they are very soluble.

Identifying boiler deposits.

1. *Carbonates*. Dissolved bicarbonates of calcium and magnesium, under heat in the boiler, break down, giving off carbon dioxide and forming insoluble carbonates. These carbonates are generally granular and porous. When an acid solution is dropped on carbonate deposits, bubbles of carbon dioxide are formed, effervescing like champagne when just opened.
2. *Sulfates*. These are much harder than carbonates, brittle, and acid will not cause them to effervesce.
3. *Silica*. This is a very hard deposit, very brittle, resembling porcelain. It will not dissolve in hydrochloric acid.
4. *Iron*. This is dark colored, usually magnetic, and soluble in hot acid.
5. *Phosphates*. These form a sludge that is easily removable. They are a result of a phosphate water treatment and are usually soft, brown or gray deposits.

Problems with deposits and sludge. Deposits and sludge can cause plugging or partial obstruction of tubes, resulting in tube overheating and failure. Corrosion underneath the deposits may occur. The boiler operating efficiency is reduced, sometimes drastically, and unscheduled shutdown may occur.

Corrosion and Pitting Prevention

Oxygen (O_2) and carbon dioxide (CO_2). The most common type of corrosion is that caused by oxygen dissolved in the water and by low pH caused from the presence of carbon dioxide. CO_2 originates from the feedwater directly or from the decomposition of carbonates in the feedwater, as follows:

$$HCO_3^- \rightarrow OH^- + CO_2 \uparrow$$

$$CO_3^= + H_2O \rightarrow 2OH^- + CO_2 \uparrow$$

In the condensate, CO_2 combines with water to form carbonic acid:

$$CO_2 + H_2O \rightarrow H_2CO_3$$

Then, the carbonic acid ionizes, producing hydrogen and bicarbonate ions:

$$H_2CO_3 \rightarrow H^+ + HCO_3^-$$

When iron corrosion occurs as a result of low condensate pH, the iron goes into solution at the anodic sites of the corrosion cells:

$$Fe^\circ \rightarrow Fe^{++} + 2e^-$$

Reduction of hydrogen ions occurs at cathodic sites of corrosion cells in two sequential steps:

$$2H^+ + 2e^- \rightarrow 2H^\circ \rightarrow H_2 \uparrow$$

When these anodic/cathodic reactions occur, the ferrous ions produced will combine with bicarbonate ions to form ferrous bicarbonate:

$$Fe^{++} + 2HCO_3^- \rightarrow Fe(HCO_3)_2$$

The bicarbonate can decompose to form ferrous carbonate or ferrous hydroxide, according to the following chemical reactions:

$$Fe(HCO_3)_2 \rightarrow FeCO_3 + H_2O + CO_2 \uparrow$$

$$Fe(HCO_3)_2 \rightarrow Fe(OH)_2 + 2CO_2 \uparrow$$

These reactions release CO_2 as stated, allowing the cycle to be repeated.

Low-pH corrosion caused by carbonic acid eats away the metal, which is especially troublesome in threaded-pipe sections and stressed areas.

Dissolved oxygen in steam and condensate, due either to air inleakage or inadequate oxygen removal by deaeration, can lead to general and localized corrosion by either or both of the following reactions:

$$H_2 + 1/2\ O_2 \rightarrow H_2O$$

$$1/2\ O_2 + 2e^- + H_2O \rightarrow 2OH^-$$

As long as the preceding reduction reactions occur, the corrosion cells will continue to be active.

Ammonia and hydrogen sulfide. Attack on copper alloys is increased when these gases are present in the water.

Chelates. Chelates are normally used in water treatment; however, excessive chelate residual—in excess of 20 ppm (measured as Ca CO_3)—may produce corrosion also.

Problems with corrosion and pitting. High temperatures and stresses in the boiler metal accelerate the corrosion process. Corrosion fatigue may also contribute to boiler problems. Cyclic stresses caused by rapid heating and cooling concentrate at points weakened by corrosion and pitting, and tube failure may occur. Replacement of the corroded equipment is expensive. These problems can readily be solved when feedwater is deaerated properly. Acids can be neutralized with alkalies.

Prevention of Embrittlement

High caustic concentration causes embrittlement. It occurs when, in the normal conditions under which a boiler is operated, caustic concentrates along drum seams, under rivets, and at tube ends where water flashes to steam. At least a trace of silica must be present for embrittlement to occur. The use of tannin, lignin, or sodium nitrate can solve this problem.

Prevention of Carryover

Foaming is the basic cause of carryover of moisture, oil, treatment chemicals, dissolved salts, etc. Certain substances such as alkalies, oils, fats, grease, organic matter, and suspended solids are known to cause foaming.

Table 9-1
Common Impurities in Water and Possible Effect When Used for Boiler Feed Purposes

Constituent	Chemical Composition	Principal Source of Contaminating Substance	Possible Effect When Present in Boiler Waters
Calcium bicarbonate	$Ca(HCO_3)_2$	Mineral deposits	Scale
Calcium carbonate	$CaCO_3$	Mineral deposits	Scale
Calcium chloride	$CaCl_2$	Mineral deposits	Scale
Calcium sulfate	$CaSO_4$	Mineral deposits	Scale and corrosion
Carbonic acid	H_2CO_3	Absorption from the atmosphere, mineral deposits, decomposition of organic matter	Corrosion
Free acids	HCl, H_2SO_4	Mine drainage and industrial wastes	Corrosion
Magnesium bicarbonate	$Mg(HCO_3)_2$	Mineral deposits	Scale
Magnesium carbonate	$MgCO_3$	Mineral deposits	Scale
Magnesium chloride	$MgCl_2$	Mineral deposits	Scale and corrosion
Oil and grease	—	Industrial wastes	Corrosion, deposits, priming and foaming
Organic matter and sewage	—	Domestic and industrial wastes	Corrosion, deposits, priming and foaming
Oxygen	O_2	From atmosphere	Corrosion
Silica	SiO_2	Mineral deposits	Scale
Sodium bicarbonate	$NaHCO_3$	Mineral deposits	Priming, foaming and embrittlement
Sodium carbonate	Na_2CO_3	Mineral deposits	Priming, foaming and embrittlement
Sodium chloride	$NaCl$	Sewage, industrial wastes, and mineral deposits	Inert, but may be corrosive under some conditions
Suspended solids	—	Surface drainage and industrial wastes	Priming, foaming, sludge or scale

Source: PB-265 713, *Measuring and Improving the Efficiency of Boilers—A Manual for Determining Energy Conservation in Steam Generating Power Plants,* Engineering Extension Service, Auburn University, Alabama. FEA Contract FEA-CO-04-50100-00, Report No. FEA/D-77/132, November 1976.

Priming is a sudden surge of boiler water caused by a sudden change in boiler load. Too high a water load can also cause priming.

The dissolved and suspended solids in the boiler water can cause foaming in the steam and contaminate the system (superheaters and turbines), causing overheating and corrosion, including failure of superheater tubes. Blowdown and antifoam agents can help combat carryover. Proper boiler design and operation can reduce carryover, too.

Table 9-1 summarizes the common impurities found in boiler water.

Raw-Water Treatment

Figure 9-1 outlines the water treatment steps, from raw-water treatment through the internal water treatments usually employed.

For convenience, a raw-water treatment includes mostly physical or mechanical purification of the raw water rather than a chemical treatment that includes chemical reactions. Most of the water impurities are taken out during the physical purification.

Thus, raw-water treatment includes:

Clarification. The large suspended particles in the feedwater settle out readily. Settling basins and filters are used for this type of suspended matter. Sand filters are commonly used for clarification. It is reasonable to expect suspended solids to be maintained under 5 ppm.

Coagulation and precipitation. Raw water also contains suspended matter such as oil, which must be removed (that under 1 ppm) to prevent foaming and boiler deposits, the particles of which are so small that they do not settle out and pass through filters. To remove these finely divided particles (usually colloidal substances), coagulants are used. Coagulants charge and neutralize the finely divided and colloidal particles into masses that can settle quickly and then be filtered. This occurs because colloidal particles have negative electrical charges and therefore repel each other until the coagulant neutralizes them and they adhere together, or flocculate.

The common coagulants used are iron and aluminum salts such as ferric chloride, ferric sulphate, aluminum sulphate (also known as alum), and sodium aluminate.

Organic polymers and special types of clays are also used in the coagulation process.

Deaeration. Dissolved gases (oxygen, carbon dioxide, and hydrogen sulfide) can be removed physically by heating the water with steam in a deaerator. Part of the steam is vented, carrying with it most of the dissolved

Table 9-2
External Treatment of Boiler Feedwaters

Method	Final Hardness as ppm $CaCO_3$	Alkalinity Reduction	Total Solids Reduction	CO_2 in Steam	Remarks
1. sodium ion exchanger	0 – 2	none	increase	high only if alkalinity high	silica not reduced
2. sodium ion exchanger plus acid treatment	0 – 2	all of it possible	none	all of it possible	silica not reduced
3. sodium and hydrogen ion exchanger	0 – 2	all of it possible	reduced with alkalinity	all of it possible	silica not reduced
4. cold lime-soda process	34–85	reduced	reduced	usually low	silica content is reduced
5. cold lime plus ion exchanger	0 – 2	reduced	reduced	usually low	silica content is reduced
6. hot lime-soda	17–25	reduced	reduced	usually low	silica content is reduced
7. hot lime-soda plus hot phosphate treatment	0 – 2	reduced	reduced	usually low	silica content is reduced
8. hot lime-soda plus ion exchangers	0 – 2	reduced	reduced	usually low	silica content is reduced
9. hot phosphates only	0 – 2	a little	a little	usually low	silica not reduced
10. deionization	0 – 2	practically all of it	practically all of it	all of it possible	silica can be reduced to 0–2 ppm

Source: Power Handbook, McGraw-Hill, Inc.

gases. The physical method is usually employed in conjunction with a chemical treatment to maintain low gas concentrations:

Oxygen: <0.005 ppm for high-pressure boilers

Carbon dioxide: ratio of alkalinity to $CO_2 > 3:1$

Hydrogen sulfide: <5 ppm

Reverse osmosis. In this process, a semipermeable membrane allows ions to pass from a more concentrated solution to a less concentrated solution without allowing the reverse to occur. This way, the dissolved solids in the raw water are reduced significantly. Installation and operating costs are high.

External Feedwater Treatment

Chemical pretreatments include the following types, summarized in Table 9-2.

Chemical Precipitation

Precipitation methods are used to reduce dissolved hardness, alkalinity, and in some cases silica. In these methods, certain chemicals are added to react with dissolved minerals in the water to produce a relatively insoluble reaction product. The following process is one of these methods.

Lime-soda softening. The hydrated lime (calcium hydroxide) reacts with the soluble calcium and magnesium bicarbonates to form insoluble precipitates:

$$Ca\,(OH)_2 \text{ (lime or calcium hydroxide)} + Ca\,(HCO_3)\text{(calcium bicarbonate)} \rightarrow 2\,CaCO_3 \text{ (calcium carbonate)} + 2\,H_2O \text{ (water)}$$

$$2\,Ca\,(OH)_2 + Mg\,(HCO_3)_2 \text{ (magnesium bicarbonate)} \rightarrow Mg\,(OH)_2 \text{ (magnesium hydroxide)} + 2\,CaCO_3 + H_2O$$

The soda ash (sodium carbonate) is used to reduce the nonbicarbonate hardness known as "permanent hardness:"

$$Na_2CO_3 \text{ (sodium carbonate)} + CaSO_4 \text{ (calcium sulfate)} \rightarrow CaCO_3 + Na_2SO_4 \text{ (sodium sulfate)}$$

$$Na_2CO_3 + CaCl_2 \text{ (calcium chloride)} \rightarrow CaCO_3 + 2\,NaCl \text{ (sodium chloride)}$$

The resulting sludge from the lime-soda reactions can be removed by settling and filtration. The resulting sodium sulfate and chloride are very soluble and do not form scale.

Coagulants are used in the lime-soda process to speed up settling of sludge by 25%–50%.

Ion-Exchange Processes

Dissolved minerals in water form electrically charged particles called ions. Thus, calcium bicarbonate ionizes as follows:

1. A calcium ion Ca^{++} with two positive charges (cation).
2. A bicarbonate ion HCO_3^- with one negative charge (anion).

Certain materials, including natural and synthetic zeolites, can remove mineral ions from water in exchange for others. Thus, using cation-exchange resins, the Ca^{++} and Mg^{++} cations are adsorbed and the Na^+ cations are released.

Using anion-exchange resins, certain anions such as HCO_3^- (bicarbonate ions) and $SO_4^=$ (sulfate ions) are adsorbed and replaced with OH^- (hydroxide ions). Strong-base anion resins will also reduce silica content.

The ion exchangers are regenerated with sodium chloride (NaCl) or acid (H_2SO_4 or HCl) for cation exchangers, and with caustic (NaOH) or ammonium hydroxide (NH_4OH) for anion exchangers, depending on the process used.

Deaeration and Chemical Treatment

Dissolved oxygen is removed down to 0.007 ppm by heating the water with steam and adding sodium sulfite (Na_2SO_3):

$$2Na_2SO_3 + O_2 \rightarrow 2Na_2SO_4$$

or hydrazine.

A special catalyzed sodium sulfite is often used to assure a faster reaction and that oxygen is eliminated in less than 30 seconds. The uncatalyzed sulfite takes longer than 10 minutes to eliminate 30% of the oxygen present. The recommended levels of oxygen scavenger are shown in Table 9-3.

Internal Boiler-Water Treatment

The internal water treatment complements the raw-water and external water treatments, since even small amounts of dissolved and suspended

Table 9-3
Recommended Levels of Oxygen Scavengers[1,2]

Boiler Pressure	Residual of SO_3, ppm	Residual of N_2H_4, ppm
0–150	30–60	.1
150–300	30–40	.1
300–600	20–30	.08
600–900	10–15	.06
900–1200	5–10	.03
1200–1500	3–7	.01
above 1500	use N_2H_4	.01

water impurities (iron, hardness, oxygen, or silica) can cause serious problems in the boiler. In cases where large amounts of condensed water are available, low-pressure boilers may use an internal water treatment only. Specifically, the internal water treatment provides:

1. Chemicals to react with the feedwater hardness and prevent it from precipitating on the boiler metal as scale.
2. A natural or synthetic sludge conditioner to prevent suspended solids from sticking to the boiler metal.
3. Antifoam chemicals to prevent carryover, while allowing a certain concentration of dissolved and suspended solids to remain in the boiler water.
4. Oxygen scavengers to prevent corrosion.
5. Chemicals to maintain enough alkalinity to minimize corrosion.

Therefore, improper internal boiler-water treatment can result in:

1. *Corrosion*, which can cause tube failure and costly repairs.
2. *Scale*, which will reduce boiler efficiency.
3. *Carryover*, which can cause turbine deposits and corrosion.

The chemical reactions in the boiler water involve the following:

1. *Water hardness*. The water hardness not eliminated during the previous treatments breaks down as follows in the boiler:

   calcium bicarbonate → insoluble calcium carbonate

   sodium carbonate → sodium hydroxide + carbon dioxide.

2. *Phosphates*. To minimize the preceding reactions, phosphates are added to the boiler water as follows:

   sodium phosphate + calcium bicarbonate →

   calcium phosphate + sodium carbonate

   The following sodium-phosphate compounds are used: NaH_2PO_4, $Na_2H\ PO_4$, $Na_3\ PO_4$, and $Na\ PO_3$, yielding the calcium phosphate, or $Ca_3(PO_4)_2$, that precipitates in the form of a sludge.
   In the presence of caustic, the following reaction occurs:

   sodium hydroxide + magnesium bicarbonate →

   magnesium hydroxide, or magnesium silicate if silica is present. The latter compound also forms a sludge. The sludge is conditioned to keep it from depositing as scale on the boiler metal. Sludge conditioners include tannins, lignins, starch, and reactive colloids. The sludge is removed from the boiler by blowdown.
3. *Chelates and polymers*. When chelates are used for internal treatment, they react with magnesium and calcium salts to form soluble complexes. Chelates used in conjunction with polymers assure that the reaction products will be maintained in suspension. The sludge is removed from the boiler by blowdown.
4. *Sulfates*. The residual calcium and magnesium sulfates present in small amounts in the boiler feedwater concentrate in the boiler, becoming insoluble and precipitating. Phosphate or chelate/polymer will prevent the sulphates from precipitating in a manner similar to that described for water hardness.
5. *Silica*. Silica tends to precipitate as scale on hot spots in the boiler. This problem can be resolved by:
   - Keeping high alkalinity to solubilize silica.
   - Adding certain organic materials to prevent the adherence of silica to the metal.
   - Treating with special synthetic polymers.

If sufficient magnesium is present in the water, some of the silica will precipitate as sludge and not as scale.

Treatment of Steam Condensate

Corrosion in the steam condensate is caused by carbon dioxide (which forms carbonic acid) and oxygen carried into the system with the steam. In addition to the water treatment in the deaerator, volatile anions or volatile

Table 9-4
Maximum Impurities in Water for Boilers[1,2]*

Feedwater	Softness	less than 1 ppm†
Feedwater	Oxygen	less than 20 ppb**
Boiler water	Hardness	less than 1 ppm
Boiler water	pH	9.5–11
Boiler water	TDS	less than 3500 ppm
Boiler water	Sulfite	30–60 ppm
Boiler water	Alkalinity	less than 800 ppm
Boiler water	Phosphate	20–40 ppm
Condensate	pH	7.5
Condensate	TDS††	less than 20 ppm

*for boilers under 200 psig
†parts per million
**parts per billion
††total dissolved solids

filming amines are used to protect the condensate system. Filming amines usually give more economical protection in systems with high make-up water, air inleakage, and high alkalinities.[1,2]

How Chemicals Are Added

Chemicals are added using solution tanks and proportioning pumps. In general, the chemicals used for internal treatment of the boiler water are added directly to the feedwater right after deaeration but before entering the boiler drum. Continuous feeding of chemicals is preferred to intermittent applications.

Dosages are calculated on the basis of chemical analysis and controls.

Water Analysis and Controls

Boiler Blowdown

The dissolved and suspended solids in the boiler water are controlled by regulating boiler blowdown. Since an equivalent amount of fresh water is fed into the boiler (through automatic controls), the boiler water is diluted. This way, the appropriate conditions for the boiler water are maintained, depending on the boiler pressure, as shown in Tables 9-4, 9-5, and 9-6.

Blowdown is usually controlled by measuring the conductivity of the water. Chloride tests are also used to control blowdown, since chlorides are not affected by chemical treatments. Thus:

$$\text{blowdown (\% of make-up water)} = \frac{\text{chlorides in feedwater}}{\text{chlorides in boiler water}} (100)$$

Table 9-5
Target Values for Boiler Water

Boiler-Water Limits Vary with Operating Pressure					
Boiler Pressure at Outlet, psig	**Total Solids, ppm**	**Alkalinity (Total) ppm**	**Suspended Solids, ppm**	**Silica, ppm**	**Specific Conductance (micromhos/cm)**
0–300	3500	700	300	125	7000
301–450	3000	600	250	90	6000
451–600	2500	500	150	50	5000
601–750	2000	400	100	35	4000
751–900	1500	300	60	20	3000
901–1000	1250	250	40	8	2000
1001–1500	1000	200	20	2.5	150
1501–2000	750	150	10	1.0	100

Maximum limits for boiler-water concentration in units with a steam drum. Silica limits based on limiting silica in steam to 0.02–0.03 ppm.
Source: PB-262 577, *Industrial Boiler Users' Manual,* Volume II, KVB Inc. FEA Contract C-04-50085-00, Report No. FEA/D-77/026, January 1977.

Table 9-6
Maximum Limits for Boiler Water[1,2]

Drum Pressure, psig	**Iron, ppm Fe**	**Copper, ppm Cu**	**Total Hardness, ppm $CaCO_3$**
0–300	0.100	0.050	0.300
301–450	0.050	0.025	0.300
451–600	0.030	0.020	0.200
601–750	0.025	0.020	0.200
751–900	0.020	0.015	0.100
901–1000	0.020	0.015	0.050
1001–1500	0.010	0.010	ND*
1501–2000	0.010	0.010	ND*

*ND = None detectable

If solids concentration is too high, required blowdown becomes excessive and improved water pretreatment may be necessary to lower the solids concentration in the make-up water.

Sludge blowdown is done at low points in the boiler; this is the intermittent blowdown. The continuous blowdown is done from the top (just below the water level in the steam valve) to control solids concentration.

Table 9-7
General Terms Used in Water Analysis

Symbol	Means	Represents
H	total hardness	total Ca and Mg content
Ca	calcium	total calcium hardness
M	methyl orange alkalinity	bicarbonates and carbonates, or carbonates and hydrates (hydrates and bicarbonates cannot exist together)
P	phenolphthalein alkalinity	1/2 carbonates and all hydrates
O = 2P − M	hydrates (OH)	calcium, magnesium, or sodium hydroxides
S = M − H	sodium alkalinity	sodium hydroxide, carbonate, and phosphate
C	total carbonates (CO_3)	calcium, magnesium, and sodium
N = S − O	sodium carbonate	sodium carbonate
[O] = M − 2P	bicarbonates (HCO_3)	calcium, magnesium, and sodium bicarbonates
[S] = H − M	sulfate hardness (SO_4)	calcium and magnesium-sulfate hardness
[N] = O − S	free lime, $Ca(OH)_2$	calcium hydroxide
TDS	total dissolved solids	all soluble materials except gases
PO_4	phosphates	sodium phosphate
SiO_2	silica	soluble silica
μmho	conductivity	ability of water to conduct electricity

Boiler-Water Tests

Routine control tests vary, depending on the chemical treatment used. Routine tests include those for alkalinity, pH, color, total dissolved solids, phosphate, chelate, and hydrazine. Occasional tests for contaminants include those for iron, silica, and oil. The most common units to express concentrations are in parts per million (ppm) and grains per gallon (gpg, where 1 gpg = 17.1 ppm).

In water analysis, calculations are facilitated if the results are expressed as "equivalent to ppm $CaCO_3$." The general terms in Table 9-7 are used in water analysis. The symbols in Table 9-7 are used only to simplify water analysis and calculations; do not confuse phenolphthalein alkalinity with the element P (phosphorous), or S (sodium alkalinity) for sulfur.

Calculations using symbols in Table 9-7.

1. If P = O, M is all HCO_3.
2. If P = M, M is all OH.

3. If $P = 1/2\ M$, $2P = CO_3$.
4. If $2P > M$, $2(M - P) = CO_3$ and $2P - M = OH$.
5. If $2P < M$, $2P = CO_3$ and $M - 2P = HCO_3$.

Also, when no hydrate alkalinity is present:

1. If $M > H$:

   $M - H$ represents sodium bicarbonate
   $H - Ca$ represents magnesium bicarbonate
   Ca represents calcium bicarbonate

2. If $H > M$ and Ca is also $> M$:

   $H - Ca$ represents magnesium sulphate
   $Ca - M$ represents calcium sulfate
   M represents calcium bicarbonate

3. If $H > M$ and $Ca < M$:

   $H - M$ represents magnesium sulphate
   $M - Ca$ represents magnesium bicarbonate
   Ca represents calcium bicarbonate

and when hydrate alkalinity is present:

4. If $S > O$:

   $M - Ca$ represents magnesium carbonate
   Ca represents calcium carbonate
   O represents sodium hydrate
   N represents sodium carbonate

5. When $S < O$:

   $H - \boxed{N}$ represents calcium carbonate
   S represents sodium hydrate
   $\boxed{N}$ represents calcium hydrate (free lime)
   $\boxed{S}$ represents sulfate hardness
   O represents calcium and sodium hydrate

Glossary of Water-Treatment Terms

Backwashing. Reversing the normal flow of water through a softener or filter to reactivate or clean it.

Brine. A solution of sodium chloride.

Cation exchange. The exchange of Ca and Mg ions for sodium ions during water softening.

Chelants. Also known as chelating agents. Nitrilo triacetic acid (NTA) and ethylenediaminetetraacetic acid (EDTA) are chelants used alone or together to keep potential scale formers in suspension.

Dynamic water pressure. Water pressure at inlet to softener when water is flowing.

Effluent. Water or solution from a water softener.

Flow rate. Quantity of water or brine, usually measured in liters or gallons per minute.

Grains per gallon (gpg). Used only in the U.S. and Canada to report water analysis. One gpg equals 17.1 ppm or milligrams per liter.

Head loss. Pressure drop, expressed as lb/in^2 or Kg/cm^2.

Ion exchange. To exchange ions in solution.

pH. This number denotes alkalinity or acidity. The pH scale goes from 0 to 14. 7.0 is neutral; below 7.0 indicates acidity.

Soft water. Water with less than 1 gpg hardness.

References

1. *Betz Handbook of Industrial Water Conditioning*, Betz Laboratories, Inc., Trevose, Pennsylvania, 19047, 1980.
2. *The Nalco Water Handbook*, Nalco Chemical Company, 2901 Butterfield Road, Oak Brook, Illinois, 60521, 1979.

10

Applied Statistics and Errors in Controls and Measurements

This chapter covers the sources of errors in control systems and the errors inherent in the measurements, whether inexpensive or expensive instruments were used. Also, basic statistics will be applied, giving examples on how to estimate the accuracy and precision of the measurements.

Sources Of Errors in Control Systems

Whether the air/fuel ratio is controlled manually or automatically, several errors are present in the controls:

Fuel variation. Fuel properties and flow will vary according to pressure, temperature, heat content, and specific gravity. See Tables 8-1, 8-9, 8-10, and 8-13 for ranges in the quality of typical fuels. In addition, the fuel-metering system wears out, becomes partially plugged up, or does not mechanically function at its peak.

Combustion air variations. Changes in pressure, temperature, and humidity will have a definite effect, as discussed in Chapter 3. Likewise, the equipment can introduce mechanical variations, such as the air dampers getting stuck in a certain position.

Exhaust gas variations. Boiler fouling and air infiltration will cause changes in the composition of the flue gases.

Boiler controls. These vary with type of control, mechanical wear, improper calibration, etc. Dukelow[1] has estimated that these errors are responsible for boilers running with 19%–44% excess air using different controls. The estimated error for oxygen was 10% excess air.

Errors in Measurements and in Fuel Composition

When the combustion efficiency is obtained by the indirect method, the only significant variables are % CO_2 and stack temperature. Any errors in CO_2 determination and in the stack temperature will cause errors in the combustion efficiency estimate. Error analysis has been reported in the literature, plotting the % error in one measurement vs. the % error in efficiency.[2]

Another kind of error is introduced when estimating combustion efficiency using slide rules and tables based on fuels with different chemical compositions. To test the significance of this type of error, combustion efficiencies were estimated for a certain set of conditions, using three different fuel sources for the three different types of fuels: natural gas, fuel oil No. 2, and fuel oil No. 6. The data are shown in Table 10-1. The following observations are made:

1. The combustion efficiencies were more sensitive to gas composition than to fuel oils.
2. The combustion efficiencies estimated for the fuel oils were much more consistent than those for natural gas.
3. The tables used from Reference 3 consistently gave the highest % increase in efficiency, followed by the Bacharach slide rule calculations. The lowest results were obtained with the Cleaver-Brooks tables.

It is concluded that for a quick and practical estimate, the Bacharach slide rule or any of the two published tables can be used. If more accurate data are needed, a fuel analysis and somewhat more involved calculations are required. Since the important thing is the *difference* in combustion efficiency under two different sets of conditions, it is apparent that using the same slide rule or the same table will give more accurate results. In other words, do not get one efficiency with one slide rule and another efficiency with one of the tables published by one source or the other in order to obtain the increase or decrease in efficiency.

Applying Statistics[4]

In calculating the efficiencies of a boiler, the precision of the results may need to be estimated. Some errors in the analysis include:

Table 10-1
Accuracy in Calculating Efficiencies and Fuel Savings Using Different Fuel Compositions[1,2,3,4]

Fuel	% CO_2	Stack Minus Ambient Temperature, °F	Efficiency			% Increase in Efficiency††		
			Bacharach Slide Rule*	Cleaver-Brooks Table†	Ref. 4†,**	Bacharach	CB	Ref. 4
Natural gas	8	400	79.25	79.1	78.3	—	—	—
	8.5	300	81.75	81.0	81.8	2.50	2.1	3.5
No. 2	8	400	79.25	79.3	79.6	—	—	—
	8.5	300	83.00	83.5	83.9	4.25	4.2	4.3
No. 6	8	400	79.25	79.6	79.4	—	—	—
	8.5	300	83.00	83.6	84.2	4.25	4.0	4.8

*Natural gas with 1060–1190 Btu/ft^3; fuel oils with 0.11–0.19 H_2/C ratios.
†Natural gas with 1000 Btu/ft^3; No. 2 oil with 140,000 Btu/gal; No. 6 oil with 150,000 Btu/gal.
**Btu content or composition of fuels not specified.
††The % efficiency increase in going from 8% CO_2 at 400°F to 8.5% CO_2 at 300°F.

1. Changes in fuel quality.
2. Steam-demand changes while taking the flue-gas samples.
3. Errors dependent on both the instrument and the training of the operator analyzing the gas.

Other errors include weak chemical solutions in the Orsat, gas "pockets" in the flue gas, the sample taken too close to an elbow or pipe turn, etc.

The use of simple statistics will tell how precise the data really are and whether more samples should be taken or taken more carefully. To do this, the following definitions are needed:

$$\text{arithmetic mean} = \bar{x} = \Sigma\, x/n$$

in which x is the individual readings, and n the number of readings.

$$\text{mean deviation} = \frac{\Sigma\, |x - \bar{x}|}{n}$$

$$\text{standard deviation} = s(x) = \sqrt{\Sigma\,(x - \bar{x})^2/n - 1}$$

$$\text{confidence limit for the true mean} = s(\bar{x}) = s(x)/\sqrt{n}$$

$$\text{precision at 95\% confidence level} = \bar{X} = x \pm 1.74\,\bar{d}$$

in which $\bar{d}$ is the difference between duplicates.

Statistical terminology includes the following definitions:

1. *Accuracy:* A measure of how close the results are to the truth.
2. *Precision:* A measure of how close the check results are to each other.
3. *Probability:* 0.05 probability means 5% probable or 95% improbable.

Example 10-1

Your quality-control laboratory reports the following sulfur content in the fuel oil No. 2:

0.57, 0.75, 0.67, 0.63, 0.70%

Determine:

1. Arithmetic mean (usually called "average"):

$$\bar{x} = \Sigma\, x/n = 0.66$$

2. Medium: 0.67 (the middle value)

3. Mean deviation: $\dfrac{\Sigma |x - \bar{x}|}{n} = \pm 0.05$

Discussion. For a quick and dirty statistical analysis, the medium is most convenient and easiest to determine. The mean deviation gives an idea of how accurate the data are.

Example 10-2

After determining $\%CO_2$ and stack temperature in three different analysis, calculate the following combustion efficiencies:

76.4, 77.2, 76.8%

Determine:

1. Arithmetic mean: $\bar{x} = \Sigma x/n = 76.8\%$

2. Standard deviation: $s(x) = \sqrt{\Sigma (x - \bar{x})^2/n-1} = 0.326$

3. Confidence limit for the *true* mean:

$$s(x) = s(x)/\sqrt{n} = \frac{0.326}{1.73} = 0.188$$

Therefore, there is a 95% chance that the true mean is:

$76.8 \pm (4.30)\,(0.188) = 76.8 \pm 0.80$

or

76.0%-77.6%

Discussion. The value 4.30 is obtained from probability Table 10-2 at $n - 1$. The standard deviation is the most useful way to show the spread of the data. Thus, the three readings can be reported as ± 0.326.

There is a strong probability (95%) that the *true mean* is between 76.0% and 77.6% efficiency, i.e., 76.8 ± 0.80.

Table 10-2
***t* Values to Estimate Probabilities**

	t at Given Probability Level				
$n-1$	0.30	0.20	0.05	0.01	0.001
1	1.96	3.08	12.71	63.66	636.62
2	1.39	1.89	4.30	9.92	31.60
3	1.25	1.64	3.18	5.84	12.94
4	1.19	1.53	2.78	4.60	8.61
5	1.16	1.48	2.57	4.03	6.86
6	1.13	1.44	2.45	3.71	5.96
7	1.12	1.42	2.36	3.50	5.40
8	1.11	1.40	2.31	3.36	5.04
9	1.10	1.38	2.26	3.25	4.78
10	1.09	1.37	2.23	3.17	4.59
∞	1.04	1.28	1.96	2.58	3.29

Note: This table is also known as "double-sided test" or "two-tailed test." The *t* values for the single-sided or one-tail test are half of those shown above, although no example of this is given here.
Source: Garcia-Borras, Thomas, "Some Examples of Basic Statistics," *Hydrocarbon Processing and Petroleum Refiner,* April 1962, p. 163.

Example 10-3

Assume that the difference between CO duplicate samples in the stack has been determined to be 1.0 ppm. A single determination gives 3 ppm CO. Determine the CO content at the 95% confidence level:

$$\bar{X} = x \pm 1.74\bar{d}$$
$$= 3 \pm (1.74)\,(1.0)$$
$$= 3 \pm 1.74 \text{ ppm}$$

Discussion. Once you have determined the spread in the analysis of duplicate samples, you may save time by obtaining only one analysis and determining its precision based on past data.

Common sense must prevail. Lots of figures and miles of computer sheets do not do much when the basic input data are doubtful or when many of them are not even necessary. For instance, since we know that for every 40°F decrease in stack temperature the combustion efficiency increases by 1%, a temperature difference of 5°F means very little, since it is equivalent to a 0.125% increase (or decrease) in efficiency. On the other hand, by *decreasing* the temperature by 5°F, say from 400 to 395°F, we know we are going in the right direction. An inexpensive thermometer may be able to give a 5°F difference. If this is the case, it becomes redundant to obtain a ±0.1°F reading or to require accuracy within 1°F–2°F.

Therefore, in determining the combustion efficiency, it is necessary to:

1. Minimize obvious errors. For instance, the strength of the Orsat or Bacharach solutions should be high enough to give reliable results, the boiler should be run under steady conditions, etc.
2. Use instruments that are reliable for the accuracy desired.
3. Use statistical analysis as a tool to see how precise measurements are or should be. The precision of the analysis includes the uncontrollable variation in fuel quality, steam demand changes, however small, etc.

References

1. Dukelow, S. G., "Combustion Controls Save Money," *Instruments and Control Systems*, November 1977.
2. PB-265 713. *Measuring and Improving the Efficiency of Boilers—A Manual for Determining Energy Conservation in Steam Generating Power Plants*, Engineering Extension Service, Auburn University, Alabama. FEA Contract FEA-CO-04-50100-00, Report No. FEA/D-77/132, November 1976.
3. Dyer, David F., and Glennon, Maples, "Boiler Efficiency Improvement," Boiler Efficiency Institute, Auburn, Alabama, 1981.
4. Garcia-Borras, Thomas, "Some Examples of Basic Statistics," *Hydrocarbon Processing and Petroleum Refiner*," p 163, April 1962.

11

Instrumentation and Boiler Controls

This chapter covers the instruments that measure and control the efficiency of a boiler:

1. *Manual instruments*. These are relatively inexpensive and practical, and require little or no maintenance.
2. *Portable electronic instruments*. These instruments do the same thing the manual instruments do, but they are more expensive and require some sort of calibration and the maintenance services of the supplier or a trained technician.
3. *Boiler controls*. These include the standard jackshaft and sophisticated automatic controls where CO, O_2, or both are measured in the flue gas.

It is beyond the scope of this manual to present a complete list of all the available instrumentation and their suppliers. The emphasis will be on effective, relatively inexpensive manual instruments, and on the principles of automatic control where CO, O_2, or both are measured in the flue gas.

The addresses of some instrument suppliers are given in Table 11-1.

Manual Instruments

1. *Flue-gas kit analyzers*. The Bacharach and Dwyer kits are commonly used to measure CO_2, the smoke quality, temperature (with a dial

(text continued on page 137)

Table 11-1
Instrumentation and Its Sources

Type of Measurement	Instrument	1983 Prices (Approx.)	Sources
Flue-Gas Composition	Orsat for measuring CO_2 and O_2	$450	Burrell Corporation 2223 Fifth Ave. Pittsburgh, PA 15219 Fisher Scientific Co. 711 Forbes Ave. Pittsburgh, PA 15219
Flue-Gas Composition and Stack Analysis	Kits—complete kits to measure CO_2, draft, temperature, and smoke quality. Kits include slide rule to estimate efficiency.	250	United Technologies—Bacharach 625 Alpha Drive Pittsburgh, PA 15238
	Kits—complete kits to include all of the above plus means to measure O_2 and CO.	500	Dwyer Instruments, Inc. P.O. Box 373 Michigan City, IN 46360 Goodway Tools Corporation 404 West Ave. Stamford, CT 06902
Flue-Gas Composition	CO indicator only	$175	United Technologies—Bacharach Instruments Co. Mine Safety Appliances 408 Penn Center Blvd. Pittsburgh, PA 15235
Flue-Gas-Appearance	Smoke spot tester	$ 55	United Technologies—Bacharach Dwyer Instruments, Inc.

Table 11-1 (continued)

Type of Measurement	Instrument	1983 Prices (Approx.)	Sources
Flue-Gas Composition	Electronic O_2 and combustibles indicator	\$ 2500	Teledyne Analytical Instruments 333 W. Mission Drive San Gabriel, CA 91776 Thermox Instruments (Division of AMETEK Co.) 6592 Hamilton Ave Pittsburgh, PA 15238 Mine Safety Appliances Westinghouse 200 Beta Dr. Pittsburgh, PA 15238
Temperature	Thermometers (for room temperature and stack gases)	\$ 6–\$40	United Technologies—Bacharach Dwyer Instruments, Inc.
Temperature	Surface thermometer (to measure heat losses from piping and process equipment)	\$ 15	Condar Company Box 6 Hiram, Ohio 44234 PTC Instruments Pacific Transourcer Corporation 2301 Federal Avenue Los Angeles, CA 90064 Arthur H. Thomas Company Vine Street at Third Philadelphia, PA 19106
Air Humidity	Hydrometer (sling psychrometer)	\$100–\$500	Arthur H. Thomas Company

(table continued on next page)

Table 11-1 (continued)

Type of Measurement	Instrument	1983 Prices (Approx.)	Sources
Pressure	Draft gauges or manometers	$100	United Technologies—Bacharach Dwyer Instruments, Inc.
Sulfuric Acid Dewpoint	Sulfuric acid dewpoint meter—model 200 is portable	$ 6,000	Land Instruments, Inc. P.O. Box 1623 Tullytown, PA 19007
SO_3/H_2SO_4 Concentration in Gas Streams	Portable meter; continuous measurement	$17,000	The Rolfite Co. 300 Broad St. Stamford, CT 06901
Gas Velocity	Gas velocity meters	$ 50–$280	United Technologies—Bacharach Dwyer Instruments Inc.
Water in Fuels	Water test kit	$404	The Perolin Company 17 Church Street Rickmansworth, Hertfordshire U.K. (WD31BZ)
Fuel Pour Point	Pour point test kit	$250	The Perolin Company 17 Church Street Rickmansworth, Hertfordshire U.K. (WD31BZ)
Specific Gravity	Specific gravity test kit	$600	The Perolin Company 17 Church Street Rickmansworth, Hertfordshire U.K. (WD31BZ)

Table 11-1 (continued)

Type of Measurement	Instrument	1983 Prices (Approx.)	Sources
Temperature	Thermocouple and digital readout	\$200–\$500	Omega Engineering, Inc. Box 4047 Stamford, CT 06907 Leeds and Northrup Co. Sumneytown Pike North Wales, PA 19454
Steam Leaks	Ultrasonic detector Esthethoscope	\$500–\$900 \$ 15	UE Systems, Inc. 1995 Broadway New York, NY 10023
Water Quality	Kits to measure hardness, sulfite, alkalinity, total dissolved solids, and amines by titration	\$ 3–\$40 per test	Hach Company Box 389 Loveland, CO 80539 La Motte Chemical Co. Box 329 Chestertown, MD 21620
Water Qualtiy	Phosphates and pH by color comparison	\$ 15 per test	Hach Company Box 389 Loveland, CO 80539 La Motte Chemical Co. Box 329 Chestertown, MD 21620
Water Quality	Conductivity meter to measure total dissolved solids	\$100–\$400	Beckman Instruments PID 2500 Harbor Blvd. Fullerton, CA 92634

(table continued on next page)

Table 11-1 (continued)

Type of Measurement	Instrument	1983 Prices (Approx.)	Sources
Viscosity	Viscometer (Shell cup viscometer)	$ 90	Norcross Corporation 255 Newtonville Avenue Newton, MA 02158
Viscosity	Brookfield viscometer	$800	Brookfield Engineering Laboratories, Inc. 240 Cushing St. Stoughton, MA 02072

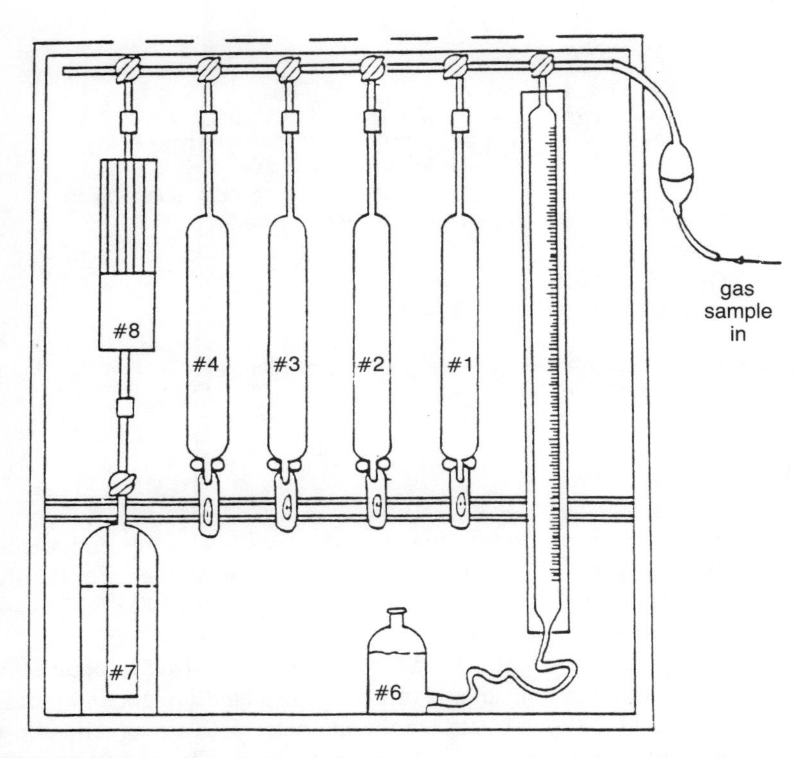

Figure 11-1. Orsat gas analyzer. A measured sample of gas is bubbled sequentially through different chemicals in absorbers 1–4. This way, the percent CO_2, O_2, CO, and illuminants are determined. A well-trained technician is required to use an Orsat. (Source: Dyer, D. F., and G. Maples, *Boiler Efficiency Improvement*, Boiler Efficiency Institute, Auburn, Alabama.)

thermometer), and draft. The kits include specially designed slide rules to calculate the combustion efficiency quickly. One such kit is depicted in Figure 7-1.

2. *Oxygen analyzer*. An Orsat, shown in Figure 11-1, or the Bacharach oxygen analyzer shown in Figure 7-1 can be used.
3. *SO_2 analyzer*. Figure 7-1 shows an easy-to-use analyzer costing less than $300 in 1983.
4. *Carbon-monoxide analyzer*. Figure 7-1 also includes a CO indicator that is especially recommended when calibrating gas boilers.
5. *Draft gauge*. Different types of draft gauges can be purchased from several suppliers. Both the Bacharach and Dwyer companies sell them. Figure 7-1 includes a draft gauge.

9318-C20 series
(shown with
9318-F50 magnetic
clip)

Cross section showing helix suspension.
Arrows indicate heat path.

Figure 11-2. Surface temperature thermometers. (Courtesy Arthur H. Thomas Company, Philadelphia, Pennsylvania.)

6. *Thermometers*.
   a. *Flue-gas temperature*: Dial thermometers and many commercial thermometers can be obtained from different sources. For the flue gas, the temperature gauge must cover 200°F–1000°F. The accuracy desired is ±5°F.
   b. *Surface temperature*: These thermometers are used to obtain heat losses through uninsulated or poorly insulated boilers and piping. Figure 11-2 shows a surface thermometer with a magnetic clip to affix it in a verticle or horizontal position. These thermometers are usually calibrated at three points and graduated in 2° intervals. The centrigrade scale is accurate within ±1°C, and the Fahrenheit scale within ±2°F.
7. *Viscometers*. The Shell Cup Viscometer from Norcross Corporation[1] has been used successfully to determine the viscosity of the fuel oil at operating conditions (see Figure 8-2). It is relatively inexpensive, and a determination can be made in less than five minutes. The following method is used to make a determination:
   a. Draw off a representative sample, say 0.5 gallons or 2 liters of fuel oil.
   b. Submerge the cup in the oil for approximately two minutes so that the temperature of the cup can reach the oil temperature.
   c. Lift the cup vertically out of the sample and record the time required for the cup to empty.
   d. From a curve provided by the suppliers, obtain the viscosity in SUS.
8. *Orsat*. An Orsat analyzes flue gas. As shown in Figure 11-1, an Orsat consists of a series of pipettes used to analyze the following components.

a. Pipette No. 1: CO_2 (use a strong potassium hydroxide solution).
b. Pipette No. 2: Illuminants, including ethylene, propylene, butylene, acetylene, benzene, and toluene.
c. Pipette No. 3: O_2 (use pyrogallol, which is a solution of pyrogallic acid and potassium hydroxide).
d. Pipette No. 4: CO (use a cuprous chloride solution).

Orsats can be ordered from Burrell and Fisher (see Table 11-1).

9. *Instruments for direct method.* The following instruments are used to estimate boiler efficiency when the direct method is employed (see Chapter 3).
   a. Flow meter: a turbine type that measures fuel and feedwater flows. It is accurate to within ±0.25% of flowrates, and is available in a 4-inch gas-turbine meter and 1 1/2-inch to 1/2-inch liquid-turbine meters. It can be ordered from ITT Barton Instruments.
   b. Pressure gauge: 0–25 and 0–160 psig gauges accurate to within ±0.5% of full scale and used to measure fuel and feedwater pressures.
   c. Thermocouple (TC): type-K chrome-alum, accurate within ±4°F, from Leeds and Northrup (model PA #8784-K). The meter used with TC is described in the following section.
10. *Steam leaks detector*. Use an ultrasonic detector from UE Systems, Inc., or a medical stethoscope.
11. *Water analysis kits*. pH meters and kits to determine hardness, sulfate content, alkalinity, total dissolved solids, and amine content by titration are available from Hach Company, La Motte Chemical Co., and others.
12. *Fuel water content measurement devices*. Perolin sells an ingenius device to determine the water content of a fuel in the range of nil to 10% in a matter of minutes.
13. *Pour point, specific gravity, compatibility, and salt-water test kits*. These can be obtained from Perolin, and were discussed in Chapter 8.

Portable Electronic Instruments

All the measurements just discussed can also be made using electronic instruments costing four to ten times as much.

1. *CO_2, O_2, and CO electronic analyzers*. Several companies have different models on the market. Some analyze CO_2, check temperature, and display the combustion efficiency. Table 11-1 shows the instruments available and the addresses of the manufacturers.

2. *Thermometers*. A digital thermometer is available from Omega Engineering, Inc., or Leeds and Northrup. A milivolt meter, accurate within ±0.35%, that reads mv output of a thermocouple, is available from Leeds and Northrup (model 914 Numotron). Infrared pyrometers costing $1,000–$2,000 are available from Omega, Mitchell, and other sources.
3. *Water-analysis device*. A conductivity meter from Beckman Instruments Co. has been used.
4. *Viscometer*. A variety of viscometers are available. One of the more versatile is the Brookfield viscometer.
5. *Dewpoint of sulfuric acid in flue gas monitor*. An instrument is available from Land Co. to actually measure the dewpoint of the acid in the flue gas.

Boiler Controls

Certain controls are needed to maintain the fuel/air ratio reasonably constant at optimum conditions. Controls are needed to:

1. Maintain optimum operating conditions.
2. Reduce boiler maintenance.
3. Improve flexibility to handle multiple fuels that vary in quality.
4. Save energy (i.e., save in operating costs).
5. Meet environmental constraints.

Figures 11-3 and 11-4 show that there is a narrow zone of maximum combustion efficiency at the point where the curves cross each other. The left curves show that when combustion air is reduced and all of the fuel is not burned, heat losses increase rapidly. The right curves show that excess air produces less drastic heat losses. In fact, the heat loss is six times faster when air is deficient than when air is in excess. Controls are needed to stay in the optimum combustion zone.

In many boiler rooms, the ''control'' is still being done by the operator who makes adjustments to control smoke, flame shape and brilliancy, carbon build-up, smell, etc. But the operator cannot tell if he is wasting 10%–30% fuel or more.

Poor boiler control also means more maintenance. For instance, flame impingement on the tubes can be caused by too much excess air, which adds velocity to the air-fuel mixture and changes the shape of the flame. This may result in more tube failures. Insufficient air will generate excessive soot, which reduces heat transfer and boiler efficiency. Widely varying temperatures because of frequent on/off ''controls'' can cause thermal shock and spalling and cracking of brick work.

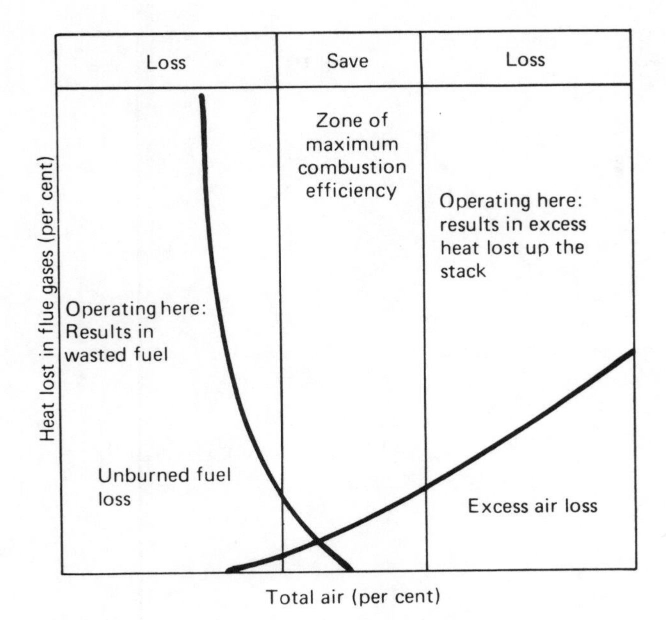

Figure 11-3. Fuel-saving methods simplified: heat loss vs. total air. (Source: PB-264 543, "Guidelines for Industrial Performance Improvement (Boiler Adjustment Procedures to Minimize Air Pollution and to Achieve Efficient Use of Fuel), by M. W. McElroy and D. E. Shore, KVB, Inc., EPA Contract 68-02-1074, Report No. EPA 600/8-77-003a, January 1977.)

It is beyond the scope of the manual to include an in-depth discussion about automatic controls. Instead, the following common controls will be covered in some detail.

Fuel-oil preheater. Strictly speaking, this is not a boiler control. Figure 11-5 is a photograph of a typical preheater, showing both the electric heating system (used on boiler startup when no steam is available) and steam heating system that takes over after steam becomes available. The boiler operating manual includes a description of the preheater and its controls. As shown in Method 4 of Chapter 3, the preheater temperature has a great influence on boiler efficiency improvement.

Combustion controls. There are two purposes of combustion control: (1) to maintain load, i.e., maintain the steam quality (pressure and temperature) as more and more steam is being used in the plant; and (2) to maintain the proper air/fuel ratio. To achieve these controls the following systems are used.

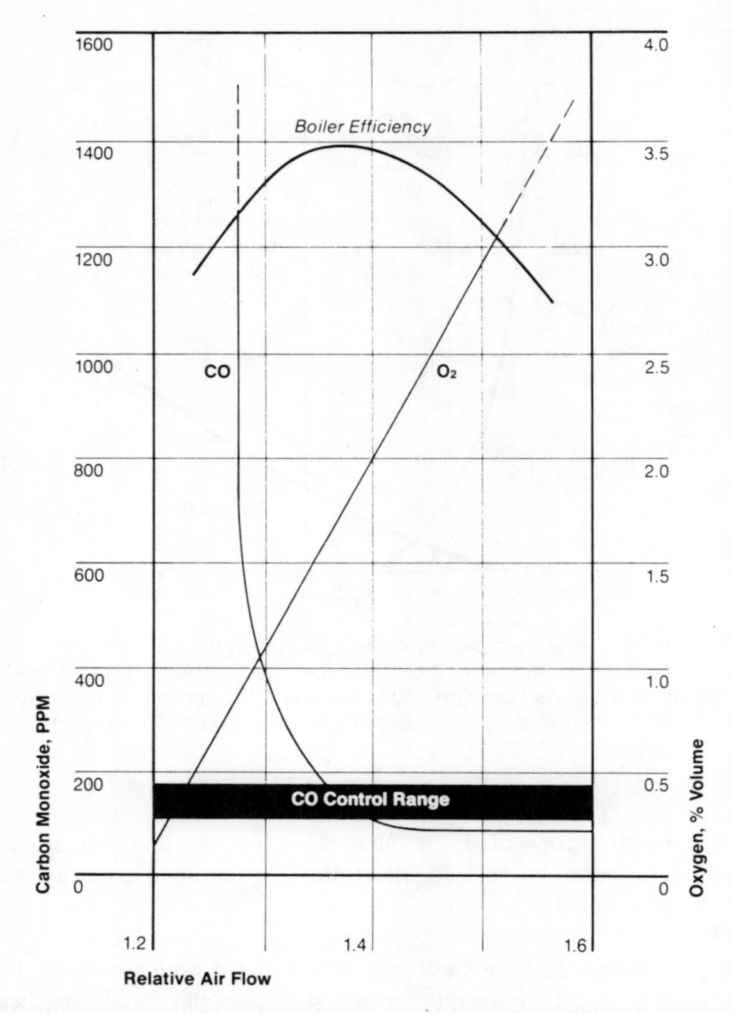

Figure 11-4. Carbon monoxide control range: CO vs. O_2 vs. air flow vs. boiler efficiency.[3]

Positioning systems. When steam pressure drops below a certain preset level, the air and fuel valves are opened. These valves close when the steam pressure rises above another preset level.

1. *High/low/off*. This system is similar to the on/off system except that there are two air- and fuel-control valve positions that give a "low" and "high" firing rate, depending on the preset valves of the steam pressure.

Figure 11-5. Fuel-oil preheater. (Courtesy Cleaver-Brooks.)

2. *Modulating positioning*. The settings on the air- and fuel-control valves are continuously varied. Settings are determined by the load. The linkage can be adjusted at one firing position.

Metering systems.

1. *One-Flow Metering System*. Either air or fuel is controlled according to load, while the other flow (fuel or air) is controlled in proportion to the position of the element controlling the first input.
2. *Full-Metering*. Either air or fuel is controlled according to load, while the second flow is measured and controlled proportionally to the flow of the first.

Control Systems

Mechanical systems. The jackshaft shown in Figure 11-6 is the simplest of the controllers. It operates as follows: As the steam pressure drops, a set point is reached. The jackshaft then moves, activating the fuel valve, and the air is proportionally increased. The fuel-valve setting is adjusted by the cam curvature, over which the fuel pin travels. Figure 11-7 shows a typical modulating cam. Each inividual cam screw must be adjusted to yield the

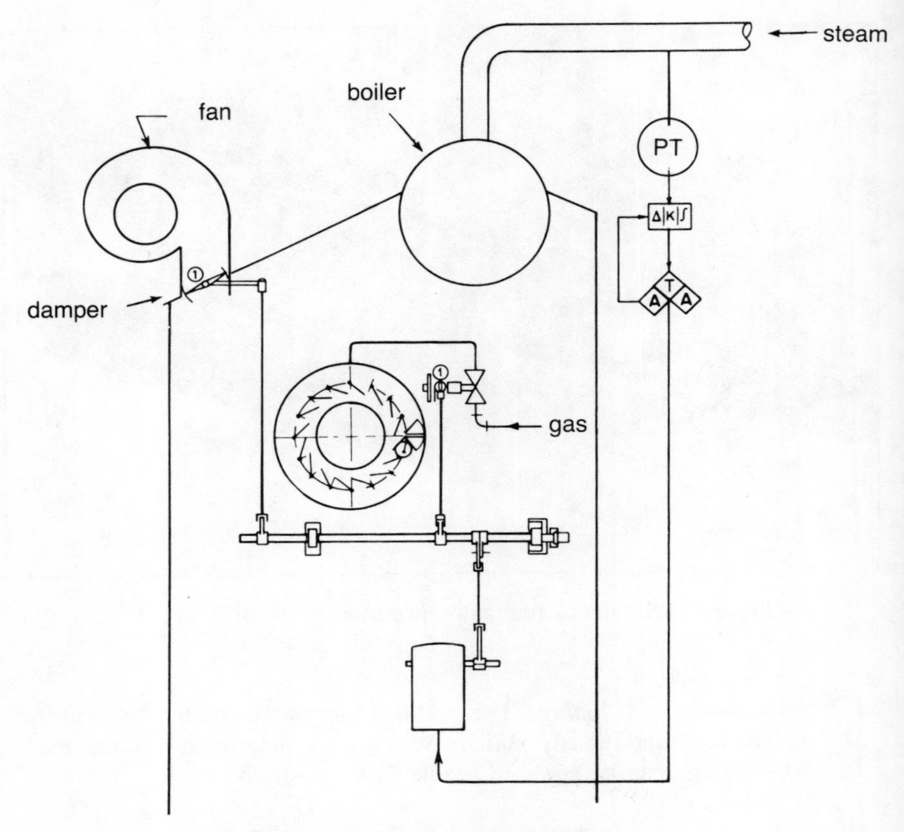

Jackshaft Combustion Control
(Positioning Type)

Figure 11-6. The simplest boiler control system.

maximum efficiency at that load. The air must likewise be adjusted to maintain the proper air/fuel ratio, say at 10%–15% excess air.[2]

Figures 11-8 and 11-9 show the oil control valve assembly for fuel oil. Figure 11-10 shows the electronic gear inside the control box. The disadvantage of the mechanical system is that the linkage wears out, eventually causing some loss of motion. Since the air/fuel ratio is fixed manually under certain conditions (atmospheric and fuel quality), any changes in content of the fuel and atmospheric conditions will not be controlled by the mechanical system. Therefore, excess air may vary widely, as discussed in Chapter 3.

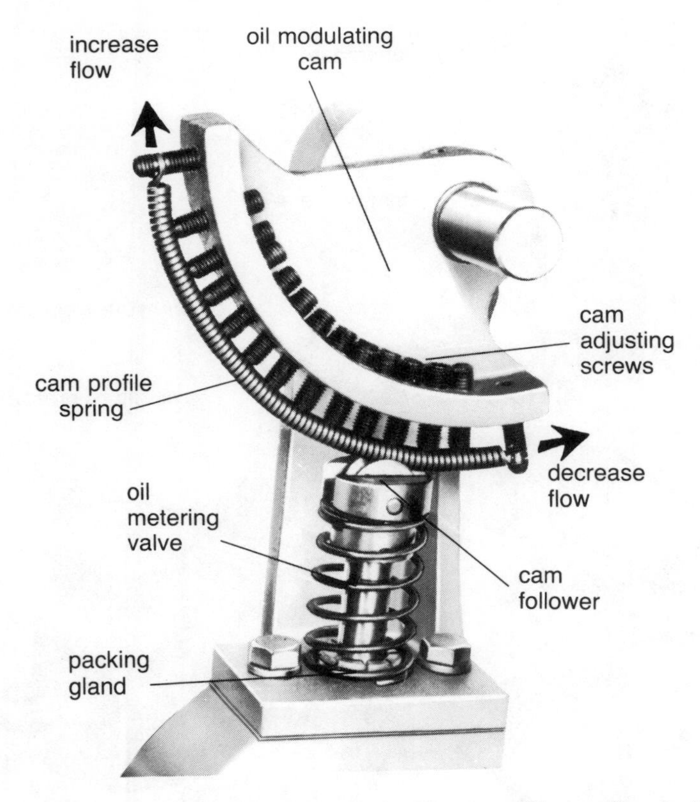

Figure 11-7. Oil modulating cam. (Courtesy Cleaver-Brooks.)

Despite these shortcomings, daily or weekly checks on combustion efficiencies and a boiler tune-up when needed can save a significant amount of fuel.

Pneumatic or electronic systems. These controls take the place of the mechanical linkage. These are not significantly better than the mechanical system, although they are in use.

Analyzers in the loop. Analyzers are available to monitor the components in the exhaust gas, including:

1. Oxygen (O_2) (use a zirconium oxide analyzer or a paramagnetic analyzer).

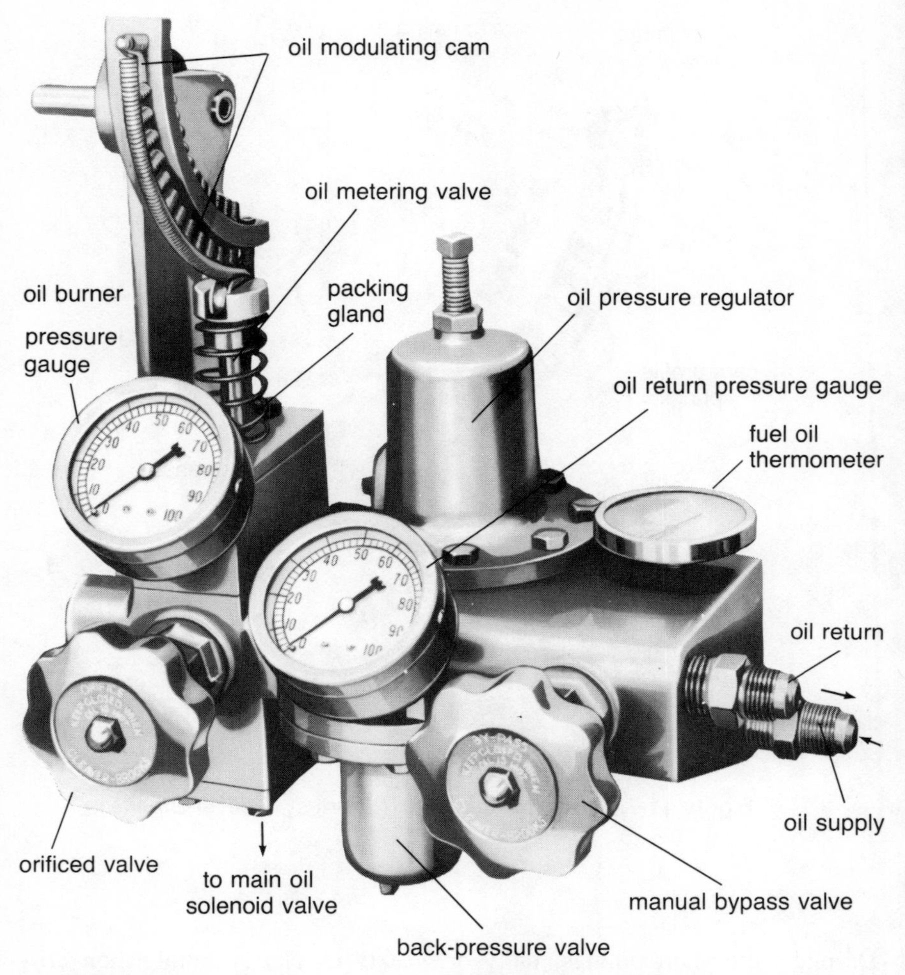

Figure 11-8. Control-valve assembly for heavy oil. (Courtesy Cleaver-Brooks.)

2. Carbon monoxide CO (infrared method).
3. Hydrocarbons (HC) (flame ionization).
4. Nitric oxide (NO) (cheminuminescence).
5. Smoke opacity (visible-light transmission).

The most popular analyzers now are the O_2 and CO analyzers, either alone or combined. Some operators also monitor CO_2, HC, and smoke. SO_2 and

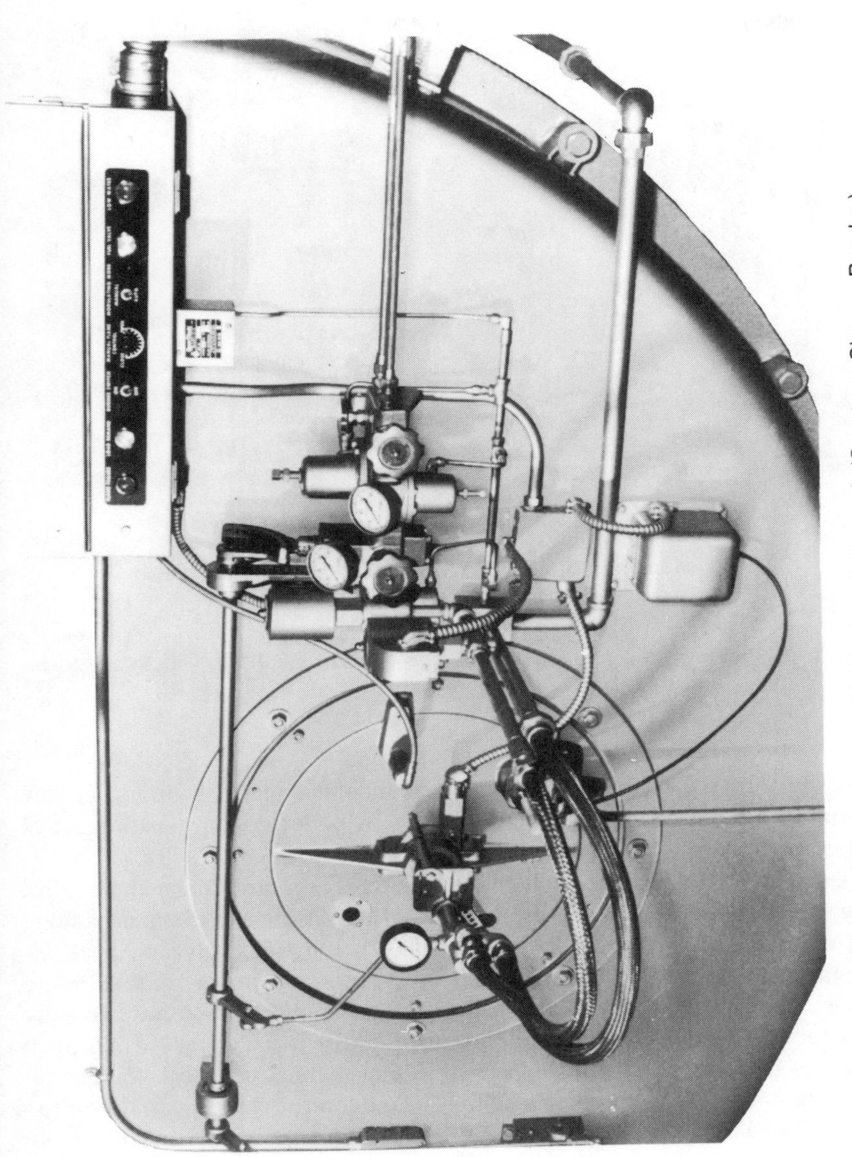

Figure 11-9. Control-valve assembly and control panel. (Courtesy Cleaver-Brooks.)

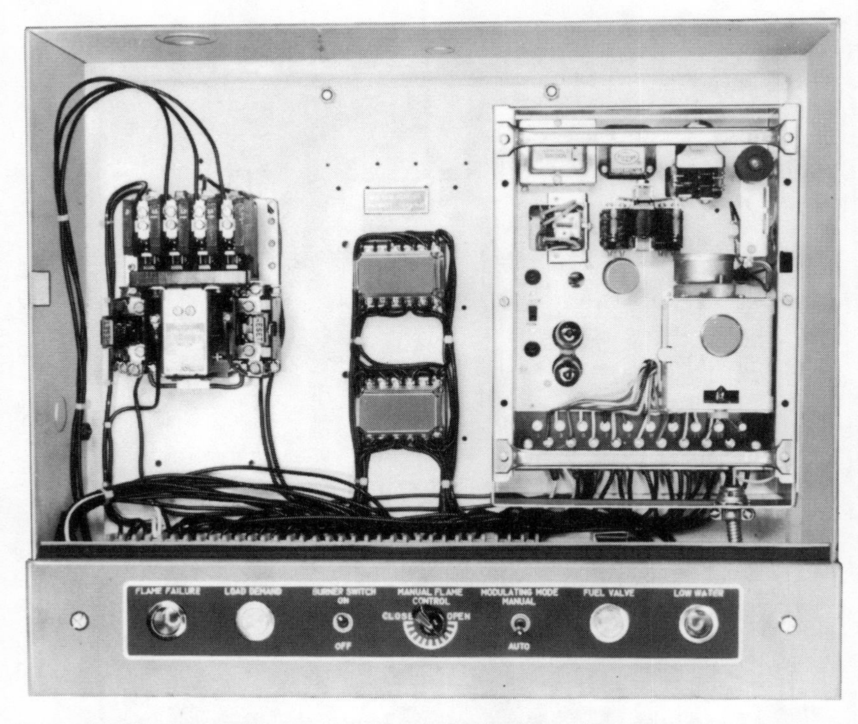

Figure 11-10. Electronic gear inside boiler control box. (Courtesy Cleaver-Brooks.)

NO_X are often measured to meet environmental emissions limits. A few operators measure SO_3. Figure 11-11 shows a boiler system where most of these components are measured.

There is no agreement on whether the O_2 or CO control alone can control any boiler at peak efficiency all the time. There is some agreement among the users and suppliers that by using both O_2 and CO analyzers, peak efficiencies are more readily obtained. However, let us emphasize that before a sophisticated control system is installed, your boiler should have already undergone an energy audit attempting to improve its efficiency using most or all of the 20 methods in Chapter 3. In other words, the best you can do must be compared to the best a sophisticated control system can do on top of it. The two most used controls for boilers will now be discussed in more detail: CO, O_2, and a combination of both.

CO controls.[3,4,5] The most precise indication of efficient boiler operation is the carbon monoxide (CO) content in the flue gas. If your boiler operates

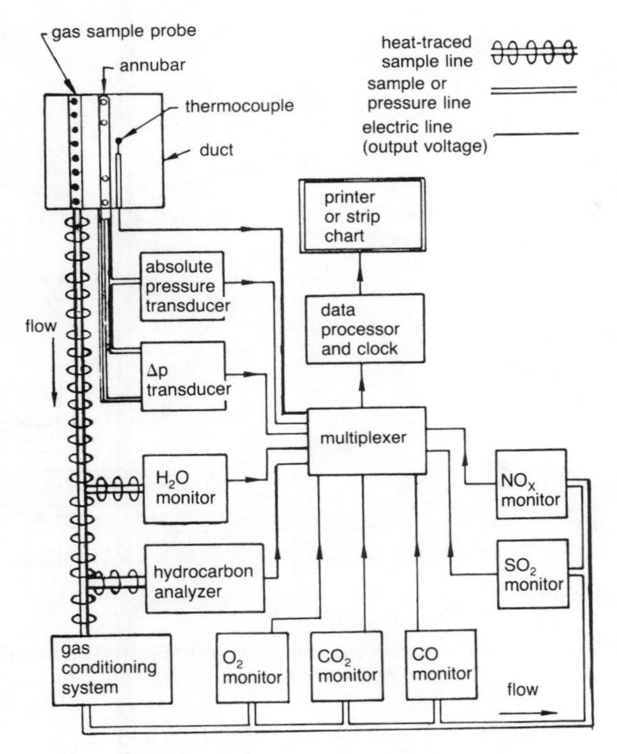

Figure 11-11. Combustion efficiency system monitoring several variables.[6]

with a flue gas of 100–350 ppm, you have done all you can do concerning *combustion* efficiency. However, there are problems in installing CO monitors:[3]

1. Reliable electronic CO monitors cost about $20,000–$30,000. This is expensive unless you have large boilers consuming large quantitites of fuel.
2. A CO monitor with automatic control costs about $10,000–$22,000 per year to maintain.
3. If your boiler is already being controlled properly (either manually or with an O_2 monitor), the CO monitor may only improve your combustion efficiency by 1%. Still, a 1% improvement is important if it is economically feasible for a particular boiler size.
4. The CO content is almost unaffected by high excess-air levels; therefore, you can have low CO content with high excess air.

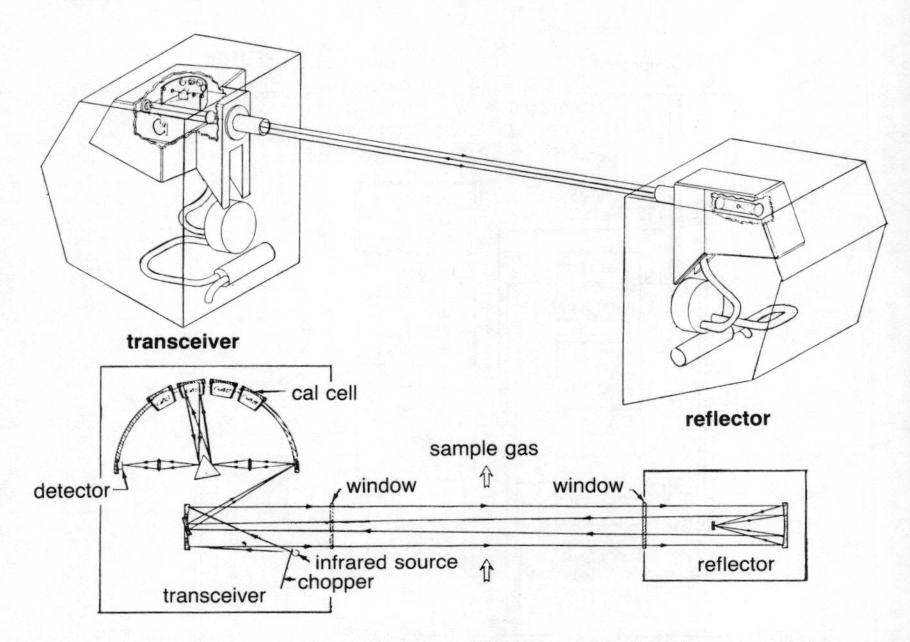

Figure 11-12. Carbon monoxide determination by infrared method installed across the stack. (Courtesy The Syconex Corporation.)

There are advantages in having CO monitors:

1. Optimum combustion efficiency is, as stated before, controlled at 100–350 ppm of CO and is not affected by the fuel, burner, or boiler design.
2. Air infiltration has practically no effect on CO concentration and thus gives a better indication of combustion efficiency.
3. As air is reduced below the stoichiometric value, i.e. the optimum air needed, the CO content rises sharply, giving a rapid response to air deficiency.
4. Accuracy is on the order of ±10 ppm.

Figure 11-12 is a schematic of a CO monitor system. The basic principle is to pass a beam of infrared radiation through the flue gas and measure the absorption at the particular frequency for carbon monoxide. To keep the windows of the instrument clean, continuous air purges are used.

O_2 controls. Maintaining an excess air of 10%–15% and monitoring the O_2 content in the flue gas can easily improve combustion efficiency 1%–

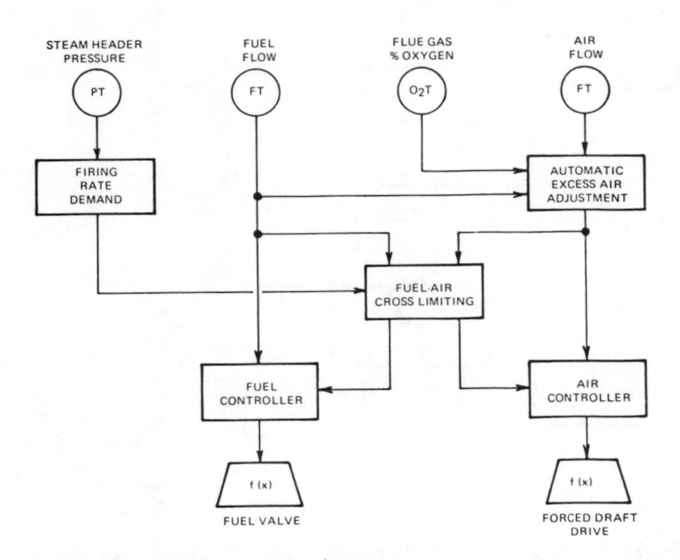

Figure 11-13. Integrated boiler controls using fuel and air flow controllers and oxygen measurement. (Courtesy Bailey Controls Company, a Division of Babcock and Wilcox.)

3%. With an automatic trim, the improved efficiency may be increased to 4%.

The advantages of an O_2 monitor are:

1. Initial cost: about $6,000 (3–5 times less expensive than a CO monitor).
2. Maintenance cost: $5000–$10,000 per year (half that of a CO monitor).
3. The oxygen content is directly proportional to excess air over the range 0%–8% (if no leakage or combustibles are present).

Figures 11-13 and 11-14 depict oxygen controllers.

The disadvantages are:

1. If infiltration air is present, the O_2 readings are high, giving an inaccurate impression of boiler control.
2. Optimum O_2 level is affected by the type of fuel, burner design, and the load of the boiler.
3. Accuracy of the readout is ±.25%–0.3% O_2 content (well below the CO-monitor accuracy of ±10 ppm).

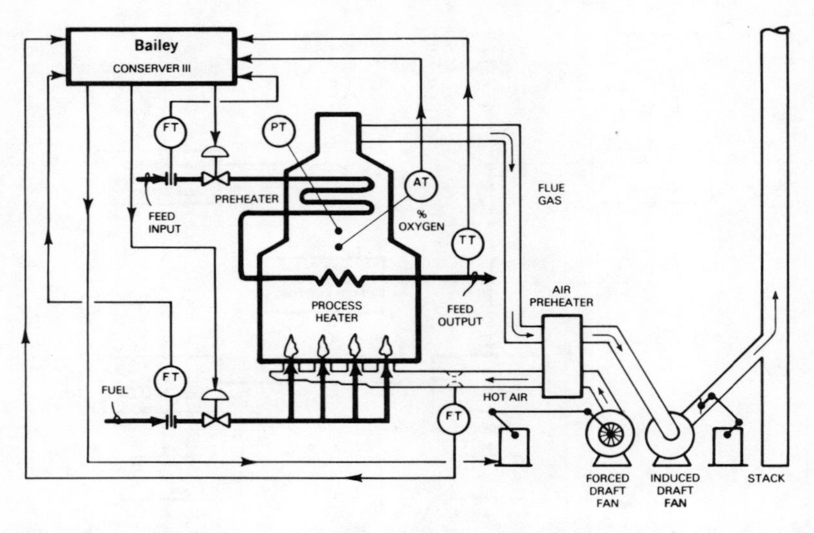

Figure 11-14. Process heater with oxygen control. (Courtesy Bailey Controls Company, a Division of Babcock and Wilcox.)

4. More excess air is needed at low loads because of the poorer mixing of the air and fuel.

The zirconium oxide cell used to control O_2 operates as follows: The cell generates a voltage that changes suddenly when the oxygen-rich atmosphere is exposed to changes in a reducing environment.

A combination O_2—CO control is often recommended to avoid the pitfalls of either one. The main pitfall, says one authority,[5] is the cost of the controllers without the benefit of an economic feasibility analysis.

O_2—CO controls. For a boiler already controlled manually with scheduled CO_2—O_2 and CO analyses (say, one check per week), the installation of an O_2—CO monitor may improve the combustion efficiency 2%–5%. These controls imply higher initial investment costs ($24,000–$36,000) and higher maintenance costs ($15,000–$32,000/year). An energy audit and an economic analysis may or may not justify the installation of an O_2—CO monitor. When the installation is justified, this dual control system should be operated as follows:

1. Couple the O_2 monitor to an automatic trim control and keep the excess air within a preset range.
2. Fine-tune the boiler with CO measurement and control.

Table 11-2
Estimated Continuous Measurement System Labor Requirements

Activity*		Estimated Labor,* Hours	
		Engineering	Technician
Acquisition	(Total)	(108–284)	(52–124)
Formulation of requirements		16–40	0
Site selection		8–16	0
Preliminary measurements for instrument scaling		4–8	8–16
System design and component selection		40–120	16–32
Equipment purchase		20–40	8–16
Instrument checkout		20–60	20–60
Installation	(Total)	(88–336)	(104–344)
Manual traverses for in-steam component location		8–16	16–40
System integration, installation and checkout		20–120	40–200
Manual traverses for in-place calibration		20–80	40–80
Writing of operational procedure		40–120	8–24
Use	(Total)	(108–440)/yr	(533–1498)/yr
System checkout (daily)		0	1–2
Data review (weekly)		1–4	1–4
Calibration check by manual traverse (monthly)		4–16	8–40
General refurbishing (annual)		8–40	20–80

*Assumes at least one gas species concentration and that the system measures total volumetric flow.

Source: EPA-600-2-76-203, *Flow and Gas Sampling Manual,* Industrial Environmental Research Laboratory, Office of Research and Development, U.S. Environmental Protection Agency, Research Traingle Park, North Carolina, 27711, July 1976.

CO and other elements. When O_2 is not monitored, other elements may be analyzed in addition to CO. The most common elements analyzed include smoke and O_2 with or without CO, and smoke, CO_2, and CO.

EPA[6] has determined the man-hours required to install and check a continuous flue-gas analyzer, as shown in Table 11-2.

References

1. Norcross Corporation, 255 Newtonville Ave., Newton, Mass., 02158.
2. Keller, R. T., ''Better Combustion Control. Who Needs It?'' Bailey Controls Company, Technical Paper TP 789, November 1978.
3. Reason, John, ''When it Pays to Monitor Flue-Gas CO,'' *Power*, p. 37, August 1981.

4. Gilbert, Lyman F., ''CO Control Heightens Furnace Efficiency,'' *Chemical Engineering*, p. 69, July 28, 1980.
5. Lord, Harry C., The Syconex Corporation, ''In-Situ Monitoring: Operational Experience Bolsters Reliability,'' p. 88, *Power*, May 18, 1982.
6. EPA-600-2-76-203, *Flow and Gas Sampling Manual*, Industrial Environmental Research Laboratory, Office of Research and Development, U.S. Environmental Protection Agency, Research Triangle Park, North Carolina 27711, July 1976.

12

Economics

In any situation where a capital investment is contemplated, a series of four steps must be followed.[1]

1. Detailed design.
2. Determination of investment cost. The cost of the new piece of equipment can be obtained from suppliers or from published data.[2,3]
3. Determination of economic feasibility. Is the new design economically sound?
4. Report. Data presentation in a concise, understandable form.

The first two steps, detailed design and investment-cost determination, are beyond the scope of this manual. An in-house engineer or an outside consultant should be able to do these jobs. The next two steps are covered here.

Economic Feasibility

The National Bureau of Standards publishes a paper entitled "Energy Conservation Program Guide for Industry and Commerce."[4] This chapter makes use of their recommendations and data.

Basic Definitions

First cost (FC): the initial cost of a capital expenditure such as a new boiler, a new preheater, or a new additive feed system. FC includes materials and labor. This cost can be obtained from the suppliers or from the literature.[2,3]

Annual operating cost (AOC): the costs of operating and maintaining the new system after its installation. For example, the cost for a new additive feed system may include the costs of electric power consumed by the feed pump and mixer, and of the labor to operate the system (loading the materials, draining the water, mixing in the tank, and cleaning the tank during shutdown periods, and supervising the feed system while in operation).

Annual fuel savings (AFS): the estimated savings received if the new system is installed. For instance, if an automatic CO or O_2 control is contemplated, the savings must be calculated on the basis of fuel savings *above* what can be done using the methods discussed in Chapter 3.

It is deceiving to imply that the new device, costing perhaps $50,000 or $150,000, will improve the combustion efficiency by 5% (i.e. saving 5% fuel per year), when the same results can be obtained easily by manually calibrating the boiler once a week.

Projected fuel price (PFP). This is a difficult item. Even in the short run, the estimated fuel price could be off. Local fuel suppliers should be contacted for their own price projections.

Estimated equipment life (EL): the period the equipment is estimated to last. Because this is used for tax purposes, the appropriate tax tables must be consulted to determine what the IRS considers reasonable equipment life.

Net annual savings (S): This is calculated as follows: (AFS × PFP) − AOC

Depreciation Charge (DC): Using the straight line depreciation for tax purposes, this would be FC/EL.

Preliminary Estimates

The preliminary estimates are simple and as such are more often used than the detailed feasibility studies. When the preliminary estimates are very attractive, a detailed study may be made merely to support what is already known: that the project should be undertaken.

On the other hand, when preliminary studies show marginal economic performance of the new equipment, a detailed study *must* be undertaken.

Payback period: PP

$$PP = FC/S$$

This estimate tells how long it will take to recover the capital investment compared to the estimated life (EL). If the payback period is less than one half of the estimated life, the investment is usually considered potentially profitable. Of course, the payback period cannot be used to compare different types of investments.

Return on investment: ROI

$$ROI = \frac{S - DC}{FC} \times 100\%$$

The rate of return should be at least as good as any other investment the company might make.

For instance, if the ROI is 8% and banks pay 15% return, the project should probably be cancelled. But other considerations may enter into the picture, which is why more detailed feasibility studies must be made.

Definitive Estimates

The time value of money is usually known as a discount rate (D). This rate represents the best internal rate on any investment opportunities.

Present worth factor (PWF): The number of years over which the net annual savings will be applied to arrive at these savings. The PWF is a function of both the discount rate and the estimated equipment life.

Table 12-1 gives the PWF at different discount rates (5%–25%) against the estimated equipment life, up to 25 years.

Present value (PV): The present value of all the future savings is estimated as follows:

$$PV = PWF \times S$$

Benefit/cost ratio (B/C): This estimate is done as follows:

$$B/C = \frac{PV}{FC} = \frac{PWV \times S}{FC}$$

When the B/C ratio is greater than one, the project is feasible.

Example 12-1

An additive firm has a boiler that burns fuel oil No. 6, and 9000 lbs/hr of flue gases leave at 580°F. Since the processing plant uses 80 gpm of hot water at 190°F for the blend tanks, it is desirable to estimate whether or not the water could be preheated economically by the flue gases. Presently, the hot water is generated by individual electric heaters on each kettle.

Table 12-1
Present Worth Factors (PWF)

Lifetime (EL)	5%	10%	15%	20%	25%
1	0.952	0.909	0.870	0.833	0.800
2	1.859	1.736	1.626	1.528	1.440
3	2.723	2.487	2.283	2.106	1.952
4	3.546	3.170	2.855	2.589	2.362
5	4.329	3.791	3.352	2.991	2.689
6	5.076	4.355	3.784	3.326	2.951
7	5.786	4.868	4.160	3.605	3.161
8	6.463	5.335	4.487	3.837	3.329
9	7.108	5.759	4.772	4.031	3.463
10	7.722	6.145	5.019	4.192	3.571
11	8.306	6.495	5.234	4.327	3.656
12	8.863	6.814	5.421	4.439	3.725
13	9.394	7.103	5.583	4.553	3.780
14	9.899	7.367	5.724	4.611	3.824
15	10.380	7.606	5.847	4.675	3.859
16	10.838	7.824	5.954	4.730	3.887
17	11.274	8.022	6.047	4.775	3.910
18	11.690	8.201	6.128	4.812	3.928
19	12.085	8.365	6.198	4.843	3.942
20	12.462	8.514	6.259	4.870	3.954
21	12.821	8.649	6.312	4.891	3.963
22	13.163	8.772	6.359	4.909	3.970
23	13.489	8.883	6.399	4.925	3.976
24	13.799	8.985	6.434	4.937	3.981
25	14.094	9.077	6.464	4.948	3.985

1. *Estimate Btu available from flue gases*. Assume that the flue gas can be cooled to 230°F. Use 0.25 Btu/lb × °F as the specific heat Cp for the gases:

$$(\text{lb gas})(\text{specific heat})(\Delta t) = (9000)(0.25)(580 - 230) = 787{,}500 \text{ Btu/hr}$$

2. *Estimate annual fuel savings*:

$$\text{AFS} = (365 \text{ days})(24 \text{ hrs/day})(787{,}500 \text{ Btu/hr}) = 6899 \times 10^6 \text{ Btu}$$

3. *Estimate water requirement*. Water comes into the plant at 72°F:

$$(\text{lb water})(\text{Cp})(190 - 72) = 787{,}500 \text{ Btu/hr}$$

In other words, the amount of heat to be picked up by the water must equal the available heat from the flue gases under the given conditions. Therefore:

$$\text{lb water} = \frac{787{,}500}{(1)(118)} = 5656 \text{ lbs/hr} = 94 \text{ lbs/min}$$

$$94 \div 8.3 = 11.3 \text{ gal/min}$$

Since 80 gpm was required, the hot water obtained from the heat of the flue gases can be used in the plant.

4. *Design*. A heat exchanger that heats 80 gpm water costs \$5000–\$7000 installed, with an estimated 5%–8% annual operating cost. The estimated life of the heat exchanger is 15 years. The company uses a 25% discount rate. The fuel price is projected to be \$2.50 per million Btu.
5. *Summary of Data:*

$$\text{FC} = \$6000 \text{ (cost of heat exchanger)}$$

$$\text{AOC} = (6.5\%)(6000) = \$390$$

$$\text{AFS} = 6899 \times 10^6 \text{ Btu}$$

$$\text{PFP} = \$2.50 \text{ per million Btu (fuel No. 6)}$$

$$\begin{aligned} \text{S} &= \text{AFS} \times \text{PFP} - \text{AOC} \\ &= (6899 \times 10^6)(\$2.50)/10^6 - 390 \\ &= \$16{,}858 \text{ per year} \end{aligned}$$

$$\text{D} = 25\%$$

6. *Preliminary economics*. The payback period is:

$$\text{PP} = \frac{\text{FC}}{\text{S}} = \frac{\$6000}{16{,}858} = 0.36 \text{ years}$$

The investment appears attractive at this point. The return on investment is:

$$\text{DC} = \frac{\text{FC}}{\text{EL}} = \frac{\$6000}{15} = \$400 \quad \text{(straight-line depreciation)}$$

$$\text{ROI} = \frac{\text{S} - \text{DC}}{\text{FC}} \times 100$$

$$= \frac{\$16{,}858 - 400}{\$6000} \times 100 = 274\% \text{ ROI}$$

Table 12-2
Cost-Estimate Classifications by Accuracy

Type	Accuracy, ± %
Order-of-magnitude (ratio estimate)	40
Study (factored estimate)	25
Preliminary (budget authorization estimate)	12
Definitive (project control)	6
Detailed (firm)	3

Undoubtedly this is a good return and far above the company's best internal rate of return.

7. *Definitive estimate*. From Table 12-1, the present worth factor is:

$$PWF = 3.859$$

The present value is:

$$\begin{aligned} PV &= S \times PWF \\ &= (\$16{,}858)(3859) \\ &= \$65{,}055 \quad \text{present value of future savings} \end{aligned}$$

The benefit cost ratio is:

$$B/C = \frac{PV}{FC} = \frac{\$65{,}055}{\$6000} = 10.84$$

8. *Report*.

Bear in mind that cost estimates have been classified by the American Association of Cost Engineers according to probable accuracies, as shown in Table 12-2. Therefore, a cost estimate must include its accuracy as part of the report. Even a firm estimate (such as quotes from suppliers and builders) must show a ±3% accuracy.

References

1. Garcia-Borras, Thomas, "Research-Project Evaluations—Parts I and II," *Hydrocarbon Processing*, p. 137, December 1976; p. 171, January 1977.
2. Guthrie, K. M., *Process Plant Estimating, Evaluation, and Control*, Craftsman Book Company of America, 1974.

3. Peters, M. S., and K. D. Timmerhaus, *Plant Design and Economics for Chemical Engineers*, McGraw-Hill Book Co. (latest edition).
4. Gatts, R., R. Massey and J. C. Robertson, ''Energy Conservation Program Guide for Industry and Commerce—Financial Evaluation Procedures,'' National Bureau of Standards, 1974, pp. 5.1–5.5.

13

Engineering Calculations

The following examples can serve as models to perform certain engineering calculations for a typical energy audit in a boiler room, or for a furnace, dryer, or steam system.

Heat Losses from a Hot Surface to Still Air

To reduce heat losses in a steam system there are two possible approaches: insulate or reduce steam pressure and temperature. In a furnace or dryer, heat losses are reduced by insulating properly. The basic equation to calculate the heat loss per unit area for an insulated pipe is given as follows:

$$\frac{Q}{A} = \frac{\Delta T}{\left(r_2 \ln \frac{r_2}{r_1}\right) \Big/ K_i + \frac{1}{h_c}}$$

in which:

r_1 = bare-pipe outer radius
r_2 = radius of outside insulation surface
K_i = thermal conductivity of insulation
h_c = combined convective/radiative heat-transfer coefficient
ΔT = difference between temperature inside pipe (stream temperature) and outside ambient temperature

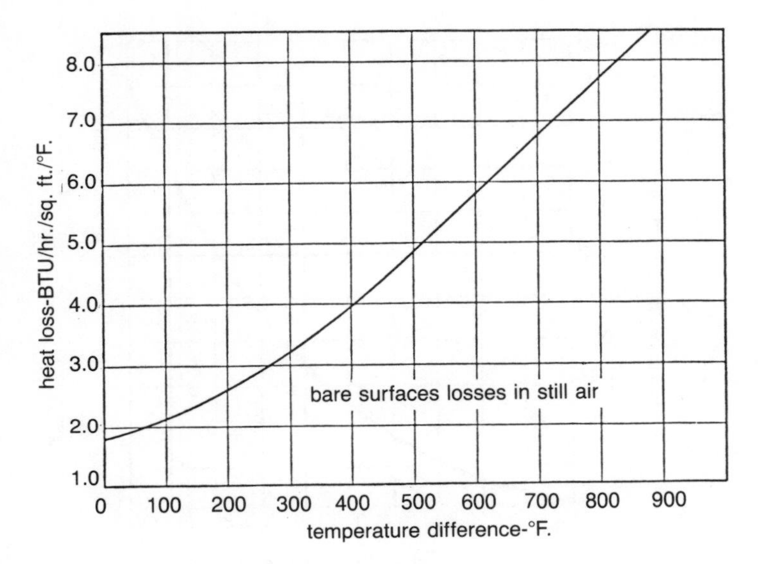

Figure 13-1. Convection-radiation heat-transfer coefficient vs. ΔT. (Courtesy Johns-Manville.)

This is a simplified equation that ignores the thermal resistance of the pipe wall and the steam-to-pipe film resistance.

For a flat surface, the simplified heat-loss equation becomes:

$$\frac{Q}{A} = \frac{\Delta T}{\dfrac{\Delta L}{K_i} + \dfrac{1}{h_c}}$$

where ΔL represents the insulation thickness.

Figure 13-1 gives the heat-transfer coefficient h_c as a function of the temperature difference ΔT. Figure 13-2 can be used to estimate heat loss directly.

Example 13-1

Steam at 290°F is transmitted through an 8-inch outside diameter pipe, with an ambient temperature of 80°F. Determine:

1. Heat loss per foot of pipe with no insulation.
2. Heat loss per foot of pipe with one-inch insulation, with a thermal conductivity K_i of 0.5 Btu-in./hr × ft^2 × °F (K_i must be obtained from the supplier)

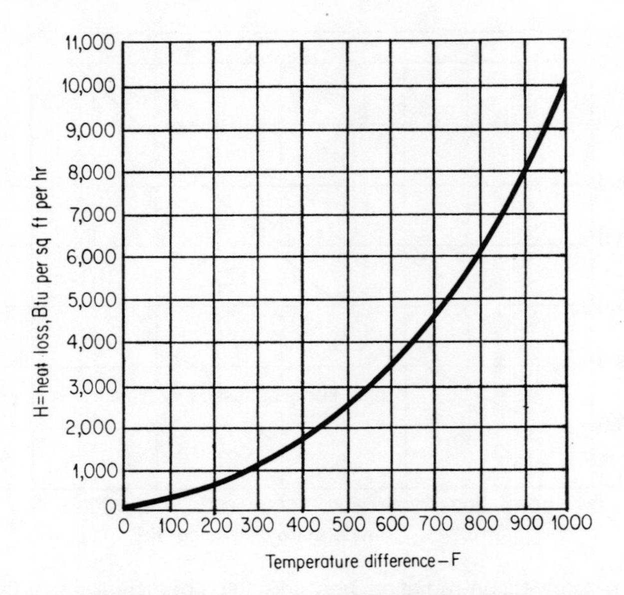

Figure 13-2. Heat loss (convection + radiation) vs. temperature difference. (Courtesy Johns-Manville and *Power* magazine.)

3. Money lost per year from 1000 ft of uninsulated pipe, if steam costs \$10.00 per million Btu.

Solution

1. From Figure 13-1, for 210°F (290° − 80°F), h_c is 2.7. Therefore:

$$Q/A = (210°)(2.7) = 567 \text{ Btu/hr} \times \text{ft}^2$$

$$A = \pi D L$$

where D is the pipe diameter. Therefore:

$$A = (8 \text{ in.})(1 \text{ ft}) \div 12 \text{ in./ft} = 2.09 \text{ ft}^2$$

Heat loss per foot of pipe is:

$$(Q/A)(A) = [567 \text{ Btu/(hr} - \text{ft}^2)][2.09 \text{ ft}^2] = 1185 \text{ Btu/hr}$$

2. Proceed as follows:

$$r_2 \ln \frac{r_2}{r_1} = 5 \ln \frac{5}{4} \doteq 1.12$$

$$\frac{Q}{A} = \frac{210°F}{\frac{1.12}{0.5} + \frac{1}{2.7}} = 80.5 \text{ Btu/hr} - \text{ft}^2$$

$$Q = (Q/A)A = (80.5)(2.09) = 168.2 \text{ Btu/hr}$$

3. The annual heat loss is:

$$Q = (1000)(1.185)(8000 \text{ hrs/year}) = 9480 \times 10^6 \text{ Btu}$$

Cost:

$$(9480 \times 10^6 \text{ Btu})\left(\frac{\$10}{10^6 \text{ Btu}}\right) = \$94{,}800/\text{year}$$

The economy of adding one or more inches of insulation is determined by balancing the cost of the insulation against the cost of the energy loss. Plotting the data, i.e. costs vs. savings, will show an optimum insulation thickness known as the "economic insulation thickness."

Therefore, the heat savings from insulation are:

$$1185 - 168.2 = 1{,}016.8 \text{ Btu/hr}$$

Example 13-2

A dryer that is 150 ft long, 13 ft high, and 15 ft wide is operated at 202°F, and the ambient temperature is 75°F. The dryer has 1.5 in. of insulation, with a thermal conductivity of 0.35 Btu-in/(hr × ft^2 × °F). Determine:

1. Present heat losses.
2. Heat losses using one additional inch of the same insulation.
3. Energy savings using additional insulation.

Solution

1. From Figure 13-1, for 127° (202° − 75°), h_c is 2.25. The present heat loss per ft^2 is:

$$\frac{Q}{A} = \frac{\Delta T}{\Delta L/K_i + 1/h_c} = \frac{127}{1.5/0.35 + 1/2.25}$$
$$= 26.85 \text{ Btu/hr} \times \text{ft}^2$$

Total area:

$$A = (2 \times 150 \times 13) + (1 \times 150 \times 15) + (2 \times 13 \times 15) = 6540 \text{ ft}^2$$

Therefore:

$$(Q/A)(A) = (26.85)(6540) = 175{,}599 \text{ Btu/hr}$$

2. $Q/A = 16.75$

   Therefore: $(16.75)(6540) = 109{,}545$ Btu/hr

3. $175{,}599 - 109{,}545 = 66{,}054$ Btu/hr

Dryer Calculations

To reduce operating costs, a dryer may be operated using higher humidities in the exhaust vent. The basic equation to make this type of calculation is:

Air Flowrate:

$$M_{air} = M_{water}/W$$

in which

M_{air} = mass flowrate of air in lbs/hr
M_{water} = mass flowrate of water in lbs/hr
W = specific humidity of air-water vapor moisture in lb water/lb air

Sensible heat content of exhaust gas, E in Btu/hr, is given by the following equation:

$$E = M_{water\ vapor} \cdot Cp \qquad \Delta T = M_{air} \cdot Cp \Delta T$$

Where

ΔT = difference between exhaust and ambient temperatures
Cp = specific heat, Btu/lb × °F

Example 13-3

Nine hundred lbs/hr of water is removed from a product in a process dryer. The exhaust temperature is 390°F. Estimate the gain obtained if the dryer were operated at 0.4 lb water/lb air instead of 0.04 lb water/lb air, using 75°F as the ambient temperature.

Solution

Find M_{air} at 0.4 lb water/lb air:

$$M_{air} = M_{water}/W$$
$$= 900 \text{ lbs}/0.4 = 2250 \text{ lb air}$$

Find M_{air} at 0.04 lb water/lb air:

$$M_{air} = 900 \text{ lbs}/0.04 = 22{,}500 \text{ lb air}$$

Estimate sensible heat content:

$$E = (900)(0.44)(390 - 75) + (2250)(0.24)(390 - 75)$$
$$= 294{,}840 \text{ Btu/hr @ 0.4 lb water/lb air}$$

$$E = (900)(0.44)(315) + (22{,}500)(0.24)(315)$$
$$= 1{,}825{,}740 \text{ Btu/hr @ 0.04 lb water/lb air}$$

Savings:

$$1{,}825{,}740 - 294{,}840 = 1{,}530{,}900 \text{ Btu/hr}$$

Assuming 8000 hrs/yr and a steam cost of $10 per million Btu:

$$(1.531 \times 10^6)\left(\frac{\$10}{10^6}\right)(8000) = \$122{,}480/\text{year saved}$$

Example 13-4

It is desirable to recover part of the heat from the exhaust gas of the dryer. Assume 0.05 lb water/lb air is used, and lower the temperature to 215°F so that no water will condense in the dryer system.

Solution

Calculate sensible heat content at 215°F:

$$E = (900)(0.44)(390 - 215) + (22{,}500)(0.24)(390 - 215)$$
$$= 69{,}300 + 945{,}000 = 875{,}700 \text{ Btu/hr}$$

Table 13-1
Data for Example 13-5

Species	Fuel Gas Analysis	Flue Gas Analysis
CO_2	1.0%	4.0%
CO	2.2	—
CH_4	63.7	—
C_2H_4	1.7	—
H_2	22.7	—
O_2	1.3	13.4
N_2	7.4	82.6
Total	100.0	100.0

Wet and dry bulb temperatures: 51°F and 70°F respectively.
Humidity of air: 0.004 molal.

Note: The molal humidity is the number of lb-mols of water vapor per lb-mol of vapor-free gas.

Savings:

$$(875{,}700)\left(\frac{\$10}{10^6}\right)(8000) = \$70{,}056/\text{year}$$

The significant annual savings suggest that the installation of a heat exchanger at the stack to recover part of the sensible heat must be looked into. An economic analysis should be undertaken.

Combustion Calculations

Many examples of material and heat balances can be found in the literature.[1,2] One example will be given in this section to show the procedure generally used in combustion calculations.

Example 13-5

Given the fuel-gas analysis and the resulting flue-gas composition shown in Table 13-1, determine:

1. Weight balance of combustion process.
2. Weight of dry flue gas.
3. Weight of air introduced for combustion.
4. Water vapor in flue gas.
5. Percentage excess air used.
6. Material balance.

Table 13-2
Solution to Weight Balance of Combustion Process, Example 13-5

Species	% Mol × Molecular Weight	=	lbs
CO_2	1.0 × 44	=	44 lbs
CO	2.2 × 28	=	61.6
CH_4	63.7 × 16	=	1,019.2
C_2H_4	1.7 × 28	=	47.6
H_2	22.7 × 2	=	45.4
O_2	1.4 × 32	=	44.8
N_2	6.4 × 28	=	179.2
100 lb-moles			1,442.0 lbs

Solution

1. *Weight balance of combustion process.* Using 100 lb-mols of dry fuel gas as a basis, the total gas burned per 100 lb-mols is shown in Table 13-2.
2. *Weight of dry fuel gas.* Make a carbon balance using 100 lb-mols of dry flue gas and 100 lb-mols of dry fuel gas as a basis.

Output: C in CO_2: 4.0 lb-atoms

Input: C in CO_2: 1.0 lb-atoms
C in CO: 2.2
C in CH_4: 63.7
C in C_2H_4: 1.7

total C in input: 68.6 lb-atoms

Moles of dry flue gas formed:

$$(68.6/4.0)(100) = 1715 \text{ lb-mols}$$

Total dry flue gas:

$$\begin{aligned} CO_2 &= 0.040 \times 1715 = 68.6 \text{ lb-mols} = 3{,}018.4 \text{ lb} \\ O_2 &= 0.134 \times 1715 = 229.8 \text{ lb-mols} = 7{,}353.9 \\ N_2 &= 0.826 \times 1715 = 1416.6 \text{ lb-mols} = 39{,}664.8 \\ \text{Total} &= 1715 \text{ lb-mols} = 50{,}037 \text{ lb} \end{aligned}$$

3. *Weight of air introduced for combustion.* Make a nitrogen balance using 100 lb-mols as a basis of dry fuel gas:

N_2 in flue gas = 0.826 × 1,715 = 1416.6 lb-mols
N_2 in fuel gas = 0.074 × 100 = 7.4 lb-mols
Therefore, N_2 from air = 1409.2

Dry air introduced:

1409.2/0.79 = 1783.8 lb-mols = 51,730 lb

(using 29 as the molecular weight of air). Since the air coming in has 0.004 molal humidity:

water introduced = 0.004 × 1783.8 = 7.135 lb-mols
= 128.4 lbs

wet air introduced = dry air + water = 1783.8 + 7.135
= 1,791 lb-mols

4. *Water vapor in flue gas*. Make a hydrogen balance using 100 lb-mols of dry fuel gas.

Input:

H_2 in CH_4: 63.7 × 2 = 127.4 lb-mols
(since one mole of CH_4 yields two moles of H_2)
H_2 in C_2H_4: 1.7 × 2 = 3.4
H_2 in H_2 = 22.7
H_2 in water vapor in combustion air = 7.1
total H_2 input = 160.6 lb-mols

Output:

In dry fuel gas: = 0.0 lb-mols
Water vapor in flue gas: = 160.6 lb-mols
Total H_2 output = 160.6 lb-mols = 2891 lbs

Therefore, total wet flue gases:

1783 + 160.6 = 1943.6 lb-mols

or

51,707 + 2.891 = 54,598 lbs

5. *Percentage excess air used.* Use 100 lb-mols of dry fuel gas.

Combustible Component		*Times*	=	*Oxygen Required*
CO:	2.2	0.5	=	1.1 lb-mols
CH_4:	63.7	2.0	=	127.4
C_2H_4:	1.7	3.0	=	5.1
H_2	22.7	0.5	=	11.4
total oxygen required			=	145.0 lb-mols
minus oxygen present in fuel			=	1.3
oxygen from air			=	143.7 lb-mols

theoretical air: 143.7/0.21 = 684.3 lb-mols

air supplied: = 1783 lb-mols

percentage excess air used:

$$(1783 - 684)/(684) \times 100 = 161\%$$

6. *Material balance.* input = output

Input:

Dry fuel gas:	1442 lbs
Dry air:	51,730 lbs
Water vapor in air:	128 lbs
Total input:	53,300 lbs

Output:

Dry flue gas:	50,037 lbs
Water vapor in flue gas:	2891 lbs
Unaccounted for:	372 lbs
Total output:	53,300 lbs

The actual weight of fuel, air, and flue gas present per time unit will depend on how many Btu per time unit used in the process furnace.

References

1. Williams, E. T., and R. C. Johnson, *Stoichiometry for Chemical Engineers,* McGraw-Hill Book Co., Inc., 1958.
2. McCormack, Harry, *The Application of Chemical Engineering,* D. Van Nostrand Company, Inc., 1949.

A

Boiler Efficiency Test Form

The test form presented in this chapter, (Figure A-1), is needed as a detailed record of boiler performance. The test form is divided into the following main sections:

1. *Boiler manufacturer's data.* This is the basic design of the boiler. It also includes the type of fuel used (natural gas, coal, fuel oil No.) and combustion controls, if any, such as CO or O_2 analyzers.
2. *Test data.* These data represent the minimum test data that should be obtained on a boiler. There are 22 individual pieces of information that must be obtained in most cases.

   The fuel and atomization pressures must follow the designer's recommendations as included in the standard-operating-procedure manual.

   Test data such as feedwater temperature and the time the boiler is on and off (if it is operated in this manner) will clearly point out areas where significant savings might be obtained.
3. *Fuel data.* An elemental fuel analysis (C, H, etc.) can be included in this section, especially if the Direct Method is used to analyze boiler efficiency. Fuel sources may change without notice. It is not uncommon to find out (through a laboratory analysis) that the No. 6 fuel oil you are now receiving has 200 ppm vanadium, when in the past, you may not even remember if your fuel ever had vanadium in significant amounts.

Boiler Efficiency Test Form

1. Company: ____________ Type of Industry ____________ Date: ________
2. Address: __________________________ Telephone no. ______________
3. Contact *(Plant Manager, Plant Engineer)* __________________________
4. Operator: ____

Equipment Data

5. Boiler no. __________ Make: __________ Year: _____ Type: __________
   ☐ Water-tube ☐ Fire-tube
6. Capacity: BTU/hr.: ________ ; HP: ________ ; lb/hr. Steam: __________
7. Atomization: ☐ Mechanical ☐ Steam ☐ Air
8. Fuel type: __________ Gal/hr. ________ 1/hr. ________ Cost: __________
9. Steam pressure controls: High __ Low __ ☐ Saturated ☐ Superheated
10. Type of Automatic Controls: ☐ CO; ☐ O_2; ☐ Other; ☐ None

Test Data

Comments:

11. Test no. ________________________________
12. % Load ________________________________
13. Time high fire on ________________________________
14. Time low fire on ________________________________
15. Fuel pressure return ________________________________
16. Fuel pressure to burner ________________________________
17. Atomization pressure ________________________________
18. Feed water temp. ________________________________
19. Fuel preheater temp. ________________________________
20. Fuel Viscosity @ preheater ________________________________
21. Air temp. ________________________________
22. Stack temperature ________________________________
23. % CO_2 ________________________________
24. % O_2 ________________________________
25. % Excess air ________________________________
26. CO ppm ________________________________
27. Smoke color no. ________________________________
28. SO_2 ________________________________
29. Position damper ________________________________
30. Draft stack ________________________________
31. Draft inlet boiler ________________________________
32. % Combustion eff. ________________________________

Figure A-1. Boiler efficiency test form.

Fuel Data

33. For natural gas: BTU/Ft³ ____________
    For fuel oil: ____________
34. API ____________
35. Sp. Gr. ____________
36. Heating value ____________
37. Water and Sediment % ____________
38. Viscosity ____________
39. Pour point ____________
40. Sulphur % ____________
41. Vanadium, ppm ____________
42. Sodium, ppm ____________
43. Nickel, ppm ____________
44. Ash % ____________
45. Fuel treatment (if any) ____________
46. Storage tank capacity ____________
47. Capacity day tank: ____________
48. Monthly fuel consumption: ____________

Water Data

49. Water softener: ☐ None; ☐ Zeolite; ☐ Other: ____________
50. Deareator: Type ____________ Pressure: ________ Temp: ______
    Exhaust plume from deareator: Velocity ________ Ft/Sec;
    Diameter: ________ Inches
51. Chemical Feed ☐ Continuous; ☐ Batch; ☐ Feed point: ________
    Water quality: *PH Hardness TDS* COND** ALKA.* SO_3 PO_4
52. Feed water ____________
53. Boiler water ____________
54. Condensate ____________
    *Total dissolved solids **Conductivity

Boiler Blowdown

55. Amount of make-up water, %: ______ ; ☐ Continuous; ☐ Batch
56. Where: ☐ Surface; ☐ Bottom

57. Blowdown heat recovery: ☐ Flash tank; ☐ Heat exchange
58. Temp. blowdown water leaving heat recovery: ____________
59. Bottom blowdown: Time each blow: ________ Sec.; Frequency per 8 hrs: ______

Additional Boiler Data

60. Steam pressure: Can it be lowered? ☐ Yes; ☐ No
61. Soot blown: Time each blow: __________ ; Frequency per 8 hrs: _____
62. No. of boilers operating at one time: __________ ; Boilers on manual: ____________ ; Boilers modulating: ____________
63. Flue gas heat recovery: ☐ Economizer; ☐ Air preheater

Boiler and Tank Condition

64. When was the last time tanks were cleaned: Outside tank(s): ______ Day tank: ________
65. When was the last time boiler was cleaned: __________ Were pictures taken? ______
66. Type of buildup found on fire-side: ___________ Sample taken? ______
67. Type of buildup found on water-side: __________ Sample taken? ______
68. How often is burner or nozzle cleaned? ☐ Daily; ☐ Weekly; ☐ Monthly

Test data witnessed by: ______________________________ Date: ________
(*signature*)

Tests conducted by: ______________________________ Date: ________
(*signature*)

Industries located in colder climates especially need to know the pour point. With this information, a certain fuel treatment can be recommended by a qualified supplier.

4. *Water data.* This information should be readily available from your own quality-control laboratory or from your water-chemicals supplier. Obviously, if the latest boiler inspection shows heavy scale and poor heat transfer, the water treatment is deficient and should have been changed a long time ago.
5. *Boiler blowdown.* These data give you useful information for increasing your boiler efficiency: could you reduce the blowdown frequency? should a heat exchanger be considered? etc.
6. *Additional boiler data and boiler and tank condition.* This is additional information that can be used to reduce fuel costs. The scale samples taken during boiler shutdown will indicate how efficient water and fuel treatments are.

When outside consultants perform these tests, it is convenient to enter the name of the company's engineer for reference purposes.

This test form does not include a complete energy audit. Thus, steam losses in leaking pipes and steam traps are everyday problems; in most cases they are the responsibility of the maintenance department. Likewise, the state of the insulation on the boiler and on the steam piping is also a responsibility of the maintenance department. However, the chief engineer may want to make sure that these checks are being carried out by the maintenance group.

Finally, this test form, when properly followed, will give three main sources of information on how to:

1. Increase boiler efficiency without additional expense.
2. Check whether water and fuel treatments are performing properly.
3. Increase boiler efficiency using additional equipment if economically feasible.

The following simplified example of a test run on a watertube boiler will give a better insight into interpreting the test data and increasing boiler efficiency. Only one boiler-load setting will be discussed in this example.

Example A-1

This particular boiler has not been analyzed in this manner since it was installed five years ago. The following test data are obtained:

1. *Fuel used:* fuel oil No. 6; no analysis was available on the last oil shipment.

2. *Atomization:* air.
3. *Fuel pressure:* the atomization pressure is found to be 15 psig under the pressure recommended by supplier.
4. *Stack damper:* fully open.
5. *% Load:* about 75%.
6. *Burner cleaning:* daily.
7. *Fuel preheater temperature:* 110°F (43°C).
8. *Fuel viscosity at preheater temperature:* 2000 SUS.
9. *Stack temperature:* 580°F (304°C).
10. *%* CO_2: 6.
11. *Smoke color:* 3.
12. *Draft:* over 0.25 in. (off-scale).
13. *Room temperature:* 78°F (25.6°C).
14. *Fuel treatment:* none.

Diagnosis of combustion efficiency test. From Figure 7-3, at 6% CO_2 there is 140% excess air. Using the slide rule shown in Figure 7-2, the combustion efficiency is 68%.

Excess air and smoke. The large amount of excess air indicates that the boiler man was using air to keep a low, visual smoke coming out of the stack. This indicates that combustion is deficient because of one or more of the following causes:

1. The fireside is probably partially insulated with soot, which causes high smoke in the flue gas. It is highly probable that if the excess air were reduced to 15%, the smoke would increase significantly, say, to No. 6 or more.
2. The atomization is deficient, depositing soot and producing smoke. This is supported by these facts:
   - The fuel oil temperature is only 110°F, yielding a very high fuel viscosity of 2000 SUS. Figure 3-3 shows that to maintain at 150–300 SUS viscosity, temperature must be in the range of at least 180°F–210°F, and in many cases even higher.
   - The lower fuel-atomization pressure also causes deficient combustion.
   - Since no fuel treatment is used in this plant, there is a good chance that the burner has deposits that contribute to poor atomization.
   - Another possibility for poor atomization is the burner itself. Is the right burner being used? Is it worn out?

Draft. The draft increased when excess air was raised to mask the smoke in the flue gas. High draft does not give enough time to the hot combustion gases to properly transfer the heat content to the water to form steam.

Combustion efficiency. The efficiency is low on account of the excess air and poor atomization present in this boiler.

Recommendations for Example A-1.

1. Shut down the boiler as soon as possible and blow out the soot.
2. Clean up and inspect the burner for build-up and defects.
3. Raise atomization pressure to design recommendations.
4. Raise oil preheater temperature, say, to 210°F.
5. Maintain 15% excess air.
6. Obtain a fuel analysis, and check especially for *V* and *S*. From the analysis and burner inspection, determine the required fuel treatment.
7. Do not reset stack damper until the preceding conditions are met.
8. Maintain stack temperature at 350°F–400°F, and lower the fuel flow-rate if necessary.

Example A-2

A second set of test data was obtained after following the previous recommendations:

1. *Stack temperature:* 400°F.
2. %CO_2: 7.0.
3. *Smoke color:* 4.
4. *Draft:* 0.20 in. water.
5. *Room temperature:* 72°F.
6. *Fuel-oil temperature:* 212°F.
7. *Excess air:* 100%.
8. *Combustion efficiency:* 79.5%.
9. *Actual measurement of fuel viscosity:* 425 SUS.

Diagnosis of combustion-efficiency test. In order to maintain a 350°F–400°F stack temperature, the fuel-oil feed rate was lowered. However, it was not yet possible to maintain a low smoke No. without a large amount of excess air. However, the combustion efficiency was improved significantly. The additional causes for smoke include:

1. Burner installation after shutdown was slightly defective.
2. Fuel was not yet being treated.
3. Fuel-oil preheatment was too low (raise temperature to 220°F).

Example A-3

A new burner was installed, and fuel was treated to minimize deposits and improve atomization. Fuel-oil preheater temperature was raised to 220°F; oil viscosity was 361 SUS. The boiler was fine-tuned, determining both CO and O_2. Combustion efficiency increased to 81.5%.

Recommendations

To maintain an optimum boiler efficiency, the following recommendations were made to this company.

1. Hold training sessions with engineers, foremen, and boiler men, and show them how to determine CO_2, CO, O_2, etc., and to estimate combustion efficiency.
2. Check boiler efficiency at least once a week (automatic CO and O_2 controls were not justified for this boiler).
3. Proceed with an energy audit of the whole steam system.

B

Scheduled Maintenance Requirements on Boilers

Table B-1
Daily Requirements

Check	Potential Problems
1. Water-level controls	1. Excessive solids or treatment. 2. Oil contamination. 3. Mechanical malfunction.
2. Pressure or temperature controls	1. Excessive steam pressure or water temperature drop may indicate excessive loading on the boiler.
3. Burner operation and controls	1. Is burner clean? 2. Are controls working properly? 3. Atomization pressure as recommended in the operating manual?
4. Motor operations	1. Are bearings wearing out?
5. Temperature exhaust gas	1. Is there a trend for higher and higher temperatures?
6. Blowdown	1. Does blowdown valve leak? 2. Any excessive automatic blowdown?
7. Records	Clear records help in solving problems and must include:

Table B-1 continued

Check	Potential Problems
	1. All of the above observations where applicable. 2. Fuel and type. 3. Any maintenance done, etc. 4. A water analysis done by boiler personnel. 5. Preheater temperature.

Table B-2
Weekly Requirements

Check	Potential Problem/Comments
1. Operation of water-level controls	1. Stop preheater pump and allow control to stop burner under normal load conditions. Do not allow water level in boiler to fall below recommended low level.
2. Blowdown water-level controls	1. Purge float level of possible sediment accumulation.
3. Safety or relief valve	1. Manually open valve with not less that 80% working pressure in the system. 2. Does valve leak?
4. Burner assembly	1. Any deposits? 2. Should it be cleaned more often?
5. Pilot and flame scanner	1. Check spark gap. 2. Check condition of electrode. 3. Check cleanliness.
6. Lubricating-oil assembly (of atomizing air compressor)	1. Oil level.
7. Gas analysis	1. CO_2. 2. Excess air. 3. Smoke. 4. Draft in stock and pressure drop through boiler. 5. Temperature.
8. Boiler operating characteristics	1. Close off fuel supply and observe flame failure and check reaction time. 2. Start boiler and observe light-off characteristics.

Table B-3
Monthly Requirements

Check	What to do/Why
1. Boiler blowdown	1. Is water being wasted? 2. Is water treatment allright?
2. Burner operation	1. Take flue-gas analysis over entire firing range, comparing these and the stack temperature with those of the previous month's.
3. Combustion air supply	1. Check air inlet to boiler room. 2. Check air inlet to boiler, and clean if fouled. 3. Clean filters on atomizing air compressor and check oil level.
4. Fuel system	1. Check pressure gauges, pumps, filters, and transfer lines. Clean filters as often as required.
5. Bolts and packing glands	1. Check belts for damage and proper tension. 2. Are packing glands leaking?

Table B-4
Annual Requirements

Check	How/Why
1. Clean fireside surfaces	1. Follow manufacturer's recommended procedures, which may include brush or water washing, vacuum cleaning to remove soot, etc.
2. Waterside cleaning	1. Follow manufacturer's recommended procedures, which may include high-pressure water hosing to push out sludge, scale, etc.
3. Repair on fireside	1. As necessary.
4. Relief valve	1. Should it be reconditioned or replaced?
5. Fuel system	1. Check fuel storage tanks for sludge and water accumulation. Any corrosion? 2. Fuel oil preheater: sludge or scale build-up? 3. Check fuel pump wear and tear.
6. Controls	1. Check electrical terminals. 2. Check electronic controls. 3. Check mercury switches.
7. Chemical feed system	1. Empty, flush, and recondition it.

C

ASTM Fuel Tests

Detail requirements. The various grades of fuel oil shall conform to the detailed requirements shown in Table C-1. It is the intent of these classifications that failure to meet any requirement of a given grade does not automatically place an oil in the next lower grade unless in fact it meets all requirements of the lower grade.

Methods of test. The requirements enumerated in these specifications shall be determined in accordance with the following methods of testing of the American Society for Testing Materials, except as may be required under the next paragraph: "Flash Point."

Flash point. The flash point, instrument, and method for determining minimum flash point shall be those legally required for the locality in which the oil is sold. In absence of legal requirements, the minimum flash point shall be determined in accordance with the standard Method of Test for Flash Point by Means of the Pensky-Martens Closed Tester, ASTM Designation D93-80.

Pour point. Standard Method of Test for Cloud and Pour Points, ASTM Designation: D97-66.

(text continued on page 188)

Table C-1
Detailed Requirements for Fuel Oils[A] (Commercial Standard CS12-48)

No.	Grade of Fuel Oil[B]	Flash Point °F Min.	Pour Point °F Max.[C]	Water and Sediment % Max.	Carbon Residue On 10% Residuum, % Max.	Ash % Max.	Distillation Temperatures, °F 10% Point Max.	90% Point Max.	End Point Max.
1.	Distillate oil intended for vaporizing pot-type burners and other burners requiring this grade.[D]	100 or legal	0	trace	0.15		420		625
2.	A distillate oil for general-purpose domestic heating, for use in burners not requiring No. 1.	100 or legal	20	0.10	0.35		(E)	675	
4.	An oil for burner installations not equipped with preheating facilities.	130 or legal	20	0.50		0.10			
5.	A residual-type oil for burner installations equipped with preheating facilities.	130 or legal		1.00		0.10			
6.	An oil for use in burners equipped with preheaters permitting a high-viscosity fuel.	150 or legal		2.00[F]					

Table C-1 continued

[A]Recognizing the necessity for low-sulfur fuel oils used in connection with heat-treatment, non-ferrous metal, glass and ceramic furnaces and other special uses, a sulfur requirement may be specified in accordance with the following:

Grade of Fuel Oil	Sulfur, Max., %
No. 1	0.5
No. 2	1.0
Nos. 4, 5, and 6	no limit

Other sulfur limits may be specified only by mutual agreement between buyer and seller.

[B]It is the intent of these classifications that failure to meet any requirement of a given grade does not automatically place an oil in the next lower grade unless in fact it meets all requirements of the lower grade.

[C]Lower or higher pour points may be specified whenever required by conditions of storage or use. However, these specifications shall not require a pour point lower than 0°F under any conditions.

[D]No. 1 oil shall be tested for corrosion in accordance with ASTM D287-39 for three hours at 122°F. The exposed copper strip shall show no gray or black deposit.

[E]The 10% point may be specified at 440°F maximum for use in other than atomizing burners.

[F]The amount of water by distillation plus the sediment by extraction shall not exceed 2.00%. The amount of sediment by extraction shall not exceed 0.50%. A deduction in quantity shall be made for all water and sediment in excess of 1.0%.

(table continued on next page)

Table C-1 (continued)

No.	Grade of Fuel Oil[B]	VISCOSITY Saybolt Universal at 100°F		Saybolt Furol at 122°F		Kinematic C.S. at 100°F		Kinematic C.S at 122°F		Gravity °API
		Max.	Min.	Max.	Min.	Max.	Min.	Max.	Min.	Min.
1.	A distillate oil intended for vaporizing pot-type burners and other burners requiring this grade.[D]					2.2	1.4			35
2.	A distillate oil for general purpose domestic heating, for use in burners not requiring No. 1.	40				(4.3)				26
4.	An oil for burner installations not equipped with preheating facilities.	125	45			(26.4)	(5.8)			
5.	A residual-type oil for burner installations equipped with preheating facilities.		150	40			(32.1)	(81)		
6.	An oil for use in burners equipped with preheaters, permitting a high-viscosity fuel.			300	45			(638)	(92)	

Table C-1 (continued)

[A] Recognizing the necessity for low-sulfur fuel oils used in connection with heat-treatment, non-ferrous metal, glass and ceramic furnaces and other special uses, a sulfur requirement may be specified in accordance with the following:

Grade of Fuel Oil	Sulfur, Max., %
No. 1	0.5
No. 2	1.0
Nos. 4, 5, and 6	no limit

Other sulfur limits may be specified only by mutual agreement between buyer and seller.

[B] It is the intent of these classifications that failure to meet any requirement of a given grade does not automatically place an oil in the next lower grade unless in fact it meets all requirements of the lower grade.

[C] Lower or higher pour points may be specified whenever required by conditions of storage or use. However, these specifications shall not require a pour point lower than 0°F under any conditions.

[D] No. 1 oil shall be tested for corrosion in accordance with ASTM D287-39 for three hours at 122°F. The exposed copper strip shall show no gray or black deposit.

[E] The 10% point may be specified at 440°F maximum for use in other than atomizing burners.

[F] The amount of water by distillation plus the sediment by extraction shall not exceed 2.00%. The amount of sediment by extraction shall not exceed 0.50%. A deduction in quantity shall be made for all water and sediment in excess of 1.0%.

Water and sediment.

Water and sediment. (For Grades 1 to 5, inclusive.) Method of Test Water and Sediment in Petroleum Products by Means of Centrifuge, ASTM Designation: D96-73.

Water by distillation. (For Grade 6.) Standard Method of Test for Water in Petroleum Products and Other Bituminous Materials, ASTM Designation: D95-70.

Sediment by extraction. (For Grade 6.) Method of Test for Sediment in Fuel Oil by Extraction, ASTM Designation: D473-81.

Carbon residue. Standard Method of Test for Carbon Residue of Petroleum Products (Ramsbottom Carbon Residue), ASTM Designation: D524-81.

Ash. Standard Method of Test for Ash Content of Petroleum Oils, ASTM Designation: D482-80.

Distillation. Distillation of Grade 1 oil shall be made in accordance with the Standard Method of Test for Distillation of Gasoline, Naphtha, Kerosene, and Similar Petroleum Products, ASTM Designation: D86-78; and of Grade 2 in accordance with the Standard Method of Test for Distillation of Gas Oil and Similar Distillate Fuel Oils, ASTM Designation: D158-41.

Viscosity.

Kinematic viscosity. (For Grade 1.) Method of Test for Kinematic viscosity, ASTM Designation: D445-79.

Saybolt viscosity. (for Grades 2, 4, 5 and 6.) Standard Method of Test for Viscosity by Means of the Saybolt Viscosimeter, ASTM Designation: D88-56.

Gravity. Standard Method of Test for Gravity of Petroleum and Petroleum Products by Means of the Hydrometer, ASTM Designation: D287-67.

Corrosion. Standard Method of Test for Detection of Free Sulfur and Corrosive Sulfur Compounds in Gasoline, ASTM Designation: D130-80 (except for interpretation of exposed copper strip).

References. Complete information regarding the procedure for making the tests specified, but not included in the preceding text, is to be found in the publication of the American Society for Testing Materials, 1916 Race Street, Philadelphia, Pennsylvania, 19103.

Significance of Tests Prescribed

Flash point. The flash point of a product may be defined as the temperature to which it must be heated in order to give off sufficient vapor to form an inflammable mixture with air. This temperature varies with the apparatus and procedure employed, and consequently both must be specified when the flash point of an oil is stated.

The minimum flash point of oils used for fuel is usually controlled by law. When there are no legal requirements, the minimum values in Table C-1 are to be employed.

Pour point. The pour point of an oil is the lowest temperature at which it will flow when cooled and tested under prescribed conditions. Pour-point specifications are included in order that oil may be secured which will not cause difficulty in handling or in use at the lowest temperatures to which it may normally be subjected.

Water and sediment. Water and sediment are impurities which are almost entirely excluded in fuel oils Nos. 1 and 2, and which are permitted in somewhat larger quantities in fuel oils Nos. 4, 5, and 6. It is difficult to eliminate them entirely from this latter group of oils, and the advantage is not sufficient to justify the cost. Water and sediment are determined together by the centrifuge, except for Grade 6.

Carbon residue. The carbon-residue test, when considered in connection with other tests and the use for which oil is intended, furnishes pertinent information and throws some light on the relative carbon-forming qualities of an oil. Values for carbon residue which are abnormally high in relation to other properties of the oil may indicate the presence of heavy residual products due to unsuitable refinery methods or contamination.

Ash. The ash test is used to determine the amount of noncombustible impurities in the oil. These impurities come principally from the natural salts present in the crude oil, or from the chemicals that may be used in refinery operations, although they may also come from scale and dirt picked up from containers and pipes. Some ash-producing impurities in fuel oils cause rapid deterioration of refractory materials in the combustion chamber, particularly at high temperatures; some are abrasive and destructive to pumps, valves, control equipment, and other burner parts. Ash specifications are included in order to minimize these operating difficulties as far as practicable.

Distillation. Laboratory distillation of a sample under prescribed conditions gives an index of the volatility of the oil. The 10% and 90% points represent, respectively, the temperatures at which 10% and 90% of the sample are distilled. The end point is the maximum temperature recorded by the distillation thermometer at the end of the distillation.

The 10% point serves as a index of the ease of ignition of the oil; the 90% point and the end point are specified to make sure that the oil will volatilize and burn completely and produce a minimum amount of carbon.

Viscosity. The viscosity of an oil is the measure of its resistance to flow. Maximum limits are placed on this property because of its effect upon the rate at which oil will flow through pipelines and upon the degree of atomization that may be secured in any given equipment.

Viscosity is measured as the time in seconds required for a definite volume of oil to pass through a small tube of specified dimensions at a definite temperature. Viscosity decreases rapidly as temperature increases, and preheating makes possible the use of oils of relatively high viscosity. The Kinematic and Saybolt universal viscosimeters are used for fuel oils of fairly low viscosity and the Saybolt furol viscosimeter for more viscous oils.

Identification

In order that purchasers of fuel oil may become familiar with the significance of grading of fuel oils and purchase fuels for the various types of burners with confidence, it is recommended that the following statement be used on invoices, contracts, sales literature, etc.:

> This fuel oil complies with all requirements for Grade _____ as specified in Commercial Standard CS12-48, developed by the trade under the procedure of the National Bureau of Standards, and issued by the United States Department of Commerce.

While the preceding is the complete standard, additional material of interest to those concerned is contained in a pamphlet entitled Fuel Oils, Commercial Standard CS12-48, issued by the Department of Commerce. This pamphlet includes, in addition to the standard, a list of acceptors, brief history of the project, membership of the Standing Committee, etc., and is available from the Superintendent of Documents, Washington, D.C., 20402.

D

Energy and Professional Resource Addresses

United States

American Boiler Manufacturers' Association
950 North Globe Road, Suite 160
Arlington, Virginia 22203

American Chemical Society (ACS)
1155 16th Street N.W.
Washington, D.C. 20036

American Institute of Chemical Engineers (AIChE)
345 E. 47th Street
New York, NY 10017

American Gas Association
1515 Wilson Blvd.
Arlington, Virginia 22209

American Society of Mechanical Engineers
345 East 47th Street
New York, NY 10017

American Society for Testing and Materials
1916 Race Street
Philadelphia, PA 19103

American Petroleum Institute
2101 L Street N.W.
Washington, D.C. 20006

American Petroleum Refiners Association
1200 18th Street N.W.
Washington, D.C. 20000

Boiler Efficiency Institute
P.O. Box 2255
Auburn, Alabama 36830

Industrial Heating Equipment Association
1901 North Moore Street
Arlington, Virginia 22209

Independent Petroleum
Association of America
1101 16th Street N.W.
Washington, D.C. 20036

Independent Refiners Association
of America
1801 K Street N.W.
Washington, D.C. 20006

Interstate Natural Gas Association
of America
1660 L Street N.W.
Washington, D.C. 20036

National Energy Information
Center
Energy Information
Administration
U.S. Department of Energy
1726 M Street N.W., Room 850
Washington, D.C. 20461

Natural Gas Supply Committee
1025 Connecticut Avenue N.W.
Washington, D.C. 20036

National Petroleum Council
1625 K Street N.W.
Washington, D.C. 20006

National Petroleum Refiners
Association
1725 DeSales Street N.W.
Washington, D.C. 20036

Petroleum Industry Research
Foundation Inc.
122 East 42nd Street
New York, NY 10017

U.S. Environmental Protection
Agency
Industrial Environmental Research
Lab-RTP
Mail Drop-64, Research Triangle
Park
North Carolina 27711

U.S. State Department
Bureau of Economic and Business
Affairs
Office of Fuels and Energy
Main State Department Building
Washington, D.C. 20520

International

Australia

Australian Institute of Petroleum
227 Collins St.
Melbourne, Victoria, Australia
3000

Belgium

Energy Directorate
Commission of European
Communities
Rue de la Loi 200
1049 Brussels, Belgium

Canada

Canadian Petroleum Association
625-404 Sixth Ave., S.W.
Calgary, Alta. T2P OR9, Canada

Department of Energy, Mines and
Resources
Mineral Development Sector
580 Booth St.
Ottawa, Ont. K1A OE4 Canada

National Energy Board
Trebla Building
473 Albert St.
Ottawa, Ont. K1A OE5 Canada

England

Department of Energy
Head Office
Thames House South
Millbank
London SW1P 40J, England

Information Services
The Institute of Petroleum
61 New Cavendish St.
London W1M 8AR, England

France

Chief of Service Documentation and Information
Direction des Carburants

Ministère de L' Industrie et de la Recherche
Poste 38-84
3 et 5, Rue Barbet-de Jouy
75 Paris 7e. France

Germany (West)

Federal Ministry for Economics
Mining and Energy Industries (Department III)
Villemombler Strasse 74-78
5300 Bonn-Duisdorf, West Germany

Italy

Énte Nazionale Idrocarburi (E.N.I.)
Piazzale Enrico Mattei, 1
00144 Rome, Italy

Japan

Managing Director
Petroleum Association of Japan
Keidanren Bldg. 1-9-4, Ohte-machi
Chiyodapku
Tokyo 100, Japan

Managing Director
Japan Gas Association
38 Shiba Kotohira-cho
Minato-ku
Tokyo 105, Japan

Latin America

ARPEL (Association of Latin American Government-Owned Oil Companies)
Piso 2°, Ofs. 206/209
Montevideo, Uruguay

Netherlands

Ministry of Economic Affairs
Bezuidenhoutseweg 30
The Hague, The Netherlands
Director-General of Energy
Deputy Director-General of Energy
Director General Energy Policy Department
Director Mining Department
Director Coal, Oil and Gas Deparment
Deputy Director Coal, Oil and Gas Department
Director Electricity and Nuclear Energy Department

OPEC

OPEC
Ring 10
1010 Vienna, Austria

United Nations

United Nations Information Center
1028 Connecticut Ave. N.W.
Washington, D.C. 20036

World Energy

World Energy Conference
34 St. James's St.
London SW1A 1HD, England

U.S. National Committee
World Energy Congress
1620 Eye St. N.W.
Suite 615
Washington, D.C. 20006

E

Unit Conversions and Data

Table E-1
Equivalent Units for Defining Boiler Output

Item	Equivalent Units
1 lb steam, from and at 212°F	970 Btu/lb
1 Boiler horsepower, (Bhp)	34.5 lbs steam/hr from and at 212°F 33,472 Btu/hr

Table E-2
Equivalent Units of Volume and Weight

U.S. Gallons	Imperial Gallons	Cubic Inches	Cubic Feet	Pounds (*)	Cubic Meters	Liters
1	.833	231	.1337	8.35	.003785	3.785
1.201	1	277.4	.1605	10.02	.004546	4.546
.00433	.00360	1	.000579	.0361		.0164
7.48	6.23	1728	1	62.4	.02832	28.32
.120	.0998	27.7	.0160	1		.454
264.18	220	61,023	35.314	2204	1	1000
.2642	.220	61.023	.0353	2.2046		1

*Based on maximum density of water at 39°F.

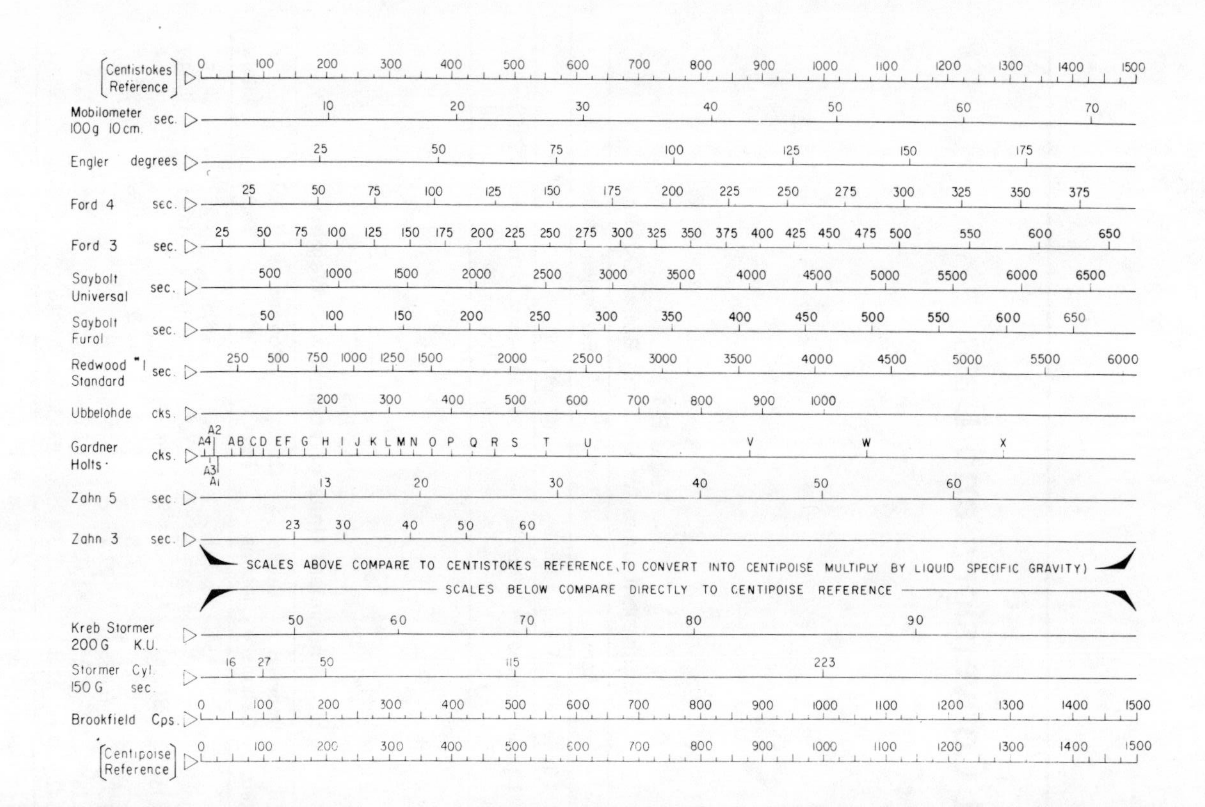

Figure E-1. Viscometer comparison chart for Newtonian liquids. This chart is intended to be an aid in comparing viscometer measurements of Newtonian liquids by referencing to absolute and kinematic viscosity. (Courtesy Brookfield Engineering Laboratories, Inc., Stoughton, Mass.)

Table E-3
Temperature Conversion

−459.4° to 0°			1° to 60°			61° to 290°			300° to 890°			900° to 3000°		
C	C/F	F	C	C/F	F	C	C/F	F	C	C/F	F	C	C/F	F
−273	−459.4		−17.2	1	33.8	16.1	61	141.8	149	300	572	482	900	1652
−268	−450		−16.7	2	35.6	16.7	62	143.6	154	310	590	488	910	1670
−262	−440		−16.1	3	37.4	17.2	63	145.4	160	320	608	493	920	1688
−257	−430		−15.6	4	39.2	17.8	64	147.2	166	330	626	499	930	1706
−251	−420		−15.0	5	41.0	18.3	65	149.0	171	340	644	504	940	1724
−246	−410		−14.4	6	42.8	18.9	66	150.8	177	350	662	510	950	1742
−240	−400		−13.9	7	44.6	19.4	67	152.6	182	360	680	516	960	1760
−234	−390		−13.3	8	46.4	20.0	68	154.4	188	370	698	521	970	1778
−229	−380		−12.8	9	48.2	20.6	69	156.2	193	380	716	527	980	1796
−223	−370		−12.2	10	50.0	21.1	70	158.0	199	390	734	532	990	1814
−218	−360		−11.7	11	51.8	21.7	71	159.8	204	400	752	538	1000	1832
−212	−350		−11.1	12	53.6	22.2	72	161.6	210	410	770	549	1020	1868
−207	−340		−10.6	13	55.4	22.8	73	163.4	216	420	788	560	1040	1904
−201	−330		−10.0	14	57.2	23.3	74	165.2	221	430	806	571	1060	1940
−196	−320		−9.4	15	59.0	23.9	75	167.0	227	440	824	582	1080	1976
−190	−310		−8.9	16	60.8	24.4	76	168.8	232	450	842	593	1100	2012
−184	−300		−8.3	17	62.6	25.0	77	170.6	238	460	860	604	1120	2048
−179	−290		−7.8	18	64.4	25.6	78	172.4	243	470	878	616	1140	2084
−173	−280		−7.2	19	66.2	26.1	79	174.2	249	480	896	627	1160	2120
−169	−273	−459.4	−6.7	20	68.0	26.7	80	176.0	254	490	914	638	1180	2156
−168	−270	−454	−6.1	21	69.8	27.2	81	177.8	260	500	932	649	1200	2192
−162	−260	−436	−5.6	22	71.6	27.8	82	179.6	266	510	950	660	1220	2228
−157	−250	−418	−5.0	23	73.4	28.3	83	181.4	271	520	968	671	1240	2264
−151	−240	−400	−4.4	24	75.2	28.9	84	183.2	277	530	986	682	1260	2300
−146	−230	−382	−3.9	25	77.0	29.4	85	185.0	282	540	1004	693	1280	2336
−140	−220	−364	−3.3	26	78.8	30.0	86	186.8	288	550	1022	704	1300	2372
−134	−210	−346	−2.8	27	80.6	30.6	87	188.6	293	560	1040	732	1350	2462
−129	−200	−328	−2.2	28	82.4	31.1	88	190.4	299	570	1058	760	1400	2552
−123	−190	−310	−1.7	29	84.2	31.7	89	192.2	304	580	1076	788	1450	2642
−118	−180	−292	−1.1	30	86.0	32.2	90	194.0	310	590	1094	816	1500	2732
−112	−170	−274	−0.6	31	87.8	32.8	91	195.8	316	600	1112	843	1550	2822
−107	−160	−256	0.0	32	89.6	33.3	92	197.6	321	610	1130	871	1600	2912
−101	−150	−238	0.6	33	91.4	33.9	93	199.4	327	620	1148	899	1650	3002
−96	−140	−220	1.1	34	93.2	34.4	94	201.2	332	630	1166	927	1700	3092
−90	−130	−202	1.7	35	95.0	35.0	95	203.0	338	640	1184	954	1750	3182
−84	−120	−184	2.2	36	96.8	35.6	96	204.8	343	650	1202	982	1800	3272
−79	−110	−166	2.8	37	98.6	36.1	97	206.6	349	660	1220	1010	1850	3362
−73	−100	−148	3.3	38	100.4	36.7	98	208.4	354	670	1238	1038	1900	3452
−68	−90	−130	3.9	39	102.2	37.2	99	210.2	360	680	1256	1066	1950	3542
−62	−80	−112	4.4	40	104.0	37.8	100	212.0	366	690	1274	1093	2000	3632
−57	−70	−94	5.0	41	105.8	43	110	230	371	700	1292	1121	2050	3722
−51	−60	−76	5.6	42	107.6	49	120	248	377	710	1310	1149	2100	3812
−46	−50	−58	6.1	43	109.4	54	130	266	382	720	1328	1177	2150	3902
−40	−40	−40	6.7	44	111.2	60	140	284	388	730	1346	1204	2200	3992
−34	−30	−22	7.2	45	113.0	66	150	302	393	740	1364	1232	2250	4082
−29	−20	−4	7.8	46	114.8	71	160	320	399	750	1382	1260	2300	4172
−23	−10	14	8.3	47	116.6	77	170	338	404	760	1400	1288	2350	4262
−17.8	0	32	8.9	48	118.4	82	180	356	410	770	1418	1316	2400	4352
			9.4	49	120.2	88	190	374	416	780	1436	1343	2450	4442
			10.0	50	122.0	93	200	392	421	790	1454	1371	2500	4532
			10.6	51	123.8	99	210	410	427	800	1472	1399	2550	4622
			11.1	52	125.6	100	212	413.6	432	810	1490	1427	2600	4712
			11.7	53	127.4	104	220	428	438	820	1508	1454	2650	4802
			12.2	54	129.2	110	230	446	443	830	1526	1482	2700	4892
			12.8	55	131.0	116	240	464	449	840	1544	1510	2750	4982
			13.3	56	132.8	121	250	482	454	850	1562	1538	2800	5072
			13.9	57	134.6	127	260	500	460	860	1580	1566	2850	5162
			14.4	58	136.4	132	270	518	466	870	1598	1593	2900	5252
			15.0	59	138.2	138	280	536	471	880	1616	1621	2950	5342
			15.6	60	140.0	143	290	554	477	890	1634	1649	3000	5432

Locate degrees F or degrees C in middle column. If degrees F, read equivalent degrees C in left hand column. If degrees C, read equivalent degrees F in right hand column.

Glossary

Air. Air composition is usually taken as 21% oxygen (O_2) and 79% nitrogen (N_2) in combustion analyses. The humidity of the air is treated separately.

Air purge. The removal of undesired matter by replacement with air.

Ash. The incombustible inorganic matter in the fuel.

ASME. American Society of Mechanical Engineers.

ASTM. American Society for Testing and Materials.

Atomization. Usually to vaporize a fuel with air or steam using a nozzle.

Atomizing media. A supplementary medium, such as steam or air, which assists in breaking the fuel oil into a fine spray.

Blowdown. The drain connection including the pipe and the valve at the lowest practical part of a boiler, or at the normal water level in the case of a surface blowdown. The amount of water that is blown down.

Boiler, high-pressure, steam or vapor. A boiler in which steam or vapor is generated at a pressure exceeding 15 psig.

Boiler, hot-water-heating. A boiler in which no steam is generated and from which hot water is circulated for heating purposes and then returned to the boiler.

Boiler, hot-water-supply. A boiler functioning as a water heater.

Boiler, low-pressure, steam or vapor. A boiler in which steam or vapor is generated at a pressure not exceeding 15 psig.

Boiling out. The boiling of a highly alkaline water in boiler pressure parts for the removal of oils, greases, etc. prior to normal operation or after major repairs.

Bourdon tube. A hollow, metallic tube, bent semicircularly, which forms the actuating medium of a pressure gauge.

Btu. British thermal unit, which is the amount of heat required to raise 1 pound of water through 1 degree Fahrenheit.

Bunker C oil. Residual fuel oil (No. 6 fuel oil) of high viscosity, commonly used in marine and stationary steam power plants.

Burner. A device that positions a flame in the desired location by delivering fuel and air to the location in such a manner that continuous ignition is accomplished. Some burners include atomizing, mixing, proportioning, piloting, and flame-monitoring devices.

Calorie. The mean calorie is 1/100 of the heat required to raise the temperature of 1 g of water from 0°C to 100°C at a constant atmospheric pressure. It is about equal to the quantity of heat required to raise 1 g of water 1°C (4.184 J).

Caking. Property of certain coals to become plastic when heated, and to form large masses of coke.

Carbon dioxide. CO_2; a product of combustion between a fuel and the oxygen of the air.

Carbon monoxide. CO; incomplete fuel combustion, usually found in very small amounts when excess air is present.

Carryover. The moisture and entrained solids forming the film of steam bubbles; a result of foaming in a boiler. Carryover is caused by a faulty boiler-water condition. See foaming.

Catalytic. Describes a chemical reaction that is enhanced by the presence of a catalyst, a substance that is not consumed in the reaction.

cfh. Cubic feet/hour.

cfm. Cubic feet/minute.

Combustion. Chemical combination of the combustible (that part which will burn) in a fuel with the oxygen in the air supplied for the process. Temperatures may range from 1850°F to over 3000°F.

Conduction. The transmission of heat through matter.

Conductivity. The amount of heat (Btu) transmitted in 1 hr through 1 ft^2 of a homogeneous material 1 in. thick, for a difference in temperature of 1°F between the two surfaces of the material.

Convection. The transmission of heat by the circulation of a liquid or a gas such as air. Convection may be natural or forced.

Damper. A device for introducing a variable for regulating the volumetric flow of gas or air.

1. *Butterfly type.* A single-blade damper pivoted about its center.
2. *Curtain type.* A damper consisting of one or more blades, each pivoted about one edge.
3. *Flap type.* A damper consisting of one or more blades, each pivoted about one edge.
4. *Louver type.* A damper consisting of several blades, each pivoted about its center and linked together for simultaneous operation.
5. *Slide type.* A damper consisting of a single blade which moves normal to the flow.

Deaerating heater. A type of feedwater heater operating with water and steam in direct contact. It is designed to heat the water and to drive off oxygen.

Descale. Removal of rust and scale from metal.

Design pressure. The pressure used in the design of a boiler for the purpose of determining the minimum permissible thickness or physical characteristics of the different parts of the boiler.

Desuperheater. Apparatus for reducing and controlling the temperature of a superheated vapor.

Dewpoint. The temperature at which a vapor becomes a liquid (condensation). Thus, the condensation of sulfur trioxide (SO_3) and water (H_2O) yields sulfuric acid (H_2SO_4).

Distillate oil. Light fraction of oil which has been separated from crude oil by fractional distillation. See *fuel oil.*

Downcomer. A tube or pipe in a boiler or waterwall circulating system through which fluid flows downward between headers.

Draft. A difference of pressure that causes a flow of air or gases through a furnace, boiler, or chimney.

Dry steam. Steam containing no moisture, or commercially dry steam containing not more than 0.5% moisture.

Economizer. A series of tubes located in the path of the flue gases. Feedwater is pumped through these tubes on its way to the boiler in order to absorb waste heat from the flue gas.

Embrittlement. An intercrystalline corrosion of the boiler plate occurring in highly stressed zones. Cracking may result.

Endothermic. Describes a chemical reaction that absorbs heat.

Enthalphy. A thermal property of a fluid which is a function of state and is defined as the sum of stored mechanical potential energy and internal energy. It is generally expressed in Btu per pound of fluid (joules per kilogram).

Entrainment. The conveyance of particles of water or solids from the boiler water by the steam.

EPA. Environmental Protection Agency.

Exothermic. Describes a chemical reaction that gives off heat.

Firebox. Also known as the combustion chamber in a boiler.

Firetube. A tube in a boiler, having water on the outside and carrying the products of combustion on the inside.

Fixed carbon. The carbonaceous residue less the ash remaining in the test container after the volatile matter has been driven off in making the proximate analysis of a solid fuel.

Flame detector. A device which indicates whether fuel (liquid, gaseous, or pulverized) is burning, or whether ignition has been lost. The indication may be transmitted to a signal or to a control system.

Flue gas. The gaseous products of combustion in the flue to the stack.

Fly ash. Suspended ash particles carried in the flue gas.

Foaming. Formation of steam bubbles on the surface of boiler water due to high surface tension of the water. See *carryover*.

Forced-draft fan. A fan supplying air under pressure to the fuel-burning equipment.

Fouling. Condensation of fuel constituents, such as sodium sulfate, on flyash particles and on boiler tubes in areas where temperatures are such that the constituents remain in the liquid state. Fouling is easily cleaned by soot blowers.

fps. Feet per second.

Fuel oil. A petroleum product requiring comparatively minor refinement that is used as a combustible for steam boilers.

Furnace. A high-temperature heat-holding enclosure; a process furnace.

Furnace pressure. The gauge pressure that exists within a furnace combustion chamber. The furnace pressure is said to be positive if it is greater than atmospheric pressure, negative if it is less than atmospheric pressure, and neutral if it is equal to atmospheric pressure.

Gauge pressure. The pressure above that of the atmosphere (14.7 psi at sea level); absolute pressure minus 14.7 at sea level.

Hardness. A measure of the amount of calcium and magnesium salts in water. Usually expressed as grains per gallon or parts per million as $CaCO_3$.

High-heat value or higher heating value. The total heat obtained from the combustion of a specified amount of fuel at 60°F before the quantity of heat released is measured.

Igniter. A burner smaller than the main burner, which is ignited by a spark or other independent and stable ignition source, and which provides proven ignition energy required to light off the main burner immediately.

Ignition. A system in which the fuel to a main burner or gas or oil pilot is ignited directly, either by an automatically energized spark or glow coil or by a gas or oil pilot.

Ignition temperature. Lowest temperature of a fuel at which combustion becomes self-sustaining.

IHEA. Industrial Heating Equipment Association.

Inhibitor. A substance which selectively retards a chemical action.

Ion. A charge atom or radical which may be positive or negative.

Ion exchange. A reversible process by which ions are interchanged between solids and a liquid.

kcal/Nm. Kilocalories per normal cubic meter.

LNG. Liquefied natural gas.

Low-heat value. The high heating value minus the latent heat of vaporization of the water formed by burning the hydrogen in the fuel.

LPG. Liquefied petroleum gas (propane and/or butane).

Mechanical-atomizing oil burner. A burner which uses the pressure of the oil for atomizing.

Mud or lower drum. A drum- or header-type pressure chamber located at the lower extremity of a watertube-boiler convection bank which is normally provided with a blowoff valve for periodically blowing off sediment collecting in the bottom of the drum.

Nitrogen. N_2; nitrogen does not burn, and in combustion analysis it is considered inert. Very small amounts of nitrogen oxides (NO_x) are usually formed, depending on combustion conditions.

OSHA. Occupational Safety and Health Administration.

Oven. A low-temperature heat-holding enclosure.

Perfect or stoichiometric combustion. The complete oxidation of all the combustible constituents of a fuel, utilizing all the oxygen supplied.

pH. The hydrogen-ion concentration of a water to denote acidity or alkalinity. A pH of 7 is neutral. A pH above 7 denotes alkalinity, while one below 7 denotes acidity.

Pilot. A small burner which is used to light off the main burner.

ppm. parts per million.

Preheating. Heating prior to the process.

Pressure. Absolute pressure; the pressure above a perfect vacuum; gauge pressure plus 14.7, at sea level.

Primary air. Air introduced with the fuel at the burners.

psi. Pounds per square inch (pressure).

psig. Pounds per square inch gauge.

Pyrolysis. Chemical decomposition caused by heat.

Quenching. Rapid cooling after heat treatment.

Recuperator. A heat exchanger used to recover waste heat.

Reducing atmosphere. A furnace atmosphere that tends to remove oxygen form substances placed in the furnace. It may be produced by supplying inadequate air to the burners, thus intentionally making the combustion incomplete.

Scale. A deposit of medium to extreme hardness occurring on water-heating surfaces of a boiler because of an undesirable condition in the boiler water.

SCFH. Standard cubic feet per hour.

SCFM. Standard cubic feet per minute.

Slagging. Deposition of molten or partially fused particles (noncombustible fuel constituents) on boiler-tube surfaces. Slag is a residue deposited by ash particles that have attained their softening temperature (1900°F–2700°F), depending on their composition. Slag may be plastic and viscous when hot. It hardens and is rather porous and brittle when cool. Slagging is a major problem for equipment used for cleaning.

Soot blower. A tube from which jets of steam or compressed air are blown, used for cleaning the fireside of tubes or other parts of a boiler.

Specific gravity. The ratio of the weight of a unit volume of a material to the weight of the same unit volume of water at a certain temperature.

Specific heat. The quantity of heat, expressed in Btu (joule), required to raise the temperature of 1 lb (kilogram) of a substance 1°F(C).

Spontaneous combustion. Ignition of combustible material following slow oxidation without the application of high temperature from an external source.

Stack. Chimney.

Steam quality. The percentage by weight of vapor in a steam-and-water mixture.

Sulfuric acid. H_2SO_4; formed during fuel combustion in the presence of excess air when sulfur (S) is present in the fuel.

Superheater. A series of tubes exposed to high-temperature gases or to radiant heat. Steam from the boiler passes through these tubes to attain a higher temperature than would be possible otherwise.

Sulfur oxide. SO_2; the first reaction with sulfur (S) and oxygen (O_2) during fuel combustion when sulfur is present in the fuel.

Sulfur trioxide. SO_3; SO_2 reacts with additional O_2 to yield SO_3. Usually 3%–5% of the SO_2 becomes SO_3 when excess air is present during combustion.

Therm. One therm equals 100,000 Btu.

Turndown. The ratio of maximum to minimum input rates.

Turnkey. Total project responsibility, start to finish.

UL. Underwriters Laboratories, Inc.

Ultimate analysis. A fuel analysis where the various chemical components are shown.

WC. Water column.

Watertube boiler. A boiler in which the water flows through the tubes, and the products of combustion surround the tubes.

Bibliography

Government Reports on Boiler Fuel Additives

PB-264 065. *Experimental Evaluation of Fuel Oil Additives for Reducing Emissions and Increasing Efficiency of Boilers,* R. D. Giammar, A. E. Weller, D. W. Locklin, and H. H. Krause, of Battelle-Columbus Laboratories, Columbus, OH, for the EPA, Industrial Environmental Research Laboratory, Research Triangle Park, NC, and the FEA, Washington, D.C., January 1977.

PB-264 068. *Combustion Additives for Pollution Control—A State-of-the-Art Review,* H. H. Krause, L. J. Hillenbrand, A. E. Weller, and D. W. Locklin, of Battelle-Columbus Laboratories, Columbus, OH, for the U.S. Environmental Protection Agency, Washington, D.C., January 1977.

Government Reports on Increasing Boiler Efficiency

PB-262 577. *Industrial Boiler Users' Manual*—Volume II. KVB, Inc. FEA Contract C-04-50085-00, Report No. FEA/D-77/026, January 1977.

PB-264 543. *Guidelines for Industrial Boiler Performance Improvement (Boiler Adjustment Procedures to Minimize Air Pollution and to Achieve Efficient Use of Fuel),* M. W. McElroy and D. E. Shore, KVB, Inc. EPA Contract 68-02-1074, Report No. EPA-600/8-77-003a, January 1977.

PB-265 713. *Measuring and Improving the Efficiency of Boilers—A Manual for Determining Energy Conservation in Steam Generating Power Plants,*

Engineering Extension Service, Auburn University, Alabama. FEA Contract FEA-CO-04-50100-00, Report No. FEA/D-77/132, November 1976.

TID-27600. *Maintenance and Adjustment Manual for Natural Gas and No. 2 Fuel Oil Burners,* Bureau of Natural Gas, Federal Power Commission, and U.S. Department of Energy, 1977.

Non-Government Publications

Betz Handbook of Industrial Water Conditioning, Betz Laboratory, Inc., Trevose, Pennsylvania, 19047, 1980.

Boiler Care Handbook, Cleaver-Brooks, Division of Aqua-Chem Inc., P.O. Box 421, Milwaukee, Wisconsin, 53201, 1978.

Combustion Technology Manual, Industrial Heating Equipment Association, 1901 North Moore Street, Arlington, Virginia, 22209, 1980.

Dyer, David F., and Glennon Maples, *Boiler Efficiency Improvement,* Boiler Efficiency Institute, P.O. Box 2255, Auburn, Alabama, 36830, 1981.

Eliot, R. C., *Boiler Fuel Additives for Pollution Reduction and Energy Saving,* Noyes Data Corporation, Noyes Building Park Ridge, New Jersey, 07656, 1978.

Fundamentals of Boiler Efficiency, Exxon Company U.S.A., P.O. Box 2180, Houston, Texas, 77001, 1976.

McCormack, Harry, *The Applications of Chemical Engineering,* D. Van Nostrand Company, Inc., New York, N.Y., 1949.

''Oil Burner Combustion Testing,'' Bulletin 4011, Bacharach Instrument Company, 625 Alpha Drive, Pittsburgh, Pennsylvania, 15238, 1975.

Power Handbook, McGraw Hill, Inc., 1221 Avenue of the Americas, New York, N.Y., 10020.

Shaw, D. E., and M. W. McElroy, ''Tuning Industrial Boilers,'' Parts I, II, and III, *Power,* April, May, June, 1977.

The Nalco Water Handbook, Nalco Chemiccal Company, 2901 Butterfield Road, Oak Brook, Illinois, 60521, 1979.

Index

TO LIVE IN PARADISE

Renée Roosevelt Denis' adventures began at a very early age, when she was almost carried off by an albatross from the deck of a ship taking her to Bali with her family. Renée's grandfather, André Roosevelt, his daughter Leila, son-in-law, Armand Denis, and their children, Renée, Armand and David, along with a tortoise sacred to the Balinese, spent a year in Bali while making the exotic, romantic film, *Goona Goona*. Leila and Armand Denis went on to make a series of dramatic films during their expeditions to primitive Africa and Asia. Renée spent years in boarding schools but on vacations she joined her brothers at the Anthropoid Ape Foundation established by Armand Denis in Florida.

When Renée was eighteen she went to live with her grandfather André in Haiti, and there she met Gustav Dalla Valle, a wild Italian dubbed "Tarzan of the Caribbean." Despite her grandfather's disapproval and her own misgivings, she fell in love with the handsome adventurer. Their carefree life in Haiti ended when "Tarzan" decided to go to the States and become a big businessman. This was disastrous for their romance and after futile attempts to salvage their marriage, Renée finally ran away to French Polynesia. There she met and wed a charming Tahitian, but this marriage also did not last. Renée was on her own, and that was when her most exciting experiences began.

Moorea, with its beautiful bays, sharp peaks and deep valleys, is considered one of the most spectacular islands in the world. The story *Bali Hai* was based on this island.

To Live In Paradise

Renée Roosevelt Denis

Lost Coast Press
Fort Bragg, California

To Live in Paradise

For more information, or to order additional copies, please contact the publisher:

Lost Coast Press
155 Cypress Street
Fort Bragg CA 95437
1-800-773-7782

Library of Congress Publisher's Cataloging-in-Publication Data
Denis, Renée Roosevelt.
To live in paradise / Renée Roosevelt Denis.
p. cm.
ISBN: 1-882897-07-2
1. Denis, Renée Roosevelt. 2. Tahiti — Biography. 3. Bali (Indonesia : Province) — Biography. 4. Haiti — Biography.
I. Title.
CT2990.D46A3 1996
996.2'11 — dc20.
[B] 96-42545
CIP

Book production by Cypress House

Manufactured in the U.S.A.

First edition

10 9 8 7 6 5 4 3 2 1

This book is dedicated to my brothers,
David Roosevelt and Armand Roosevelt Denis,
who, although we were often oceans apart,
always remained closest to me.

Table of Contents

To Live In Paradise

Prologue

Haiti

"You're just asking for trouble, girl!" my grandfather André shouted. "Haven't I told you enough about that man?" But as he glared at me, his face red with exasperation, I saw the slight shift in his eyes. Yes, he had said plenty about Gustav in that amused way he always spoke about him. André, too, had been an irresistible ladies' man all his life. He loved women, all women, and they adored him. André was still handsome, although now he had lost most of his hair. Six feet tall and broad shouldered, cantankerous when he didn't hear well, he could look frightening. But he was genial and charming, and I loved him.

We were having breakfast on the terrace of the hotel. It was going to be another beautiful day. Happily and naïvely, I told André that Gustav was going to take me out tonight. I had expected a kindly, paternal warning, but, to my astonishment, André slammed down his coffee cup and lit into me.

It was the first time I had seen him angry. I said with dignity, "I'm not going to be alone with Gustav. We've been invited to a dinner party."

André didn't hear me. "I thought you had more sense, Renée. The least of his intentions is to use you for camouflage."

"Camouflage?"

"Yes, innocent child. Don't you know about that woman at the British Consulate, Evelyn Burns, whose husband is always away?"

"Yes, of course," I said coolly. "She may be crazy about him but Gustav never takes her out."

"Naturally he doesn't take her out. That would be too blatant. So he just happens to show up wherever she goes. And at night he slips through the garden and up onto her balcony like a thief."

"Hardly a thief if she lets him," I protested, thinking, just like Errol Flynn in those old romantic movies. I didn't want to hear any more about Evelyn. "Well, he couldn't be all that much in love with her. What about that actress last month? He takes out other women —" I stopped short but André caught me.

"And you'd like to be one of them! Do you think he just takes them out to dinner? You're seen with him, my girl, and your reputation is shot."

That's right, I thought crossly, a man can do anything he wants but a woman no more than steps out the door with a man and she gets a bad name. Since when did André worry about reputations, particularly there in Haiti? We lived in the hotel high above Port au Prince and never went anywhere.

"Damn," I muttered. I didn't want André to be angry with me. He was my only family now. Of course there was my father and mother and my brothers — more than a thousand and five thousand miles away. None of them wondered what would happen to me when they all left. André was the one who went to the States and got me. Well, maybe he didn't go for that reason, but he cared enough to bring me here to live with him.

I tried to smile as I pushed back my chair and stood up. I reached for a red hibiscus and tucked it into my hair. André watched me as he put his cigarettes into his shirt pocket and got up too. He came around the table and put his hands on my shoulders, blue eyes smiling kindly.

"That's it, my girl, use your good sense. Don't get mixed up with Gustav. He's too sophisticated and too old for you. He's past thirty and you're what, eighteen? You're still wet behind the ears. Oh, you're pretty enough — in fact you've got a damned beautiful body, I'm proud to say. Don't give it away to a rake like that." He lifted my dark hair away from my face and studied me a moment. "You have the kind of looks that will grow more beautiful as you mature. With those high cheek bones, don't worry, you'll have plenty of men falling in love with you when you are a woman."

He said the wrong thing. I was already a woman. I had enough of that boarding school innocence. Years of nights in study hall and lights out at ten, while other girls were dating boys and having a good time. I knew my grandfather wanted to protect me, but no one had taken me anywhere since I'd been in Haiti. Of course I liked helping him in the hotel and I had met lots of people, but I wanted to go out. I wanted to go out to dinners and shows, and go dancing. With a man.

"Oh," I wailed, "I just want to go to the party with him. He isn't going to rape me — the granddaughter of his good friend? Besides, it wasn't he who asked me. It was Corinne."

"Rubbish! Gustav told her to." André fumbled for his glasses in his other pocket, angry again. "He knows I wouldn't approve. Good God, the man's married! He left his wife, lived with another woman, and I saw the insane female he sent packing back to Madrid or wherever he got her. You think it will be just an exciting party but you'll be hooked, little girl. You're going to get hurt." Seeing my disappointment, he hesitated a moment. But, shrugging his shoulders, he turned away, and caught the cooks, Dieu Merci and La Petite, hiding behind the fan palm, eavesdropping.

"Get out of there, you black Jezebels!" he roared as the two women ran giggling back into the kitchen. Dieu Merci was beautiful, and everyone knew that André had slept with her. She paid no attention to his noise and ran her kitchen as she liked. Seeing her, I thought resentfully that André was in no position to tell me what I shouldn't do.

Not that my grandmother was there, that year of 1950. They had been divorced for thirty-five years; separated, my grandmother said, because she refused to acknowledge the Mexican divorce. She was still sitting up there in Connecticut, her mouth primly pursed, and who could imagine that she had been quite wild, back in the days when she was a young artist in Europe, the darling of the avant-garde. Not surprisingly, she and André were drawn to each other. He was as handsome as she was beautiful, and even more attractive to her was his illustrious family name.

André took his name for granted. He could have held important positions despite his lack of business sense, but he didn't give a damn about prestige. He was an adventurer. He traveled around the world, and wherever he found an exotic land with sensuous women, he stayed to become happily familiar with the country and its people.

What's funny is André playing the heavy father now with me instead of Mother. I thought, maybe he feels guilty about leaving her when she was a small girl. André just wasn't a family man. There was too much to see and to experience in the world. He left and didn't return for twelve years. He soon left again, for Bali and elsewhere no one knew, for another seventeen years. Mother too had travel in her blood. No wonder she became an important part of Father's expeditions. And now I was on my first trip. Where did it all begin?

Theodore Roosevelt. He was the first explorer in the family. André's father and Theodore were cousins, and although André's family lived in France, the Roosevelt clan traveled often between continents and Theodore was like an uncle to the boy. André was thrilled when Theodore, always delighted to have an eager, young audience, recounted his marvelous adventures in the Badlands of the Dakotas and the wild country of the Indian nations. Teddy Roosevelt was president when André turned twenty-one and came to the United States for the first time. For several months André lived at the White House, which was very informal in those days, overflowing with Theodore's several children, relatives and guests.

André had to complete his education, but as soon as he had graduated from Harvard he was off to the Far West and then the Orient. He received an allowance for several years and then came into his inheritance. André had no feeling for business. He was always surprised when his investments did not prosper, and he had to find a job. As he was dynamic and charming, superb in public relations, promoting travel in particular, he was in demand by the growing industry of travel services wherever he was in the world. But he rarely stayed long in one place. His marriage had lasted only four years. André had soon realized that his wife was a self-centered, ambitious snob. He asked her to divorce him. She refused, so he went to Mexico and divorced her. And kept on going. I did not know André until he went to the States two months ago, to promote his hotel. He was surprised to find all of the family gone except me.

Somewhere on the hill below the hotel a donkey brayed, raucous and ridiculously plaintive. I loved the sound and it drew me to the railing overlooking the valley. I thought of when I first heard the donkeys in the cool dawn of that enchanted morning when I woke up and realized I was in Haiti. Delighted by the crowing of the roosters and that rusty complaint, I ran out onto the terrace in my pajamas and hung over the railing to see the farm. There was none, only tiny thatched houses clinging to the hillside, chickens scratching in the manure, still warm and steamy and earthy smelling, how I loved it, and a donkey with droopy ears nibbling at the sparse grass. The sun was rising beyond the rim of the ocean in the east, and I watched as its pale golden light touched like a wand upon the whitecaps in the bay, the sail of a pirogue just leaving the port, the roof of the cathedral, the president's white palace, and then all the houses of the town.

I was giddy with happiness. A rush of gratitude for my grandfather filled my heart. Carefree at last, I had a world to discover, a whole lifetime to enjoy it.

Meeting Gustav was the most exciting thing that had happened to me, already. Despite what I had heard, I did not expect him to be so fantastically handsome. From the Italian Tyrol, blond with blue eyes that were startling in his deeply tanned, rugged face, he looked more like an Austrian alpinist than any Italian I had ever seen. He had lived in Venice, Rome and Paris, had produced a film, and was an artist and a champion spearfisherman — and I'd been in boarding school practically all of my life.

I suddenly realized the enormous difference between us. Whatever could I talk about to him? André's right. Gustav is too sophisticated for me. Sure, he goes around in old sandals, his shirt half-buttoned, but the shirt is made of the finest linen, and the sandals are probably custom-made in Italy. The day I met him, I had gone looking for André in the bar. There was no mistaking the man who was with him. The two of them were arguing loudly, something about paintings, which was as much as I understood with my school French. André saw me and paused long enough to introduce us. Gustav glanced at me curiously, but then they both ignored me. I felt like a ninny. I thought they would not notice if I left. Just in case, I breezily waved good-bye. But Gustav turned to me and said, in his unique English, "If liking, come on the boat to-morrow. André and you always welcome." That was what Gustav was doing in Haiti, taking people out sailing and spearfishing. I was pleased that he had noticed me.

The next morning I went happily down to the pier. Gustav greeted me, a Haitian crewman helped me on board, and we soon set sail.

Evelyn was on the boat. She was with some friends, but I knew which one was Evelyn because she sat by the tiller. Gustav was standing on the bulkhead, steering with one bare foot. He looked seven feet tall, and now and then Evelyn casually touched his leg. Really! I thought. And couldn't take my eyes away.

How gauche I must have looked. Gustav was just being nice, inviting me on the boat. He probably saw me watching him. One time when he was laughing, so hugely careless, and put his arm across Evelyn's shoulders, my stomach lurched with envy.

I came back to the present. André's right, I convinced myself. I'm just asking to be hurt. Better to have nothing to do with the man. I sighed as I turned away from the view and walked back across the terrace.

"Renée!" The girl who worked at the reception desk called me, beckoning excitedly. I ran to the door. "Telephone for you — it's Gustav!"

"Oh!" My heart started to beat wildly. I tried to compose myself before picking up the phone. "Hello?" I said calmly.

"Renée?" Just the sound of his voice—

"Yes, it's Renée."

"Ah. Is Gustav." I thought his English was delightful.

"Yes, Gustav."

"Ah. I call for reminder tonight. The dinner, Corinne's house."

All reason flew out of my head. "Yes?" I said weakly.

"I come get you at eight. Is all right?"

Could I have said no? Impossible. "Yes. Yes, Gustav."

"Good. I see you then. *Au'voir.*"

"*Au revoir,*" I replied softly, as I put the receiver down.

1

Child of My Parents

Child of my parents? Not when I was a child. I don't blame them, as they didn't know what it was to raise children. My mother, Leila, was only twenty-one when she already had three children. She was born in 1906 and was named after rich, elderly Aunt Leila. When Aunt Leila died, little Leila's mother, Heidi, received a settlement that was invested wisely and allowed Heidi to travel and indulge in her passion for art, while Leila was put away in a Connecticut boarding school. Apparently unaffected by her parents' neglect, Leila grew up to be a lovely and charming woman. When she was eighteen, she met a brilliant, young scientist and inventor from Belgium, whose name was Armand Denis.

Armand, born in 1895, was the son of a judge of the Belgian Supreme Court, and he was educated, as were all the sons of well-off, bourgeois, Catholic families, at a strict Jesuit seminary. During the holidays the family traveled around Europe, staying in fashionable resorts and hotels. The following story is often quoted in articles about Armand Denis:

> "Judge Denis took his family for a holiday to the elegant resort of Deauville on the French Atlantic shore. In the party was his son Armand, who insisted upon taking with him his colony of pet lizards and snakes. The reptiles were kept in a pillowcase under the boy's bed. One morning Armand carelessly tied the string that was to secure the pillowcase, before the family went down to breakfast. It was not long before the hotel erupted in an uproar as the lizards and snakes appeared, scuttling and slithering out the door and into the hallways. To the accompaniment of shrieks, ringing bells and slamming doors, the Denis family hurriedly packed their bags and headed for the railway station."

When Armand was fourteen he persuaded his father to let him go off by himself for the holidays. His father gave him enough money to reach his destination, pay his expenses and return. Each summer Armand traveled until he had used up all of his allowance, then cabled home for more money to buy his return ticket. When he was sixteen and tried the same thing again, his father lost his patience and cabled back to Armand to find his own way home. With no money left, Armand was spending a few nights in a Carthusian monastery. The superior was dismayed when Armand frankly admitted that he did not believe in God, so he undertook to convince the young man of God's existence. After much discussion Armand agreed to become a novice for six months. If he was persuaded that God did indeed exist he would join the brotherhood. The fight for Armand's soul was strenuous but unsuccessful, and the superior agreed that he did not have the vocation to be a monk.

The First World War began. Armand joined the army and was almost immediately caught by the Germans. He managed to escape and made his way to England, where he got a job washing bottles for a chemistry professor at Oxford University. As he had been a brilliant student and was fascinated by chemistry, his employer sponsored his admittance to the university. He graduated a Rhodes Scholar. Armand came to the United States in the mid-twenties, spent a year at Harvard University, then bought a car for $100 and drove across the country to the West Coast. At the California Institute of Technology he began to experiment with radio and cinematography.

On a holiday trip to Long Island my remarkable father met my lovely mother and they fell in love. They were married in the garden at Sagamore Hill, the Roosevelt estate at Oyster Bay, Long Island, in 1926, and within two years they had three children, my twin brothers, Armand and David, and me.

Meanwhile, in the early twenties, our grandfather André, finding himself running out of money again, accepted a job with Raymond & Whitcomb Travel as program director on a posh cruise ship making a world tour. The ship broke down in Singapore. Faced with a three week delay, what was André to do with the passengers?

Bali, then as now, was a legendary romantic island, almost unknown. André and a few of the more adventurous passengers dared to take the smelly, local cargo boat called the Pig Express, to make the trip to Bali, stay three days, and return.

André was enchanted by the island. Bali was the

tropical paradise, with coconut palms swaying over white sand beaches, forested mountains rising beyond terraces of rich green rice fields, beautiful temples and peaceful villages. Best of all were the most exquisite, exotic women André had ever seen. At the end of three days, André politely put his charges back on the boat, saying, "I am so glad to have known you, but I am staying here in Bali."

Mr. A. Morzer Bryns, Dutch Consul in Djakarta, Indonesia, wrote about his own arrival in Bali in 1924. "The Dutch Steam Company, KPM, offered me the job of establishing a tourist office in Bali. Without an office there on the island they could not compete with the Roosevelt-Minus Tourist Office which represented American Express, Thomas Cook, and Raymond & Whitcomb. My instructions were to fight them tooth and nail. In those days it took three days by ship from Djakarta to Buleleng, the port of north Bali. The ship arrived at 6:00 a.m., and before anyone from KLM showed up, this very nice man came up to me on deck and said, 'Good morning! My name is André Roosevelt. Welcome to Bali. Can I help you in any way?' From that moment on André and I became fast friends. His name is still a legend in Bali."

Four years later André came to the United States to make a promotional tour. He arrived in the middle of winter at a Brooklyn dock, in the snow and slush — miserable. He honored his tour engagement up to a point, but Chicago's icy winds were too much. He wired my parents, "I have to get out of here! Let's go back to Bali and make a film."

Yes, it was just about as casual as that. I mean, André coming back like that after all those years. Mother was thrilled, and my father and André liked each other immediately. The two men were remarkably alike. They were both adventurers and shared a passion for cinematography. My father needed no persuasion to leave for Bali, and they were soon calculating what they would need: movie equipment, film, developing solutions and heat-proof containers. Mother had to purchase everything necessary for three adults and three babies for a two-month voyage on a freighter, plus an unknown length of time on a tropical island on the other side of the world.

I loved to hear the story about how Mother almost lost me on the trip across the Pacific. She left me alone one day on a mat on deck while she went below for a moment. I was too small to crawl but I was strong and active, kicking my feet energetically in the sunshine. Mother came back on deck just in time to see a tremendous bird, an albatross with a ten-foot wing span, swooping down to catch my small, wiggling body. Terrified, screaming, she raced across the deck and reached me just ahead of the albatross.

An unusual passenger and member of our party was Jake, a fifty-pound Galapagos tortoise, given to my father by scientists returning from South America. A very active tortoise, he had a tremendous appetite for vegetables and fruit. My father and the ship's cook argued constantly over storage space for huge quantities of fresh greens and perishable fruit for Jake. The family often went without salad so that the tortoise could satisfy his voracious appetite. My brothers, a year old, loved to be held on the hump of Jake's shell and be taken for a ride. Jake surprisingly became a very important contributor to the success of the trip.

Our arrival in Bali was celebrated by a host of André's friends, and we were treated with more than ordinary ceremony. People came from miles around to see us and we were showered with flowers and gifts. André finally found out why.

Our family was considered favored by the gods. My brothers were twins and that was a great blessing. Twin girls would have been mildly unfortunate, but a boy and girl would have been a tragedy. They could not be raised in the same village, nor could their mother ever see them because she had condoned their living in sin, together in her womb for nine months. More important than the twins, we also had a god with us. Most Balinese were Hindus who believe that the earth rests on the back of a tortoise. The Balinese, however, had never seen a live tortoise. Jake was worshipped and adorned daily with flowers (which he ate) and he almost exploded from the copious sacred offerings of fruit.

The great advantage for my parents was the enthusiastic cooperation of the Balinese in making the film, called *Kriss*, or *Goona Goona* (meaning "love powder"). It was a poetic Balinese drama with remarkably talented actors. It was a simple story of a prince who, when riding through the countryside, sees and desires a lovely peasant girl, but she is married to a peasant boy. The prince persists in seeing her and succeeds in seducing her. Her husband, returning from the rice fields, catches the pair in *flagrant délit*. The climax of the film is a fight between the prince and the peasant. This scene had to be violent and dramatic. It was difficult to get the Balinese, who are very peaceful, to understand what they should do. The fight was with the kriss daggers, and finally, after several rehearsals, the men enjoyed the mock battle so much that they had to be separated before they really injured each other.

In those days of early photography, exposed film had to be processed as quickly as possible because the image would quickly fade. A dark room was built of bamboo and thatch, remarkably well by the Balinese, who could not understand why these people wanted a house completely sealed off from light. The Balinese woodworkers also made the developing tanks, out of mahogany, and they were works of art, so well done that there were no leaks. But, despite careful control, the tropical heat caused the film density to vary so that some parts were too light and other parts too dark. To correct this in the processed film, Father designed an electric circuit which controlled the light in the copy projector. When the negative was too dark the light became brighter, and when it was too light the projector light dimmed, thus balancing the overall tone of the film. Later, Father was able to convert his electronic light control to invent what is now called the A.V.C. circuit, the automatic volume control that is used in every radio, TV set and similar electronic devises made today.

The success of the film, *Goona Goona*, and my father's love of animals, attracted the attention of the well-known, wild animal collector and movie producer, Frank Buck. He needed a cameraman for the production of his next film, *Wild Cargo*, in Malaysia. The man must have no fear of animals. My father accepted without hesitation.

"Wonderful!" Mother said. "When do we leave?"

"Leila, you can't go."

"What do you mean, I can't go?"

"This is Frank Buck's company, his expedition. I can't just take a wife along."

"I'm not just a wife! I worked hard in Bali. I spent hours and hours developing film for you and working with the actors, making sure that the sequences matched—"

"I know, I know—"

"Well, did you tell him that?"

"Yes, I did—"

"What did he say?"

"Leila, he thinks you are a fine woman, but he will not permit a woman on the expedition."

"Why, why—!" Mother sputtered.

"Leila, don't try to insist. You can't go."

Father was shocked and disgusted to find out that all of Frank Buck's films, every one of them, were done in a compound outside the town of Singapore. Nothing was done in the bush. The compound was very large, planted with trees, vines, and tropical bush to resemble the jungle, and in the walls and trees there were platforms for the camera men to film the action below.

Frank Buck was a good actor and really looked the part of an intrepid animal hunter, but he was an alcoholic, and spent most of his time at the bar of the Raffles Hotel. As for his love of animals, he didn't hesitate to sacrifice them to obtain the dramatic sequences he was famous for.

In the film, *Wild Cargo*, there is a gripping scene of a fight between a panther and a python. This could never have occurred in the jungle because the two animals' paths never crossed, or they simply avoided each other. Frank Buck bought his animals from the Singapore zoo, and the "true life" captures were to be filmed in the compound. Now, how to get a python and a panther to fight?

The panther was put in a cage with a gate opening into the wall of the compound, and the python was let loose in front of the gate. In order to have the action in complete detail, there were half a dozen movie cameras zeroed in on the scene. It was expected that as soon as the gate was opened, the panther would leap out, and, as he would be immediately confronted with the python, the two would certainly fight. All was ready, and the gate was jerked open. The panther took one look at what was outside, and said, "No way, I'm not going out there." The men started yelling and prodding the panther, but he refused to leave the cage. Buck was determined to have this "extraordinary, unexpected jungle experience" in his movie. He had the men build a fire at the back of the cage. When the flames shot up, the panther leaped out. The python then did what was natural to him — he struck out and caught the panther in mid-air, by the nose, and immediately threw his constricting coils around him. That was the end of the fight. It was over in a matter of seconds, hardly time to see what had happened. It was only by using the footage taken from several angles, and working the shots back and forth, that the sequence was drawn out to a suspenseful minute. The panther never had a chance.

Another episode was to be the capture of a panther which had climbed to the high branches of a tree. The idea was for Frank Buck to shoot the branch it was crouched upon, and the panther would fall down into the net spread on the ground. They didn't need Frank Buck for the actual shooting, as he was generally inebriated and would miss the branch in any case, so there was a sharpshooter for the actual shot. A close-up of Frank Buck aiming the gun would be spliced in. But when the scene was set up with the panther high in the tree, and the sharpshooter ready to aim, Buck drove into the compound, stumbled out of his car, and declared that he would shoot the branch out from under the panther. All right, if he wanted to try, let him, because he was sure to miss. He fired and did miss the branch — but shot the panther right between the eyes.

There was the exciting scene of Frank Buck being attacked by a python. The python bit Buck's arm, which was of course protected by a metal shield under his shirt sleeve. When the python struck and started to wrap his coils around Buck's body, Buck saved himself by shooting the python. This time his aim was true, he couldn't have missed at arm's length, but he shot too soon. The sequence was redeemed by tying the python's head to Frank Buck's arm, while the crew manipulated the long body of the snake out of range of the camera, agitating and twisting it so that it appeared to be about to coil and crush its victim.

Another episode was the capture of a tiger by trapping it in a concealed, deep pit. The sequence was to show Frank Buck being lowered into the pit and presumed to somehow tie up the tiger. They had the tiger and the pit was dug. They managed to direct the tiger along the path in the "jungle" and down he went, into the pit. Unfortunately, there was a monsoon rainstorm the night before, the pit was half full of water, and the tiger drowned. Not to lose the sequence, the tiger was pulled up and dropped into another pit, Frank Buck was let down, and he struggled energetically with the dead tiger, manipulated like a puppet, to create Frank Buck's brave and daring capture.

It happened that the local rajah, curious to see how

movie making was done, came to the location. Of course he would have been shocked to see the dead tiger, so he was refrained from approaching the pit "because of the risk of great danger."

Father, disgusted with it all, left the employ of Frank Buck.

2

Children of Travelers

Mother had had her first taste of travel and adventure on the trip to Bali, and she did not give up easily. She was furious that she could not go on the Frank Buck expedition simply because she was a woman. It didn't take her long to decide how she would show up the men. She put an ad in the New York papers for a woman mechanic who would like to make a trip around the world. She found her mechanic, as good as any man and fully capable of handling any emergency. Mother purchased a second-hand, white Ford panel truck; she painted a map of the world on the side, with a line tracing the route they would follow, and the legend, "New York to New York Around the World." She pinned $1000 to her bra for expenses.

And that is the way they went. Across the Atlantic by ship, across Europe, Asia Minor, Persia, Baluchistan and

India, across mountain ranges and desert, through jungle, mud and war zones (with a Red Cross flag over the top of the truck). They followed the centuries old spice route of the camel caravans, Marco Polo and Ghengis Khan, across the Kyber Pass. They were received by presidents, rajahs and kings. By the time they arrived in Singapore to meet my father they were far more famous than Frank Buck. They went on to cross the Australian outback alone, and from Sidney, world wide news reporters accompanied them on their triumphant return across the Pacific and the United States to New York City.

I clearly remember our reunion with Mother. We children, who had stayed in England while our parents were gone, came back shortly after Mother reached New York. I remember looking down from the ship's rail and seeing thousands of people on the pier, surrounding the white Ford truck. They had come to see Mother greet her children. I remember us riding on top of the truck as it was driven slowly through the crowds of cheering people. We were thrilled by all of the attention. Mother was so beautiful and we were her children! She was the darling of New York; even the fashion houses climbed onto the bandwagon. I collected all the glamorous pictures of her in elegant dresses, her dark hair brushed back in soft waves framing her perfect features. It was all so very romantic for a small daughter.

I adored my mother. I was so proud of her fame, but I wished she could stay home with us. We didn't even think about Father; we didn't know that fathers had anything to do with children.

Mother had taken us to Belgium when she started on her trip around the world, but my father's parents were too old to take in three rambunctious children four and five years old. So she took us to England and put us in a boarding school.

Until that time, we had always been together, in the same playpen when we were babies and sleeping in the same room. I was never without my brothers. At the boarding school I was separated from my brothers for the first time in my life. It was as though I had lost two-thirds of myself and I was desperately unhappy. Put in with a group of about twenty girls, we spent most of our days in a large room that was both classroom and playroom. Each girl had a toy box, but Mother did not have time to buy toys for my brothers and me before she left. That didn't bother me but the other girls would not let me touch their toys, nor would they play with me. I was different, from America, a place that chose to have no part of England so they would have no part of me. I don't know how those children became aware of that historical dissension but it was quite enough for them to scorn and ignore a newcomer with a funny accent. I don't know why there were so many small children in a boarding school (I always thought it was an orphanage) but perhaps it was the English method of education in those days. Mother loved us, surely, but she never seemed to realize that children need affection and caring. Did the parents of the other children not realize it either? We were all too small to be living away from our parents, in the care of institutional personnel who had not the time or inclination to give a hug, much less comfort a very lonely little girl.

I escaped into a dream world. The fantasy was always in the setting of an imaginary house and garden; perhaps because we had moved around so much the house was important. Most of all, it was being with my brothers, and Mother coming back. What was so wonderful about my dream world was knowing it would all come true.

Long after, Mother protested, "But you never complained about the school. You never said anything about being unhappy. You talked about nothing but the seashore."

"Probably because we were there last," I said, "when Jeannette came to England to get us, and took us to the seashore to wait until the ship was to leave." Jeannette was our nurse when we were little, and came back, more like a housekeeper, whenever my parents returned. She was wonderful, always kind and cheerful. If only we could have stayed with her every time they left.

It was 1934 when we returned from England to find Mother in New York and Father too, back from Asia. We were all together, just like in my dream world, for two years. Father bought a farmhouse in Connecticut, with a big barn and a pond, in sixty-three acres of fields and woods. What a wonderful place it was! In the fields there were huge boulders left from the glacial age, and these were castles, fortresses and ships for my brothers and me. We went to day school and came home again every day. Mother played games with us. I remember her running "home free" when we played hide-and-seek, laughing gaily as we chased her across the lawn. In the house there was a playroom with lots of toys and even an old piano. The living room was for adults. Our favorite time of day was the late afternoon, when guests came to the house for cocktails. We children were allowed to join the grownups for a little while. We stopped at the door of the living room, waiting for permission to enter. We made the rounds of the guests, formally kissing them, then, sitting silently on the floor, we watched our parents. We were really waiting for the canapés of crackers, cheese and olives, given to us with a glass of milk. In a short time we asked to be excused and walked quietly to the door, to rush happily outside to play before the sun went down.

Our parents went to Africa in late 1935 when we were seven and eight years old. This time we were left in Connecticut, in the care of our grandmother Heidi. Heidi (as she insisted we call her) had no patience with us. She took us to the same school that Mother had attended for twelve years. It was now a finishing school so there were no children. There were several cottages on the property and Miss Lowell was pleased to provide one for her former student's children. The school grounds were spacious and there was a pond with ducks, and we thought it would be fun to stay there. Miss Lowell was very old and Heidi said that someone else would be taking care of us. Jeannette! "No," Heidi said, "Jeannette isn't coming back. You know she is getting married and moving away."

It was Miss Hewland who came to take care of us. How could our grandmother have left us with such a terrible woman? Heidi lived just a few miles away and came to take us for an hour's ride every Sunday. Why didn't we complain to her? Perhaps we said something about how mean Miss Hewland was, but we had always accepted without question that we must stay somewhere when our parents left, and we didn't know that it was much different from the way other children were raised.

Miss Hewland kept a chart on the wall, listing our names. Whenever we did anything bad, such as making too much noise, getting dirty, or spilling something, she put a mark after the name of the bad child. Three marks meant that he would get a whipping that evening, with switches we had to cut ourselves. My brothers were whipped so often they finally got used to it. But I was so terrified that I was often sick by evening, even when I had only one mark. It was enough to hear my brothers be punished; I was so careful, so good. Not that it made her like me any better. Somehow we got through the two years.

How angry Mother was about it when she returned. She said to me, long after, when we were able to talk about those times, "I didn't mean to neglect you. I never forgave Heidi for putting you in the hands of that dreadful woman. But it was my fault because I never should have left you."

"Mother, forget about that," I said. "We are fine. Plenty of children who are neglected or mistreated turn out to be better people than other children who are spoiled or have perfect parents. We knew that you loved us. When you came back we forgot all about those times."

Our parents came back from Africa and the most wonderful part of our childhood began. We returned to the New England farm house and our beloved woods and fields.

Every time Father went on a trip he collected animals. From Africa he brought two magnificent, tame cheetahs. These streamlined members of the cat family were used by the rajahs of India for hunting. The cheetahs were trained to ride with the riders on the backs of horses. When the horses got within range of the quarry, the cheetahs were released in a burst of speed of up to sixty miles an hour to bring down the prey. Our cheetahs were kept on long wires stretched between the trees, so they could run. We children took them with us out in the fields where we would let the cheetahs stalk one of us through the high grass. They were on long chains but we risked being jerked off our feet and dragged behind the charging animals, who knew it was all a game and happily licked the face of the victim. The cheetahs' greatest sport was to stalk the cows that a farmer kept in our large pastures. The unsuspecting bovines would graze near the stone wall behind which the cheetahs were crouched, and when the cows came close the cheetahs would flash from cover. The panic-stricken cows plunged away, bellowing, their udders swinging wildly. The farmer complained that it soured the milk.

We collected frogs, snakes and turtles and turned the dairy shed into a vivarium. We had three monkeys, much more fun than dogs, but they were a bane to the maid. If the pantry door was left open for a few minutes, the sharp-eyed rascals were in and out before she could catch them, strewing cracker boxes and banana peels behind them.

We always had three or four domestic, plain and fancy, cats. The most remarkable was a pure gray Maltese who had traveled with Father and Mother on the African expedition, by truck from Brussels, Belgium — where King Leopold, sponsor of the expedition, gravely noted the addition to the party of travelers — to Capetown, South Africa. Pooka (from pu-ss ca-t) quickly adapted to the mobile life and got lost only once, in North Africa. When the trucks were ready to leave, my parents realized that Pooka was missing. A frantic search was organized and soon all the Arabs as well as the French population were aware of the disappearance of Pooka. Finally my parents had to give up the search and start out across the

Sahara Desert. Two days later they saw a small plane approach and then circle directly overhead. Something was dropped from the plane by parachute. To their astonishment they saw it was a small cage, and in the cage was Pooka. He had been found and proudly delivered to them by the French Foreign Legion.

The Sahara was almost more than Pooka could cope with. Let out upon this immeasurable sandbox, he held up the expedition while he dug holes until he was quite desperate, unable to decide upon a hole that was satisfactory. The greatest danger Pooka encountered was in the Congo, when he went beyond the perimeter of the camp one night. Mother and Father were preparing for bed when they heard Pooka screeching. They threw open the tent flap and Pooka came flying in with a lion hard behind him, so fast that the lion couldn't stop and he tore right through the back of the tent on his way out. It happened so suddenly that no one had time to be frightened, except Pooka and the lion.

During the Connecticut winter the wild animals were kept indoors. The cheetahs were excellent house pets and they often stayed in the large, downstairs bathroom. It amused us children enormously to politely indicate to new guests where the bathroom was, and then wait for the expected scream of fright as the friendly cheetahs rose to greet them.

Every night we children hurried through our baths, put on our pajamas and robes, and sat down around Mother's chair near the fire. The cheetahs were there also, lying on the hearth, watching the flames with half-closed eyes, purring like domestic cats. Mother read wonderful adventure stories to us, like "Gulliver's Travels" and the "Swiss Family Robinson." She read with great expression, bringing the characters to vibrant life, and while she read, she knitted. She had taught herself to knit on the African veldt, to have something to do while Father waited for hours, cameras poised, for the rare leopard or rhinoceros to come within range.

Father was usually up in the studio, editing film. The barn had been renovated and the loft became his workshop. The walls were lined with great bins of film strips, and the long table down the center held splicing machines and projectors. When Father was not at home he was in New York, integrating sound and music, and finally arranging for the release of the film. The result, released in 1938, was the award-winning, full-length, fascinating documentary on primitive Africa, *Dark Rapture.*

The film was a tremendous success. *Life Magazine* gave *Dark Rapture* the greatest coverage they had ever done on any subject, with eleven pages of photographs and story. Now we had publicity agents and photographers coming to the farmhouse. It was very exciting for us children as the visitors always made a fuss over us and the animals. We were part of the charming picture of Armand Denis and Leila Roosevelt; the twin boys always dressed in matching clothes, blue for Armand and red for David, and a rather plain but nice little girl with dark hair and eyes like her mother. Photographers taking pictures of our parents would say, "Now let's get some shots of you with the children." We were kept ready, neat but not dressed up. The pictures were to be informal, showing how we lived at home; a touch of typical family life, though I wore a quaint, mirror-encrusted jacket from North Africa and David happened to be holding his pet snake.

3

Growing Up

And the Anthropoid Ape Foundation

Father and Mother left again in 1939, this time for India. We were now old enough, eleven and twelve, to go to regular boarding schools. We were separated again and I went first to Low Heywood in Stamford, Connecticut and later to Oak Grove in Vassalboro, Maine. I didn't get along with the other girls. At home, my brothers had been my companions and I wanted no others. I remembered too well the boarding school in England and I didn't trust girls. But I envied them.

Mother no doubt thought that the school uniforms, designed by a French couturier, were all the clothes I needed, but the other girls had closets filled with pretty dresses and shoes, and they wore their own clothes at every opportunity. They went to New York to spend the weekend with their families or were invited out to lunch with friends. I went to New York every week to have the braces on my teeth adjusted, and every week I wore my Sunday best uniform. Mother would not have hesitated to buy me clothes if she knew, but she was far away in India.

I became a tomboy and scorned the girls' silly talk of parties, clothes, and weekends at home. I knew their homes and families were very dull compared to mine. Fortunately there was a riding school adjacent to the boarding school, and I spent hours in the stables helping the old man who fed and curried the horses. We seldom talked. I wanted only to dream, to remember and elaborate upon my dreams about the wonderful life of my family.

My parents returned the next year and we were together again, but our home was no longer the same. I didn't know that for several years Mother and Father had not been the romantic couple I had believed them to be, that they had remained together to perpetuate their image before the public as well as for us. I clung to that image, not wanting to realize that the cocktail hour was starting earlier and earlier in the day, that there were often only men friends of my mother, and I didn't like them. Father was rarely present. He didn't work any more in the studio up in the barn; he worked on the new film in New York.

Perhaps that was the year, when I was twelve, that I started to become a separate person. We were no longer children, my brothers and I. They were learning about things that did not interest me. They built a workshop down in the basement and spent hours there without me. And there was Billy Pope. How I hated him. The son of a friend of Mother, he often came to our house. It was he who laughed at Armand and David for allowing their sister to play with them. Boys did not play with girls; that was sissy. I was very hurt and didn't know how to fight Billy, although I tried swaggering and talking tough. He looked at me pityingly and walked away. I could see that Armand and David were embarrassed but they followed him. I should have turned to Mother but something had happened to my loving and caring image of her. I went out across the fields by myself. In the far meadow I found an oak tree with a low branch that had grown horizontally and then up, like the back and neck of a horse. I didn't have a saddle but I invented one, as I invented to perfect all my dreams.

I hardly minded going back to school that fall, particularly because of the stables and horses. Father had promised me riding lessons. I was so excited that nothing else mattered. Riding across the beautiful Connecticut countryside, or in the covered ring when cold weather and snow prevented us from going out, I concentrated upon improving my riding skill and lavished my love upon the horses.

I did fairly well in classes, particularly in English literature. I was thrilled by Shakespeare and joined the theatre club. Being one of the tallest girls, I usually played the role of a man in the twice-yearly productions. In my biggest role, the part of Hamlet in our condensed version, I was a surprising success. The praise of the other girls showed me at last that I could make friends. I went out for competitive sports and played field hockey and basketball. The school competed with other boarding schools, and the students were divided into two teams to encourage everyone's participation in the games. I became popular with my team and was eventually elected captain. I finally liked boarding school.

The mistress of the school fancied herself as being quite progressive, but she offered little guidance to help the students discover any talents we may have had, or to pursue any interests. The girls were expected to go to college and choose a career. We had little idea of what women could do, other than teaching or nursing, which would be temporary in any case as we all expected to get married. We thought only about which college would be the most enjoyable.

I decided I would study veterinary medicine because my brother David planned to, and we could be at a university near home. I certainly wasn't interested in medicine.

My father might have helped me find another field of animal studies. He might have suggested a career in the theatre or film industry. He had chosen the school I attended because it offered the education that prepared a girl for marriage. This encompassed a general knowledge of the world, an appreciation of the arts, and social graces that would catch the attention of an intelligent, imminently successful, young man. The important thing for a woman was security and fulfilling her function as wife and mother, as it was in Belgium.

How could my father, Armand Denis, inventor and adventurer, have been so conservative about women and family! That may have been why he paid so little attention to us; he felt that we were Mother's responsibility. He had no patience with children. When I was eight I had attended a ballet class, and then of course wanted to practice at home. David started to play the piano. Father soon stopped us, saying he couldn't stand the noise. He said we would probably never do well so it wasn't worth trying. He greatly admired talented people, but he was illogically disparaging of anyone who did not immediately show potential proficiency. Of course we were not aware of that; we believed he knew we had no talent. Father knew everything and he was never wrong.

As I grew up, Father paid more attention to me. At sixteen I talked intelligently and I had lost my lanky tomboy appearance. It may have amused him that despite his neglect of us, I admired and idolized him.

I spent my sixteenth summer in New York with Father and it was one of the happiest times of my life. We lived in an apartment that Father kept for years, on 8th Street in Greenwich Village. In the morning I made breakfast, put the apartment in order, and then went over to Washington Square to stroll by the artists' paintings displayed along the walkways, pet all the dogs, and chat with their owners. Father went uptown to the film studios, and at noontime I took the subway and rode up to 52nd Street to join him for lunch. We usually went to one of the smart restaurants around there, often with Father's business associates or friends. Father was known everywhere and it was fun to be treated like important people. Sometimes we walked up to Central Park to the zoo. Father always liked to walk. Unusually tall and distinctive, striding along as though he were crossing the African veldt, someone would recognize him and call, "Hello, Mr. Denis!" Sometimes he would tell me to wait for him in the Village and we would go down to Chinatown by way of Fulton Street, stopping outside the fish market to eat delicious oysters and clams sprinkled with lemon juice. Father took me to the museums, and his knowledge of archaeology, history, and ancient art made him a fascinating teacher. We went to one or two art galleries; however his criticism of modern art was uncompromising and pungent. He didn't like ballet but he loved the theatre and we went to see all the best Broadway shows.

Father was a wonderful story-teller. He was always the center of attention wherever we went, relating his adventures, and the misadventures, with his very British wit. He was my ideal of sophistication — brilliant, charismatic and fun.

The wonderful summer ended. Although I am sure that Father enjoyed it as much as I, he apparently forgot me after I left. I was disappointed but accustomed to it. I returned to boarding school, and it was all right, but that year seemed endless. The family had moved to Florida, and home there was going to be more exciting than ever.

Father had set up the Anthropoid Ape Research Foundation in 1942, in Dania, a small town south of Fort Lauderdale. He had long wanted to establish a holding station for animals, monkeys and chimpanzees in particular, to be used in medical research. Johns Hopkins and other foundations were extremely interested. Father went back to Africa to collect the animals. This time Mother stayed at home, at the Chimpanzee Farm, as it soon came to be known. Shipments started arriving, chimpanzees and several hundred monkeys at a time, with an assortment of other animals that Father could not resist. Father had decided that my brothers should stay in Florida to help with the animals. It was not good for their schooling but they didn't care, though David would have liked to go out for football. It wasn't possible; the year was 1943 and men had gone to war. My brothers became responsible young men very quickly. It was often they who drove non-stop to New York to meet the ships and bring back the animals in the loaded truck to Florida as fast as possible. It was necessary to tame and train the chimpanzees so they could be handled by the scientists who came to study their near human behavior, as well as take blood samples or test medications destined for human care.

Johns Hopkins University learned about the Chimpanzee Farm, and Father's purpose of taming and training animals to actually enjoy participating in experiments for scientific research. Being the closest relative to man, chimpanzees were a source of valuable information for both medical and psychological studies.

I didn't understand at first; the animals would not only permit, but also enjoy participating in medical research? David explained,

"Take the example of Kip—"

"The one who likes to be dressed and walks so straight?"

"Yes. He is a one hundred and fifteen pound chimpanzee but he is so gentle and — happy, is the word — he can always be taken out without a chain or any restraint. We were asked to go up to Johns Hopkins in Baltimore, as they wanted to see if chimpanzees really would let scientists work safely with them. So Armand and I went up to Baltimore with Kip. When we arrived at Johns Hopkins, Armand waited outside with Kip while I went in to find Dr. Schultz, who was expecting us. When I spoke to his receptionist she said I would find Dr. Schultz in the lab. I could hear a chimpanzee screaming in fear and anger. I went into the lab and saw the doctor, his assistant and a nurse, trying to control a chimpanzee weighing no more than thirty pounds. They wanted to hold him down to run a cardiogram, and he had already bitten the doctor, struggling to get loose. I introduced myself to Dr. Schultz while the nurse bandaged his hand. I asked, 'How can you manage to get a normal cardiogram when the chimpanzee is panicked and struggling to get away?'

" 'How else can we get any results?' he replied. That was my cue.

" 'I'll show you.' I went out and called to Armand to bring Kip. They came in; Kip, in his jeans and T shirt, walked over to the startled receptionist, shook her hand, and then walked into the lab. I said, 'Kip, get up on the table.' He hopped up immediately. 'Lie down.' He lay flat on his back. 'Now you can do what you want,' I said.

"Dr. Schultz was astonished, and still not convinced; he had just been bitten by a much smaller chimp. He said, 'But we have to shave some of that hair off his arms.'

" 'Go ahead.' To reassure the nurse, I held Kip's hand while she hesitantly shaved some hair off, and Kip watched interestedly. When Kip held out his other hand, she laughed and took it without hesitation, attached the straps, and within a few minutes Dr. Schultz had his cardiogram.

" 'That is how you can work with a chimpanzee and, as you see, he enjoys it.'

" 'You said that your chimps will even let you draw blood?' Dr. Schultz asked.

" 'That's right. In fact, they'll let you do anything for an ice cream cone. They see us come by the cages and put their arms out, hoping we have a needle and ice cream. One of them doesn't even care about the treat. We had to do some dental work on him and gave him a dose of Nembutal. Now he always sticks out his arm, hoping to get another buzz."

Chimpanzees were easy to tame when they were young, but even those became untrustworthy and dangerous when they became adults. Napoleon had been trained to a certain extent, but he was now fully grown, one hundred and fifty pounds, sly and aggressive. He loved to stamp his feet, make the thick hair all over his body stand on end, and then slam from one side of his cage to the other, whooping and hollering, exciting the rest of the colony and creating a total uproar. We kept the big males for breeding.

One day Napoleon broke out of his cage. No one saw him until he suddenly appeared outside the office at the main entrance. Father, in the office, heard people screaming. He ran out and saw Napoleon, his hair on end, pounding on the wooden porch and apparently ready to charge. Father didn't hesitate a moment. He went straight up to the big animal, held out his hands as though he expected Napoleon to play. The chimp was surprised enough to take his hands and continue jumping up and down, undecided about what else to do. David and Armand arrived at that point, but before they could plan anything, Napoleon pulled away from Father and ran across the lawn to the open tool shed. David thought he had gone into the cottage next to it; he raced up to the door, and looked straight into the barrel of a '45. It seemed that every man who heard the shouting had a gun. David yelled at the top of his voice, "Don't anyone shoot! We'll get him, so don't shoot!" At that, Napoleon tore out of hiding, but, confused by the noise and getting frantic, he charged at the first person he saw, who was Armand. He recognized Armand and stopped short, but then he raised himself up, hair bristling, and continued to advance. Armand had a gun but had no intention of using it unless Napoleon attacked. Finally backed up against a car, he had no choice, but yelled, "David, think I should shoot?"

"You've got to, Armand. Shoot!" He pulled the trigger, and it just went, click. Whatever stopped Napoleon, perhaps just the gun aimed at him, at the click he whirled and raced off in another direction. David and Armand were after him again as Napoleon headed into the angle of the main building. At that moment the local fire engine showed up. David yelled to the fireman to give him the hose. "Is there plenty of pressure?" he cried, and the fireman, high on his seat but scared stiff, nodded that there was. David held the nozzle like a stick and commanded, "Napoleon, go back to your cage!" At a younger age, Napoleon would have obeyed, but now, he turned aggressively on David and David snapped the nozzle open. Only a few drops of water spurted in Napoleon's face; the fireman was so terrified that he had forgotten to open the valve. David shoved the nozzle at Napoleon, who grabbed it out of his hand, and with a swipe of his arm he slammed David against the wall. David's glasses flew off, and that made him lose his temper. Half blinded, he attacked Napoleon. Napoleon was so astonished that he turned and fled, David hard after him. Suddenly a jeep came roaring up behind him, with Armand at the wheel. His intentions were good, the noise was frightening Napoleon even more, but if Napoleon had stumbled, David would have fallen on top of him, and the jeep would have wiped out both of them. Napoleon knew, though, that David was mad enough to kill him, and he raced as fast as his legs and arms would take him, straight back into his cage. When the door closed on him he was safe and happy.

When I went to Florida that summer, I was put in charge of the babies. When baby chimpanzees are born in captivity, the mothers often neglect them, so we had to get them away as soon as possible. We raised them like human babies, with diapers and bottle feeding. When they were newborn they had to be fed every two or three hours, and we managed this at night by taking turns. I would give the 10:00 feeding and put the baby in Armand's room. He gave the next feeding and put the baby in David's room, then David put him back in mine.

One time the three of us had driven to New York to pick up a shipment of animals, including a chimpanzee who had unexpectedly given birth just before the ship from Africa had docked. The man who had accompanied the animals had taken good care of the baby so he was fine, but the mother, a rare and valuable pigmy chimp, was ill and would never have survived the truck ride back to Florida. We went by plane, David going with me to care for the mother, required to travel as cargo, at the two unavoidable stopovers — and I carried the baby, diapered and well concealed in a blanket, in the passenger section. People looked at us sympathetically, a very young couple with their tiny baby, and the stewardess was glad to heat the bottle of milk David took to her. David came back to his seat and shortly after the stewardess arrived with the warm bottle, leaning across to give it to me. Before I could take it a tiny, black and hairy arm shot out of the blanket and grabbed the bottle. The stewardess gasped in shock and horror, as the determined baby pushed his head out to latch on to the nipple, his black eyes gleaming at her. Realizing he was not human, she protested he was not allowed in the passenger section — but the plane was not going to land then to put us off.

We all had our favorite animals and of course I had a horse. I had hung around a ranch out west of town, watching the black cowboys work, until the rancher noticed me. He asked if I could ride, and then, if I wanted to take a broken but untrained horse for the summer. Did I! I was so excited — a real cow pony. We drove out to the grazing lands and the rancher separated a young pinto from the herd. Back at the ranch the pinto was saddled and bridled. He let me mount him easily and I started off in style at a nice canter. Good cow ponies stop on a dime, but this one had come in from a couple of years on the range, so I figured I could just pull gently on the reins. He stopped so fast I went right over his head.

I generally rode across the sandy pine woods into the abandoned naval air station where we could gallop the length of the airstrips. Armand told me that during the war the Chimpanzee Farm was a vital landmark for the naval planes coming in to land at the base, because from twenty miles away the pilots could see the reflection of the sun off the squares of sheet metal that were the roofs of the cages. It was reassuring for them — we were very close to the Bermuda Triangle.

Seminole Indians lived down by the canal behind the Chimpanzee Farm. Their homes were simply open platforms several feet off the ground, roofed with thatch. The Indian women still wore their traditional dresses of bright colored strips and squares of fabric sewn in rows of intricate patterns. When one young woman was very pregnant, Mother told her to let us know when she was ready to go to the hospital. A few days later she came to the house with her newborn baby in her arms and said she was ready to go.

The woman's young husband, Johnny Tiger, was an

alligator wrestler. The alligators were not only in the Everglades; when we captured several that came up the canal, David and Armand learned to wrestle young alligators too. Then they organized a show for visitors to the Chimp Farm. One day Armand was emceeing the program while Johnny was in the pit with an eight-foot alligator. The highlight of the show was when Johnny pried open the alligator's jaws and put his head in the beast's mouth. Armand was still facing the crowd when the people gasped and suddenly began to scream. Armand whirled around to see Johnny's head clamped in the alligator's jaws and the giant lizard already whipping over to break Johnny's neck. Another turn and it would have been done. Without hesitation, Armand leaped over the wall and fell down upon the alligator. With the strength of fear and desperation, he managed to hold the animal down and pry his jaws open again, freeing Johnny, badly bitten but alive.

4

Going Home

How I hated going back to boarding school at the end of the summer. That year seemed to take forever until, at last, graduation. My suitcases were packed, I hugged my friends good-bye, knowing I would never see them again, and took the first train for New York. I went directly from Grand Central to Penn Station, to board the Silver Meteor that would take me to Florida and home. I didn't mind the long journey because it was for the last time, and I happily watched the scenery fly by. That night I woke up about three o'clock, and saw the moonlight silhouetting the stands of thin, straight pine trees across the flat and sandy terrain dotted with clumps of fanned, corrugated leaves that were palmettos, and I knew we were in Georgia. I watched for the first real palm trees and caught glimpses of the ocean as we sped down the coast of Florida.

As the train drew in to Fort Lauderdale I pressed my face against the window to see who had come to meet me. I quickly saw them, my brothers, of course. Armand, tall and thin, with curly, blond hair, searching for me with his kind, blue eyes, and David, a little heavier, his hair dark and straight as mine, his sharp eyes a deep brown. Who would believe they were twins? I scrambled down the steps and ran into their open arms. They laughed as I broke away again and danced a jig of happiness. "I'm home! I'm home! Home for good!"

"Stop it," Armand grinned, "You're making a spectacle of yourself."

"I know it!" I cried, dancing along beside them as they picked up my luggage. "I want everyone to know I'm home for good!"

"Shut up and get into the car," David growled, giving me a whack on the seat.

"Yes, yes, whatever you say!" I cried, silly with joy. I could hardly sit still. "Tell me all the news!"

"All right," Armand said. "First of all, Father's here."

"He is? I thought he might be because I called the apartment several times before I left for New York but there was no answer. How's Mother?"

"Well—" They both started to speak and then glanced at each other.

"Well, what?" Although I could guess. She and Father were all but legally separated and Mother was always very nervous when he was at home.

"All right," David began again. "First, she's upset because she really wanted to go to your graduation, but you know how she is. Actually she has been better lately." He looked at me. "She has met someone she really likes and she wants to marry him. So she and Father are getting a divorce."

I stared at David and then at Armand. Armand was driving and concentrated on the road. He said finally, "It was bound to happen. Anyone could see it coming."

"Yeah," David said. "What difference does it make? They haven't really been together for years. I guess they were just waiting until we were grown."

"Yeah." I copied his manner unconsciously . "It doesn't make any difference." The important thing was that the three of us were together. I put my arms across their shoulders and tried to hug them both at once.

"Hey!" Armand cried. "You want to smash us up?" He smiled at me as he pulled back. I saw David looking at him.

"Why do you keep looking at each other? I'm okay. I don't care!" and I started plying them with questions about their girlfriends, which had us all teasing each other until we reached home. Mother greeted me so warmly that I forgave her for not coming to my graduation. I knew that, as distracted as she had become, she really did love us. She had never been able to cope with the responsibility for the family, and then the Anthropoid Ape Foundation, when Father returned to Africa. For the last couple of years or so it was Armand and David who had been, practically, taking care of her.

"Father had to go to Miami," Mother told me. "He said to tell you he would be back before dinner."

"Oh, then," I cried, catching Armand's hand, "we have time for a swim. I'm dying to get into the ocean!"

"Okay, let's go!" They grabbed my bags and the three of us raced upstairs to change into bathing suits, and headed for the beach. A couple of hours later we were back and Father met us at the door. He held out his arms to me with a big smile and I threw myself into his embrace. I could not be disappointed that he also did not come to my graduation. With him it was honestly "out of sight, out of mind." He was always so absorbed in his work, his travels, and the constant excitement of his life. I was simply glad to see him whenever he came back. He hugged me tightly a moment and then let go. I thought he would step back to look at me, the way he always did to see how I had grown or changed, and he would ask me about school or something. But he didn't. He put his arm around my shoulders, yes, but he looked past me at Armand and asked him if he had found the tools he wanted.

"Yes, I found the tools but couldn't get the right size containers. We can modify them, though, to fit."

"Let's see what you have," Father said, and without more than a word to me he left. I wandered into the house and back to the kitchen. Mother was there, preparing a salad. "Can I help?" I asked.

"There's not much to do," Mother said. "David is setting up the grill outside. You go and shower."

I wanted to talk but she seemed preoccupied and nervous. Perhaps she thought I didn't know yet about the divorce. I wanted to tell her it was all right but it wasn't the right moment.

A little later we were all at table. It should have been a happy occasion but even I was subdued by then. Father complimented David on the steaks and Mother asked if the salad dressing was all right, and everyone assured her it was. Then Father asked me, "Which university have you decided to attend?"

I looked at him. He didn't seem to be really interested in what he asked so I said, "I'm not sure yet." I glanced at David. "It depends upon where David wants to go. We both want to study veterinary medicine." I waited for David to confirm it, but he suddenly looked pale. And then my family totally disintegrated.

"Didn't they tell you?" Father asked. "Both David and Armand are going with me to Africa. We'll be leaving in a week or so."

I was stunned, unable to say a word. I couldn't look at my brothers. The food on my plate had turned to straw, or it was my throat that was terribly dry. I picked up my fork and forced myself to eat.

I don't remember talking to David or Armand about their leaving. Of course they were thrilled to be going to Africa and I didn't want to make them feel badly. What I should have done was kick and scream and maybe Father would have taken me too. But I was well disciplined; I had learned the hard way to accept and be agreeable.

I don't even remember talking to my mother. How could she have not realized how lost I would be? I don't know where she went; certainly to join the man she was going to marry, but couldn't she have thought at all about me? They were all uncomfortable about telling me they were leaving, so how could they leave me alone again? Did they think that because I was eighteen and had graduated, I was now able to take care of myself? A boarding school is like a convent and a girl comes out totally unprepared for living on the outside. A few days ago I was so ecstatically happy to be home with my family, and now I was lost, confused, and alone again.

Then, no more than a few weeks after everyone had left, my legendary grandfather André came into my life, as in a fairy tale. He had no wand; he just telephoned. I answered and was not surprised when a man's voice said,

"Leila?" Mother and I had voices almost exactly alike. I said,

"Leila isn't here but this is her daughter, Renée. Perhaps I can help you?"

"Good Lord," came the reply. "I thought you were still a child. This is your grandfather, André."

It was my turn to be astonished, and thrilled. "Where are you? Where are you calling from? Are you here in the States?"

"Yes, I'm at the airport in Miami. I've come to promote a hotel I have just opened in Haiti. You know, in the Caribbean. I feel as though I am talking to your mother. Where is she?"

"She's in New York right now, I think, but I don't have any phone number where you could call her. Are you staying in Miami? I can go and meet you and take you wherever you want to go. Maybe I can help you, drive you around if you like."

"That would be bully, my girl. I'll wait for you here and we'll talk then about what I am doing." How typical of him to arrive like that! And how fortunate for me.

Arriving at the airport an hour later I recognized him right away, although we had seen no pictures of him since we were all in Bali. He had a big twinkle in his eye when he saw me, probably pleased to see that his granddaughter was quite presentable. We liked each other right away. We had a wonderful time driving all over Miami to find the travel agencies for promoting his hotel. I brought André home to Dania and the Chimpanzee Foundation. He was astonished at the size of the place and the number of animals.

"Who is in charge here? You are apparently free to leave whenever you want, but are you working here?"

"I was when I came on vacations, and I loved working with the animals. Maybe Father thinks I want to stay here now. I don't know what I am going to do. I wanted to go to university, but I'd hate to have to board again. One thing is sure, I won't stay here. I don't like the man Father left as manager. He is making changes that I know Father wouldn't like — and he thinks he is in charge of me too."

André intended to spend only a few days in Miami but he stayed two weeks. We drove up to see more travel agencies in Fort Lauderdale and farther north, and generally returned to Dania by evening. The day before André was to leave he asked me if I had any more of an idea of what I wanted to do. "I don't want to leave you here like this," he said.

I felt like crying and couldn't speak. I didn't expect him to care, although it seemed as though, after these two wonderful weeks, we had always known each other.

"Renée, what do you think of coming to Haiti? You could help me in the hotel. Would you like that?"

"Oh, yes!" How wonderful, how kind of him! I was ecstatic again, so glad, so glad. It didn't take me long to pack my bags, and I was off to Haiti.

5

Tarzan of the Caribbean

Ibo Lélé is the Haitian god of the forest, and Ibo Lélé was the name of my grandfather's hotel. Built high on the mountainside above Petionville, in Mediterranean style, it gave the impression of an airy, gracious mansion. Colorful frescos of Haitian peasants were framed by the elegant entrance arches. The terrace was cantilevered over the side of the mountain like the graceful prow of a ship. One could see sixty miles across the plain, from the Bay of Gonaïve to the frontier of San Domingo; incredible sunsets filled the sky with fantastic purple and gold. Petionville was a little village of colonial houses surrounding a pretty square overflowing with pink and magenta bougainvillea and the feathery spreading branches of crimson poinciana. Further down the mountain, the picturesque town of Port au Prince sprawled from the shores of the deep blue waters of the bay into the wide plain dotted with rich green banana and sisal plantations.

My job at the hotel was to be hostess for the guests, advising them about where to go and what to see in Haiti. As the roads were very poor it wasn't possible to go much farther out of town than the beaches along the bay. In a few days I had explored all of Port au Prince.

The most fascinating place in town was the "Iron Market," which spilled out from under its vast roof and wrought iron gates into the open square. The vendors squatted beside their little stands or bits of matting heaped with fruit, vegetables and every variety of dried beans. Tiny black mushrooms cooked with rice made *ris jonjon*, conches made a delicious spicy stew, and dried beef was marinated in hot sauce and barbecued.

The Haitians wove fantastic hats for tourists and carved beautiful mahogany statues, platters and bowls. Small boys with intelligent eyes gleaming in their ebony faces, offered two or three paintings for sale. They were primitive pictures of Haitians, of the way they live and work, painted with wonderful imagination and in vivid colors. I wanted to buy them all. Interestingly, I never saw paintings of their voodoo ceremonies, although perhaps since then some of the artists have overcome their fear of the spirits.

From the hotel, after the sun set across the plain, I could hear the Haitians drumming; they gathered in clearings in the hills to play the tall, wooden goatskin-topped *tambours* and bamboo flutes, and to dance. The rhythm of the Haitian drums, intricate patterns played only with the hands, was irresistible. Rubbing the taut hide with the heel of the thumb created shivers, the sound was so throbbing and sensual. The Haitians would dance all night, and a wedding or funeral could last three days.

I had been in Haiti about two weeks when I met the man called "Tarzan of the Caribbean." It was a day when the few guests in the hotel had gone on a tour, so I had the afternoon to myself. I was lying in the sun by the pool, the smooth tiles deliciously cool against my skin, when I remembered an errand I had promised to do for my grandfather. I got up and crossed to the white stairwell leading to my room. The shutters were closed against the sun but the room was hot. I stripped off my suit, turned the shower on full and delighted in the stinging fresh spray. I reached for a towel, then stopped to look at myself in the mirror, pleased with my dark tan. Actually, it made me look quite Indian. I pulled my hair back in a smooth, thick pony tail, and slipped into a white cotton shirt and shorts.

I looked for André in the bar and saw him arguing with the big man I knew must be his friend, Gustav. No wonder the Haitians called him Tarzan. They were fascinated by his long, blond hair (shoulder length was considered long in those days, particularly compared to their kinky black caps). The Haitians were in awe of Gustav because he was not afraid of the spirits of the sea. They were terrified of going into the ocean, whereas Gustav defied the god Agbéto every day, and even attacked sharks.

Despite the difference in ages, he and André had been close friends since they had arrived in Haiti about the same time two years before, in 1948. They had stayed in the same hotel on the square in Port au Prince. The Park Hotel was a converted colonial mansion. Each of the original large rooms had been divided into two rooms, the only electric light remaining in the center of the ceiling over the partition. Inevitably, guests sharing the electric light would make each others' acquaintance, and thus André and Gustav met. They quickly found they were kindred souls.

Gustav was born in a remote town in the Italian Alps where, he told André, his grandfather, a feudal landowner, had exercised his right of "prima notte." When a young couple in his domain married, it was the custom and an honor for the seigneur to spend the first night with the bride. Gustav's father was a dour man, devoted to his wife and to accumulating wealth, so he did not take advantage of his rights. When Gustav's turn came, the times were too modern for him to revive the custom of "prima notte."

Gustav was not his real name. Born in 1917, he was christened Amerigo Angelico. What a name to be burdened with! Amerigo/Gustav was an extremely hyperactive, willful child. When he was six he appalled his younger brother by defying God. In the middle of a thunderstorm Gustav flung open a window, bared his chest and cried, "God, if you exist, strike me dead!" Secretly terrified that God might do it, he was very relieved when nothing happened. He was no less defiant of authority as he grew older. Gustav claimed that he was farmed out to distant relatives, living like a peasant, but it was more likely that he went to the best schools in Italy.

By the time Gustav went to university, the Fascist regime was established by Mussolini, and all students were automatically enrolled in the party. Gustav was totally oblivious of politics, but Italy went to war and Gustav went into the army. He was wounded, shot twice in the stomach, and for a while it was not sure he would survive. He did not like to talk about it.

After the war Gustav became passionately interested in films. He acted a bit, playing the part of a German because of his blond, Austrian features. He participated in the production of *Bitter Rice*, a film portraying the Italian peasants' struggle for survival in wartime. He invested in more films, then a play and even a ballet, but his attempts at original and meaningful productions were not successful. So Gustav tried his hand at painting, thinking that if he came to Haiti, where primitive art had just been discovered, he would be inspired to become a real artist. After a few months in Haiti he abandoned painting altogether, finding he really preferred to go fishing.

Gustav bought a ketch, then realized that in Haiti there were no boats for hire to go sailing or fishing. He bought a motorboat and spearfishing equipment and was in business. He built a glass bottom boat and the hotels sent him all of their guests. Gustav no longer needed to have money from his inheritance sent from Italy; the business was a success and he enjoyed it.

The day I went out on his boat, we sailed to Sand Cay, where the coral reef formed a plateau in the middle of the bay. It took about an hour to get there; all the way the water was so clear that I could see coral formations, large fish, and an occasional turtle gliding along over the white sand a hundred feet down. The glass bottom boat had been towed behind the ketch. When we arrived at Sand Cay, the crew pulled it over so we could step in and sit on the benches surrounding the glass windows. We could see the coral, magnified, in shades of pink, green and yellow, glowing in the sunlight that drifted many feet below the surface of the water. We moved slowly over the reef. A cloud of small fish appeared, a kaleidoscope of brilliant colors flashing in the sunlight. Larger fish came into view, scattering the little ones, but they came back and swarmed beneath the glass. The boat had stopped moving.

Suddenly the huge form of Gustav, half again life-size, appeared beneath the glass. The fish were not frightened; on the contrary, they crowded around him. Although he was using only a mask and there was no way for him to breathe under water, he was resting on the bottom, holding something out in his hand. It was the shell of a sea urchin with the spines brushed off. As we watched, Gustav broke the shell, and the fish became frantic, darting down to snatch the food from his hand. Gustav surfaced, then dived under the boat again. He broke another shell and put it between his teeth. The fish hesitated, then darted at him again, snatching bits of urchin from his mouth. The photographers on the boat were ecstatic.

Back over the shallow sand bank we all went into the water for a swim. I put on a mask and snorkel and floated on the surface. Down in the sand there were trails an inch or so wide that seemed to end nowhere. When Gustav came nearby, I asked him what caused them. "Come," he said, taking my hand. He swam slowly, leading me along one of the trails. He stopped and pointed to where the trail ended. I stared but saw nothing. Gustav let go of my hand, dived down, and dug a few inches in the sand. He brought up a beautiful, living conch, with all the colors of a sunset within its pearly shell. Gustav was amused by my delight. He took my hand again and led me over the reef. The coral seemed so close that I was afraid of getting scratched. Gustav looked back at me clinging tightly to his hand so that I would not have to kick my legs. He smiled and shook his head to say there was no danger. I was thrilled; some of the thrill was in being so close to Gustav.

I was sorry when it was time to leave. Back on the ketch, a bit sunburned and sleepy, I stretched out in the shade of the sails and dozed as we headed back to Port au Prince. Waking up a little while later, I became aware of Gustav standing beside me. He was looking at me with a pensive expression, but simply asked if I would like something to drink. I accepted, feeling warm, but not from sunburn. However Gustav did not stay with me. He told a boy to serve me and then went over to the beautiful woman sitting by the tiller, whom I knew to be Evelyn, his not-so-secret companion. Back at the pier I thanked Gustav but, although I thought he was interested in me earlier, he hardly noticed me then.

A few days later I saw Gustav again at the Ibo Lélé bar. Corinne, one of the women who had been on the boat that day, was with him. She and her husband lived in Haiti and had the American Bar in Port au Prince. Corinne had that cool, sophisticated look that I admired so much. She and Gustav were talking and laughing. Why couldn't I be casual, just friendly, like that? I stopped a moment to say hello, but used the noonday heat as an excuse to leave immediately to dive into the pool. After a few minutes I climbed out onto the far side and lay down in one of the deck chairs. I was surprised when Corinne walked over to me. She said, "Renée, we're having a party at the house Wednesday night. If you would like to come, Gustav can stop by and pick you up."

"Oh!...Yes, all right," I managed to say.

Corinne smiled. "Good. He'll be by about eight. See you then." She left as I stared after her, still not sure I had heard right. Had Corinne arranged this — for Gustav? What was there about me that had attracted his attention? I had no idea.

The next two days I could think of nothing else but Gustav. I could hardly believe he had found me attractive. Every girl dreams of such a handsome and exciting man, but men like that seldom appear in real life. When they do they are likely to be in love with someone else, or they take advantage of girls, love them and leave them. What André had not said about Gustav I learned from local gossip. He was definitely not the man for a girl with old-fashioned morals. I thought my grandfather knew I wasn't a silly girl, and going to a party with Gustav would be exciting but certainly not foolhardy. I regretted that he was angry about it but I was determined to go to the party anyway. I would have a wonderful time and felt sure that, like Cinderella, I could leave the ball before everything turned to pumpkin and mice.

After much hesitation I finally decided to wear my favorite blue skirt and a frilly white blouse, with brown and white pumps that accentuated my long legs. I still looked like a girl just out of finishing school, but thought Gustav must like me the way I am. That did not keep me from feeling nervous when I went down to the terrace. Gustav was already there, talking to André. He took my hand and kissed it. I could tell by his quick appraisal of me that he approved. André watched us but said nothing. I kissed him reassuringly and we left.

"We go with Caroline," Gustav said, as we went down the steps.

"Oh?" Who was Caroline?

"Caroline, my car. She not big but she is nice."

"Oh!"

I suddenly felt very happy. He was not so frightening after all. I thought, I really will enjoy the evening.

We drove down the mountain towards Port au Prince, then up again into a residential area overlooking the town. High walls enclosed the courtyards of large, whitewashed houses. Gustav drove the car through an open gate attended by a Haitian servant who led us out to the garden where there were several other guests. Corinne greeted us affectionately and introduced me. Of course everyone knew Gustav.

The evening was warm and beautiful under the starry, tropical sky, and the people were very friendly. Self-conscious at first, I was drawn into the group and then relaxed and enjoyed myself. Gustav kept me near him all evening. He was always the center of attention, his colorful French and mutilated English dominating the conversation.

About midnight we all went down to the new casino on the bay. Besides the gambling there was an excellent orchestra for dancing on the terrace over the water. We walked out to the edge where we could see the lights of the town reflected in the circle of the bay. Then Gustav led me to the dance floor and took me in his arms. He danced very well, and I was soon oblivious to everything but the pleasure of dancing. When the orchestra played *Le Bateau*, a rocking, sensuous Haitian rhythm, I slipped off my shoes to dance more easily. Gustav laughed and held me closer, and I was giddy with the beat of the music and the motion of our bodies. He suddenly said, "Let's go to another place."

I saw Corinne raise an eyebrow at Gustav when we went to the table to say good bye, but I left with him without hesitation, feeling beautiful, sophisticated, and in control. Maybe I had had a little too much wine.

We drove along the shore and passed the Haitian bistros with strings of Christmas colored lights, the local version of a red light district. Turning up a long winding road, we came out upon a promontory overlooking the bay. A late moon, reflected in the shimmering, steel gray sea, was more beautiful than the lights of the town below. I was completely enchanted. Behind us was a thatched roof sheltering a small bar. It seemed to be deserted, but when we walked in, a sleepy barman rose up behind the counter. There was a phonograph on a table in the corner, and Gustav told the man to put on a record. We danced slowly and I closed my eyes, giving myself up to the romance of the moment.

Did I really think the night could end on such a lovely, chaste note? Yes, I did. I was so naïve that I thought Gustav's friendship with André was too honorable; he would not take advantage of his friend's granddaughter, no matter how tempting the opportunity. I still couldn't believe that Gustav would really want to make love to me. It was thrilling enough that he wanted to spend an evening with me, and I would give it no more importance than he did.

Though he held me closely when we danced he did not caress me in any way, and after a while we went to sit at a table under the trees. Gustav picked up one of my shoes that I had slipped off again. He smoothed the white suede for a moment and then gently put the shoe down. He looked at me. It seemed to be the first time he had really looked at me.

"I like you," he said. "*Tu es fraiche et charmante.*"

I was pleased and nervous at the same time as he continued to gaze at me. He smiled and I felt more sure of myself, but was glad when he began to talk. I had had enough practice by then to understand his French, which was often as picturesque as his English. He told me how he and André had planned to operate the Hotel Ibo Lélé together; André had some experience in running a small pension in Bali. They had both lived in many hotels and they presumed that an accountant and a good cook could take care of anything they didn't know. André went to Miami on his promotional tour, leaving Gustav in charge. Gustav then found he had no patience with the details of running a hotel, so upon André's return, he left, preferring his life on the water.

"It is late," he said finally, and woke up the barman again to tell him he could close. We got into the car and drove back down to the shore road. To my dismay, Gustav turned the car towards the beach instead of the town. I said quickly,

"It's almost dawn. I must get back to the hotel."

Gustav did not reply. Why did I leave the casino with him! How could I believe he would not try to make love to me after this night together.

He drove down to the beach and turned off the motor. He got out of the car and came to my side, opened the door, and took my hand. I did not want to get out, but he insisted, taking me in his arms. I turned my head away and said firmly, "Gustav, I don't want you to kiss me."

He laughed and tried to turn my face towards him. I pulled away angrily. He was surprised.

"But why don't you want me to kiss you?" I didn't answer. He waited a moment, then shrugged his shoulders. "All right, I'll take you home."

We got back into the car. Gustav watched me while I straightened my blouse and combed my hair. He started the car and drove slowly back to the hotel, glancing at me now and then. I didn't know what he was thinking and didn't care. As exciting as it had been earlier, it was not worth the risk of falling in love with this man for whom I was apparently just another pretty girl.

We arrived at the hotel. We hadn't spoken a word on the way and we didn't then. It was already dawn and as luck would have it, my grandfather, an early riser, was just coming out to the entrance with a cup of coffee, and saw us. He stopped at the top of the steps so I was obliged to go around him. He saw my angry expression but I didn't stop to say anything. Gustav could tell him anything he wanted, I didn't care. But Gustav simply said, "*Bonjour*! " and accompanied me across the terrace and to my door. I took out my key to put it in the lock, but Gustav took my shoulders and turned me to face him. He studied me a moment while I waited silently. Then he kissed my cheek, let me go, and walked away.

I don't know what Gustav said to André, if anything, as André never mentioned it. Gustav often came back to the hotel, apparently to see André or just for a drink. Sometimes he stayed for lunch or dinner. I became used to seeing him, and he never referred to that night.

Despite my resolution, I was falling in love with Gustav. It was impossible not to, he came so often and I could not avoid seeing him. He was always pleasant and amusing, pleased to see me, courteous but not particularly attentive, so of course I was all the more attracted to him. I started going out on the boat every few days. Gustav was as friendly on the boat as he was at the hotel. Sometimes Evelyn was with him, and then I avoided looking at her or Gustav. It seemed there was no change in their relationship; I was just one of the several women

Gustav enjoyed having around him.

Then, after several weeks, I heard the rumor that Evelyn was leaving. Gustav had not come to the hotel for several days. Was he going away with her? It didn't seem likely that he would leave Haiti. The gossips said that her husband was divorcing her. Well, I didn't want to hear any more about it. If Gustav left, I said to myself firmly, it was the best thing for me. Why go on dreaming about a married man, even if he cared for me?

A few days later, I was having lunch with André and some guests, when Gustav came again to the hotel. Despite my advice to myself, my heart soared. Gustav accepted a cup of coffee while he and André talked about things as usual. Then he turned to me.

"Please come for a drive with me."

I looked at my grandfather. André looked at Gustav, then said to me, surprisingly, "Go along. You've been working more than enough lately."

We went out to the driveway and I smiled when I saw Gustav's car, Caroline. We drove to Petionville then turned onto the mountain road that led to Kenscoff. We drove for almost an hour, the air becoming much cooler as we left the tropical vegetation behind and climbed through rugged terrain until we reached a pine forest. Sunlight filtered down through the trees, touching the soft, brown carpet of needles. We left the car and walked through the trees to a clearing where there were wild strawberries. We picked them, savoring the sweet morsels while we talked. Gustav asked me about my short life, but what did I have to say? He had so much to tell; besides, I was only too happy to listen.

We drove up to the Refuge, a chalet near the summit of the mountain. The air was thin, almost cold, and misty. There was a blazing fire in the chimney place and we were served tea and pastries. It was as though we were in another country.

It was not that night, I know, because it happened on a night of the full moon again. We had gone out onto the ketch moored at the pier. The reflection of the moon touched the waves in the bay with silver. There was only the distant sound of the town and the water lapping against the hull beneath us. Gustav led me down to the cabin. A lantern was hanging from the beam overhead, glowing softly. Gustav kissed me gently, my mouth, my eyes, my throat, until I was trembling. He slowly took off my clothes and laid me down in the golden light.

Gustav came to the hotel every day. We often went to the chalet at Kenscoff because the cool air and the smell of the pines reminded Gustav of his home in the Tyrolean mountains, and because the host rented us a lovely room with red geraniums growing in the window boxes. Gustav made love to me. I had never dreamed how love could be. Afterwards, wonderfully tired, we would go in by the fire again and eat a big meal of snails and steaks with *ris jonjon*, finishing with bowls of wild strawberries and cream.

6

To America and Back

A friend of Gustav, Jim, wanted to take his yacht to Miami but he had little sailing experience and he had no crew at that time. As it wasn't a long distance from Haiti to Miami, nor was it hurricane season, two men could easily handle the boat. Jim asked Gustav to go with him, although Gustav had never made the trip. The route through the Caribbean Islands has many hidden shoals where treasure hunters still search the graveyards of ships wrecked on the coral reefs over the last two centuries. The trip was a challenge that appealed to Gustav. He acquired a book of lights and carefully studied it to be sure they would recognize the beacons and landmarks that they would be passing. He had the book with him when he came to the hotel to see me the night before they were to leave. He forgot it there when he left.

I discovered the book the next morning and rushed

down to the pier, but the yacht was gone. The men would surely turn back when they realized the book was missing. I waited all day but they didn't return. I could not believe that Gustav would be so foolhardy. I was frantic with worry. André tried to calm me, saying they would surely put in to Cuba and get another book.

It wasn't until the first night out that Gustav realized the book was not on board. He decided it was too late to turn back and he didn't want to waste time stopping in Cuba. He knew how dangerous the trip was; not only that, he had to stay at the helm all of the way because Jim became violently seasick. Gustav, never certain of where they were, kept the poor man hanging over the bow for hours, watching for any change in the color of the water that would indicate shoals or reefs. Only when he began to see fishing boats did he know they were approaching Florida. He followed them in the late afternoon, going one way, then another, until he was certain of the direction the largest number were headed, and sure enough, it was Miami.

Meanwhile, I did not leave the hotel, agonizing, waiting to hear the good news or bad. Finally it came, simply a cable. "Arrived safely. Come to Miami." I took the plane the next day.

In Miami, going to the Port Authority, I found the yacht without much trouble. I boarded it, joyful with the anticipation of seeing Gustav. He was not there. Jim was there, waiting for me, and he said,

"Gustav was taken into custody by the Immigration officers. Something about investigating his activities in Italy."

"Activities in Italy? What activities? Is he in jail?"

"I don't know. They said it was routine because he

came without a visa. They said it shouldn't take long so I don't know why he isn't back yet."

Totally panicked, I cried, "What should I do?"

"Calm down. Go to the Immigration Office and ask for him."

I rushed off, located the Immigration Office, and was gratefully relieved to find the officer was very congenial. "It is just routine," he repeated, "although it is taking longer than usual because Security is waiting for the reply from Italy. He's on the Fascist list."

"But all Italian students were enrolled in the Fascist Party. That doesn't mean they did anything!"

"Miss, I shouldn't even be talking to you about it. But I'll tell you where he is. He's in the hotel down the street."

"Hotel? Can I see him?"

"Why not? He's under guard. I'll give the guard a call." What a nice man!

I went to the hotel and asked for Gustav's room. The guard was outside the door and he was as nice as the officer. He grinned when he saw me and immediately opened the door. I threw myself into Gustav's arms. He hugged me, delighted that I had managed to see him. I asked the guard if I could stay. He said, "Sure, it's all right by me." So we spent our first night together in America in the care of the United States Government.

Gustav was cleared the next day and we returned to the boat. Poor Jim, still sick, was waiting for us. We assured him we would take care of the boat, so he flew home to Kansas or some place where he had a turkey farm. We ran the boat up to the North Shore, a beautiful place beyond the outskirts of Miami Beach, where we spent a month of lazy, happy days.

Gustav had never been to the States. He was fascinated by the American way of life, the efficiency, the friendly and helpful people, who were probably more helpful than usual because they were fascinated by Gustav. We explored Miami, from the sumptuous hotels lining the shores of Miami Beach to the dives along the docks, filled with commercial fishermen, dock workers, and ladies of pleasure from every Latin-American country. We drove all over southern Florida, through the Everglades and as far as Lake Okeechobee because Gustav was enchanted by the name. We drove across the overseas causeway from one golden island to another until we reached Key West. We were charmed by the old Creole houses with filigree balconies overlooking the narrow streets and the squat shrimp boats that lined the quays.

We went spearfishing by the reefs off shore, and one time found ourselves in the middle of a school of barracuda. I had seen one or two before; Gustav said they never attack, but these barracuda were definitely tracking us, and they did not seem to be only curious. I was terrified. I clung to Gustav's hand, shutting my eyes tightly until we reached the pier and climbed out of the water.

Most of the time, though, we stayed on the boat, and most of the time we made love. I was quite sure by then that Gustav loved me very much.

I woke up one morning as the early light of day filtered into the cabin. The air was cool, a gentle breeze coming in through the porthole. Gustav was already up. He often went spearfishing shortly after dawn. I stayed in bed that morning, curled up with my head on Gustav's pillow, while the faint lapping of the waves put me to sleep again. The caress of Gustav's hand on my face woke me up. He took me in his arms and kissed me. His body was warm.

"Mm," I murmured, "You didn't go fishing."

"No. I walked to the lighthouse and back. Renée, I have been thinking about us." Still half asleep, I ran my fingers softly over his wide, bronze chest and shoulders, never tiring of touching his smooth, muscular body. I put my arms around his waist and pulled him down, burying my face against his stomach.

"Renée, listen to me," he said. "I want you to come back to Haiti to live with me. You are my wife now."

Those words woke me up completely. We had never discussed what we would do beyond the next day or so. I had avoided it because I could see no future and had not wanted to think about it. I wanted only to take advantage of what I had now. I supposed that for Gustav it was the same. He was married. I wasn't his wife. And how long would this last? I realized then that I did not care. Or, rather, that I had no decision to make. I loved him too much and I could not separate myself from him.

"You know I cannot get a divorce in Italy," Gustav said. "It was an unfortunate marriage. She wants to be free also, but there is nothing we can do about it. Renée, for me it makes no difference; it is of no importance. I want you to stay with me."

I could only nod my head, holding him tightly. If he loved me I would stay with him as long as he wanted me, forever.

We returned to Haiti. At first we lived in a small hotel in Petionville, owned by a French couple. The wife apparently had many lovers. Her husband was furious, although it was more than obvious he detested her. They fought loudly and bitterly, and we soon knew all the details of their marriage. He would not agree to a divorce, despite her contempt and flagrant behavior. Quite suddenly he died, for no apparent reason. "Voodoo!" the servants whispered, and soon the whole town heard. The wife was taken into custody while the police investigated the death. The body was examined and no proof of foul play was found, but it was not yet permitted to be shipped to France. Having no other place to keep it in Haiti, the body was kept in the deep freeze of the hotel, until finally the wife was released for lack of evidence. There were always rumors of voodoo in Haiti.

We moved as soon as possible to a house in the quiet district of La Boule, half way up the mountain. The house was pretty but the stone walls, held together with the local cement, were fragile. Lights and paintings fell down frequently. One day we came home from a fishing trip during which Gustav had drunk a bit too much rum. As he drove around the curve of the driveway, he lightly hit the retaining wall. It cracked and then disintegrated, the stones falling like dominoes until half the wall went down the mountainside.

We had a little goat, Cabri, who loved orchids. They were common and grew wild, a lovely, delicate mauve. I transplanted some, trying to keep them out of Cabri's reach, but did not succeed. He didn't eat the plants, only the flowers as soon as they bloomed. Cabri wandered off one day and discovered a peasant's vegetable garden nearby. The man soon appeared at our door, complaining that the goat had eaten all of his vegetables. We paid him for the value of his loss, but we didn't tie up Cabri because there were no other gardens nearby. But word of the compensation spread throughout the district and we soon had peasants coming from a mile away, insisting that Cabri had eaten up their gardens too.

Our maid, Agathe, was very efficient and cheerful. Sometimes she giggled softly when she served us at table, waiting for us to ask her what was so funny. She would then collapse in a paroxysm of laughter, throwing her apron up over her head. We had to pull the apron down and calm her, so that she could tell us about a neighbor's infidelity, and how the husband caught the lovers, and all the villagers saw the naked fellow racing out of the hut with the husband hard on his heels, brandishing a machete. Or, more seriously, how another neighbor's garden had been torn up by voodoo horses, which were really human lovers who had transformed themselves into horses so that they could travel faster to meet each other. We saw them too, galloping freely along the road at night, and wondered, because horses were valuable and always kept well tied.

Our yard man, exasperatingly slow, was named Têtemarteau — hammer head. The name was probably not meant to be derogative. One boy, no less intelligent than his brothers, assured us his name was T'idiot. Other colorful names were very nice — Veritable, Mon Prince, and Bon Ami.

One of Têtemarteau's duties was to fill the shower tank, a fifty gallon drum, every morning. We did not have running water so rain was collected off the roof into a cistern. Têtemarteau carried buckets of water from the cistern back up to the roof so that by afternoon we had hot showers. The rest of the day Têtemarteau squatted on his heels with a machete, cutting each tuft of grass in the lawn individually, and never had time to do anything else. After a couple of months of trying to cope with him, I was dismayed to learn that he was the local voodoo priest. Though it seemed he never took offense when I scolded him, who knows what he might have done if he were angry, so we dismissed him as subtly as possible. He returned to his "plantation," as the Haitians called their small crops of vegetables, where he worked less strenuously, raising beans.

Beans with rice made up ninety percent of the Haitian diet. There was not much meat available, and most Haitians were too afraid of the god of the sea to go fishing. The cost of living was incredibly low; nevertheless, I felt very uncomfortable when I paid the servants their wages. Agathe's salary was nine American dollars a month, and the laundress and gardener were paid four dollars, yet wealthy Haitians complained that the foreigners paid their servants too much. Servants always had a room in the kitchen courtyard where they could keep a child or two, but they were expected to buy most of their food out of their wages.

I went shopping in Port au Prince once a week to buy household supplies but bought fresh fruit and vegetables from the peasant women every morning. They began passing on the road at early dawn, walking twenty steep miles from the mountain to the market, carrying heavy baskets of produce on their heads. Sometimes it required two women to lift the baskets in place, on a circle of twisted cloth to help balance them. A stool to sit on at the market was put on top. The women carried everything on their heads; I saw one with two trussed chickens on her head, their feet sticking up in the air. The large baskets of flowers made the most charming pictures.

Every day or so I went out to find wildflowers, the zinnias, forget-me-nots, and red lilies that grew in the hills. I followed paths that ran along the slopes, watching out for tarantulas. Sometimes they crouched in the pathway, big ones as much as five inches across, their red mandibles spread, ready to bite. It was the only danger in Haiti. Occasionally, I saw strange men whom I suspected might be zombies. They shambled instead of walked, staring at the ground. They never turned their heads to look at me and say "*Bonjou*'" the way the other peasants did.

I started going out on the boat regularly with Gustav. It was the fish and the astonishing variety of marine life that interested me. I sent for books on tropical fish and marine biology. I began explaining to the tourists on the glass bottom boat how the coral was made up of colonies of tiny animal polyps, how the octopus changes its color to match his surroundings — and thus I became a guide. It was a great help to Gustav, who became exasperated when the Americans didn't understand his English. I freely used the marvelous names Gustav invented for the fish. Some fish were particular friends of Gustav. The most famous was Anatole.

Marcel Izy-Schwartz, the well-known French explorer and photographer, came to Haiti. An excellent diver, he wanted to film sharks that were in the deep water off of Gonaïve, the island in the bay. Gustav took him out and he got excellent sequences of the sharks attacking fish that Gustav had just speared, only a few feet away.

They were surprised one day to find a shark circling around Sand Cay where we took the tourists. Gustav killed the shark, spearing him while Marcel filmed the action. When Gustav started back across the reef, he discovered a pilot fish swimming along beside him. When he stopped, the fish swam over and attached himself to Gustav's side. It was a remora, a fish that attaches itself by a suction cup on its back, to sharks and other large fish, to go along for the ride and partake of meals his host provides. When Gustav killed the shark the remora shifted over to Gustav. Gustav regretted leaving the orphan behind when he got out of the water, but when he returned to Sand Cay the next day the remora was still there, waiting for him. Every day he waited and attached himself again to Gustav as soon as he went into the water. Gustav named him Anatole and one of Marcel's most successful films told the story of Anatole and his devotion to Gustav.

It never occurred to me that Gustav could have an accident, he was so at home in the ocean, but one daring episode was nearly fatal.

We had taken a group of spearfishermen off the reef near the island of Gonaïve where the water was deep. I stayed on board while the others went into the water, followed by the Haitian boys in rowboats to collect the fish. The men scattered as usual and the boys were busy rowing from one to another. Suddenly, one of the boys shouted. He was standing and pointing to an area well beyond the reef, where something was churning the water. I shielded my eyes from the sun, then saw the surface break as a tremendous manta ray leapt out of the water. In its back there was a spear.

Only Gustav would have gone out so far. I searched for him anxiously as the boys rowed furiously to the place where the ray had disappeared again. At least two minutes passed until I finally saw him — something was very wrong. His head was just bobbing on the surface. I screamed at the boys but they did not need to be told what to do. They were immediately at his side, ripped off his mask and pulled him, unconscious, into the boat.

Even from where I was, I could see the brilliant, red blood flowing down his face. I waited in agony while the boys rowed back as fast as possible. They got him up the ladder and laid him on the deck. Trembling, I washed the blood away and saw that his face was not cut but the blood was coming from his nose. The men turned him over so he wouldn't suffocate and I put ice on the back of his neck. We were already on our way to Port au Prince.

Crouching helplessly beside Gustav, I finally saw him move and open his eyes, and I cried with relief. He gave me a wry smile. He said slowly, "Don't worry. I'll be all right." But he closed his eyes again after the effort to talk. His hand tightened on mine and I clung to the assurance his grip gave me.

The bleeding had stopped by the time we reached Port au Prince. The men carried Gustav carefully to the car and we drove to the hospital. After a thorough examination, the doctors assured me that although his body had undergone a tremendous strain he would recover completely in several weeks. A few hours later he was able to tell us what happened.

Gustav had seen the enormous ray swimming along the wall of the reef, deep enough so that it was unaware of him at the surface. Although Gustav knew perfectly well how dangerous it was, he couldn't resist the temptation to catch such a record prize. Gustav dived down within a couple of meters of the manta and shot it. The reaction was tremendous. The manta charged up with a violent surge of his wings, crashing against Gustav and sweeping the line of the speargun around his foot. The next moment, unable to do anything to save himself, Gustav was being dragged down to the bottom. He was aware of nothing more until he came to in the boat. Fortunately, the line had pulled loose as the manta again surged up to the surface, Gustav's foot was released, and buoyancy brought him up to where the boys reached him just in time.

Gustav had trouble with diving for some time after that. The doctor said his sinuses were affected. He suggested a treatment with sulfur water so Gustav collected several bottles of the stuff at a source nearby. He was instructed to heat the water and breath in the fumes. Every morning I woke up to the odor of rotten eggs. Neither of us could stand it for very long and eventually his sinuses healed by themselves.

A few months later Gustav and I took a week off to drive up to Cap Haitian on the northern coast of the island. We wanted to see the fortress of Roi Christophe, the negro slave who aided Toussaint l'Ouverture in liberating Haiti from the French in the late 1700s. A freed man, Christophe was elected governor, then declared himself emperor. He became a despot, a cruel and oppressive tyrant. Christophe conscripted thousands of men to build a fortress at the top of a mountain overlooking the ocean and the central plain. Walls ten feet thick rose from the sheer face of the towering cliff that already would have been an insurmountable barrier to any invading forces. A troop of soldiers, ordered to prove their loyalty to Roi Christophe, was marched directly off the wall of the fortress to their death. Christophe's nemesis did not come from an invading army but from his own people. In a massive insurrection they succeeded in deposing him, whereupon he committed suicide.

The road to Cap Haitian, neglected and full of potholes, discouraged venturing beyond the limits of Port au

Prince. At one time there had been grandiose projects of developing the plain, but work stopped when the regime changed, probably because the outgoing president had confiscated the treasury. Stanchions, abandoned in the middle of construction, rose like buttresses over the river beds, supporting nothing. Fortunately, the rivers were bone dry most of the year so buses, converted trucks, were generally able to lumber down the banks and up the other side. The buses, painted in bright patterns of red and blue flowers, did not have the names of their destinations, but names such as *Joie de Vivre* and *Délire des Iles*.

Gustav had regretfully traded Caroline for a more practical, sturdy jeepster. It was a rough ride with hours of bouncing and swerving, but we arrived without mishap at the little colonial town of Cap Haitian.

Gustav and Marcel Izy-Schwartz, who had flown in to the airport with all of his photographic equipment, went out to explore the reefs off Tortuga Island, with much caution because of the mako sharks that patrolled the deep waters. I was glad to have a few days to myself. Sore all over from the trip, I did little but lie in the sun. I realized that as much as I loved Gustav I needed some time apart from him, and at that particular time, I had to be by myself to think.

For the last two weeks I had not wanted to face the possibility, but now I felt sure I was pregnant. Gustav had not noticed my being late; he knew I was very careful. One time, however, in the first days we were together, we had made love without any precaution. I was frightened of the consequences.

"You said you would be careful, Gustav! Suppose I get pregnant?"

"Pregnant? Does that mean *enceinte*?

"*Enceinte*? I guess so — I don't know!" I was exasperated. "I mean I could have a baby!"

"*Enceinte*. Well, don't worry about that. I shall send you to my mother in Italy and she will be very happy."

So he didn't mind having a child. But a child being taken care of by his mother in Italy was very different from having a child with us in our home. How would Gustav react when I told him? I didn't even know if he liked children. Of course we would have a nurse. A baby didn't have to change the way we lived, and I would make sure it didn't. Would Gustav realize that or would he feel, even as I grew big with the child, that I was complicating his life?

Gustav's life was organized just the way he wanted it, a carefree existence with, as he liked to say, his perfect wife. He had bought the recording of "Porgy and Bess" just to play the song for me, "Bess, You is My Woman Now." He wanted me with him all the time and did not like it if, for any reason, I was interested in anyone else. I too, did not want anything to interfere with my freedom to be with him.

If Gustav had thought about it at all he may have supposed we would have children some time in the future, if ever the divorce in Italy was granted. His wife had initiated the proceedings but they both knew it was all but impossible to divorce, or annul a Catholic marriage.

How was I going to tell him? I could not bear it if he were angry. Of course he wouldn't send me to his mother; he wouldn't want me to leave, but would his love cool to, finally, indifference? I couldn't tell him yet, and maybe he wouldn't notice for a month or so. I felt very well. I didn't hesitate to go the day we decided to make the excursion on horseback to Roi Christophe's citadel.

We rode up the steep mountain trail on tough little horses, a two-hour trip, to arrive at the fortress. The guardian, an old man who lived alone in that sinister place, showed us around. We climbed up to walk along the wide ramparts, still littered with cannon balls, and peered over the side, trying to imagine how it must have been for those soldiers who marched off the wall, knowing they were going to die. In the interior of the fortress, gloomy corridors lead to dank and smelly dungeons. Perpetually cold in this silent and deserted tomb, it seemed as though sad ghosts of men, dead so cruelly, reached out to touch us longingly.

By the time we returned to the valley in the late afternoon, I was exhausted and depressed. Gustav was surprised by my moodiness. When we were back in the hotel, he said impatiently, "What is the matter with you? Why are you so cross?"

I burst into tears. Gustav was astonished. He took me in his arms and waited for the storm to subside. Taking comfort, I cried for several minutes, but I wondered at the same time how he could be so obtuse! How could he have not noticed I was late? Unreasonably, I wondered how he could have been unaware that something was wrong. I had carefully acted as usual, but he should have been more sensitive, more aware of me. I had to tell him, and finally, hiccuping, I said,

"It's — it's because I am pregnant!"

Gustav stopped rocking me but he didn't let me go. He muttered, "I should have known." After a moment he began to mechanically stroke my hair, and I looked up to see his face. It was without expression, his eyes unseeing. I did not dare ask what he was thinking. Finally he moved, pulling me down with him into a large armchair where he held me in his lap. I curled up against him, saying to myself over and over, he isn't angry, he isn't angry. At last he said, "It's all right, my darling. Don't worry any more, it won't make any difference. I love you and I'll love our baby."

How gentle he was that night. He teased me for being afraid to tell him, for doubting his love. I was his wife, he said, and nothing would separate us. I was completely reassured, happy again. I believed we would be together always.

When we returned to Port au Prince, Gustav went to see a lawyer and learned that a divorce obtained in Haiti was legal anywhere in the world except Italy. He knew he would probably never return to Italy to live, so he immediately started proceedings for a Haitian divorce.

I carried my baby easily and Gustav did not stop loving me even when I grew big. I went out with him on the boat most mornings as usual. Sometimes we stopped at the Ibo Lélé to have lunch with André, but we seldom went to the American Bar or other places where Gustav used to go. It was Gustav who changed, losing all interest in the parties and the social life of the foreign colony. In the early evenings we took long walks around our mountain district of La Boule and often drove up to the pine forest at Kenscoff. We started a vegetable garden, with Gustav supervising the progress of every plant. Mountain born and man of the sea, he decided he was also man of the soil. I started sewing clothes for the baby. We had brought a stereo from the States and in the evenings we listened to classical music. Friends who came to visit us were astonished to see Gustav being so domestic. Gustav was even looking forward to the birth of the baby, who he was sure was a boy and always called Guido.

A month before the baby was due I went to the States. The hospital in Port au Prince may have been adequate, but Gustav wanted the most modern and safest delivery for his son and heir. I went to New York to stay with Mother at her home in the country near Mahopac, a small town near the Hudson River. And just a month later, in April, 1952, our daughter was born.

A girl. Why had it seemed impossible that the baby would be a girl? Gustav was so sure it would be a boy that I was convinced of it also; we had simply not considered the possibility of having a girl. Our daughter, whom we named Marisa, looked just like Gustav, with the same blond hair and blue eyes, and, we were to learn, much of the same personality.

Mother had cabled Gustav as soon as Marisa was born, and later that day his answer arrived, with an enormous bouquet of flowers, saying, "I prefer a girl. Shall be there soon."

Gustav had had to stay in Haiti, waiting for his divorce to become final, and it did, exactly the day that Marisa was born. A week later Gustav flew to New York. He arrived at Idlewild airport and took the train to Mahopac.

I drove to the station to meet him. I was still apprehensive, not at all sure he really preferred a girl. The train arrived and he stepped down, looking for me. A moment later I was in his arms. Ignoring the people around us, he kissed me hungrily. I knew then that everything was still all right. We left the station with our arms tight around each other. It wasn't until we reached the house that Gustav asked about the baby.

Gustav watched as I picked her up, folded back the blanket, and showed him his daughter. He didn't want to hold her but was surprised and amused when he touched her hand and her fingers closed over his in a firm little grip. The next morning I brought her into our bed, putting her on Gustav's chest. Marisa lifted her head and struggled, kicking her feet. Gustav decided she would be a good swimmer and declared himself quite satisfied with her.

A month later we went to New York City and were married at City Hall. Mother and baby Marisa were witnesses. The next day we flew back to Haiti. The Haitians had always called Gustav "Tarzan." They called me "Madame Tarzan," and little (*ti* in Creole) Marisa became "Ti Tarzan."

7

Leaving Haiti

We had taken a group of tourists out to Sand Cay as usual, Gustav had finished his underwater show and pulled himself up beside me on the boat, when one of the passengers spoke to me.

"Renée?" she asked. "Renée, is your last name Denis?"

"Yes. That is, it was before I married." Then I recognized her — Jane, a girl who had been at boarding school with me. Jane said she was there on her honeymoon; she introduced me to her husband, Bob, and I introduced them to Gustav.

"Your husband!" Bob exclaimed, and then, embarrassed, "I mean, it's hard to imagine somebody like Gustav being married."

We laughed. Gustav's wild and independent manner made it difficult to imagine him in a home with a wife and child. The perpetual show that Gustav put on in public had become second nature to him, and it amused me because I knew how different he really was.

Gustav did not tire of our domestic life in La Boule. In fact, he did not always go out on the boat any more. Etienne, a young Haitian who often went out with us, started spearfishing and even began copying Gustav's underwater show. He became very good at it so Gustav was able to take days off without canceling the tour. Etienne was well-educated and spoke English, so he substituted for me on the boat as well.

I wanted to stay at home more because Marisa was becoming a little tyrant with the servants. Her first year was like our first year in Bali, when Mother had three nurses, one for each of my brothers and me. Marisa was never left alone and was carried almost all day long in the arms of her nurse, the cook or the laundress. So I established a normal routine of naps in her room and play time with her toys on the floor beside me. This suited her just as well, as she was very good-natured; she quickly realized that if she behaved she could go out in the car and on the boat with us.

When Marisa was several months old I became pregnant again. This time I had no qualms about telling Gustav because he loved his daughter and also wanted a son. With this pregnancy, however, I wasn't well. I was sick every morning, so I could not go out on the boat. By noon I was usually all right, and the three of us then went for walks or drove up the mountain to Kenscoff. The pines, daisies, and nasturtiums reminded me of New England. It would soon be spring in the States, and I thought nostalgically of the white dogwood flowering in the northern woods. The crocus were up, soon there would be daffodils, and the apple trees would burst into bloom. I was homesick.

It wasn't that I wanted to leave Haiti. I only wanted a change, a trip to distract me and shorten the long months of pregnancy. I begged Gustav to take a month off and go to New England while it was springtime.

"It would be more sensible to go later so that you could stay until the baby is born," Gustav said, "but, all right, we'll go now for a month or so. I suppose the earlier I decide what to do about the import business there, the better. Mario has been after me to go up and see what can be done."

Two years before, when Gustav and I went to Florida, we had visited friends of Gustav who had lived in Haiti and were living then in Miami. Mario and Kati Alborghetti were Italians. They were delighted to see us, and we stayed at their home for several days while Gustav and Mario talked about starting a business together. Gustav had been offered the representation in the Western Hemisphere for Cressi, an Italian manufacturer of spearfishing equipment. Gustav knew nothing about running a commercial business; besides, he was living in Haiti. He agreed to invest in the company but Mario would run it. The Alborghettis moved to New York, the center for importing and distribution, and rented a house with a large garage for stocking and shipping. Within a few weeks Mario received his order of spearguns, masks and fins, so he set out to sell the Cressi products.

Although the equipment was the best on the market and the retail stores wanted to buy it, Mario ran into tough competition with a company called U.S. Divers. They made aqualungs, and scuba diving was becoming popular. U.S. Divers would not sell aqualungs to stores unless they also took their guns and fins. After a year of struggling, Mario wanted to give up and get out of the business. Gustav had to find another partner or get rid of the stock and be done with the company.

We planned to leave for New York in a couple of weeks. The day before we were to go, Gustav brought our friend, Etienne, up to the house. I told the maid to set another place at table. During lunch I learned that Etienne was there for more than a farewell visit. Gustav told me, "Etienne wants to buy the glass bottom boat. His family is willing to back him. He also wants to buy the spear-fishing equipment and motorboat, so I have agreed to sell the tour business to him." I was speechless with surprise. Gustav hurried on to say, "It's easy enough for us to do the tours on another island. Jamaica, for instance. There are no boat tours there. An English island would be better for you, and for the children when they are old enough to go to school."

Gustav had misunderstood my expression. I was not dismayed at the idea of leaving Haiti. It was a beautiful island but had no particular appeal for me. We didn't own the house nor did we have special friends. It was Gustav and the way we lived that was special, and that could be on any island. I was surprised because I had no idea that Gustav was thinking of leaving Haiti. Perhaps wives were not consulted about business matters in Italy. I would not have questioned Gustav's plans, but this concerned our personal life and I thought I knew him well enough to be aware of what he was thinking. But, all right, Jamaica sounded fine to me.

"Are we going there now or are we still going to New York first?" I asked.

"New York, though it will take about two weeks to turn the business over to Etienne, so I will have to stay here until it's finished. But," he watched me, "you can still go tomorrow if you don't want to wait."

I knew he wanted me to say no, I would wait. I thought, two weeks is a short time whether I stay here or wait for him in the States. If there was any delay in the transfer, and I knew Haiti well enough to expect it, spring would be over. I said,

"When you get to New York you'll be busy with Mario so you won't have much time to go out to the country. I'll go up to Mother's house in Mahopac until you come."

"All right. You and Marisa leave tomorrow and I'll join you as soon as I can." I didn't realize how much he minded my wanting to leave instead of waiting for him.

I flew to New York with Marisa, and Mother came to meet us at the airport. She was delighted to see her granddaughter, lively, pretty, and already very feminine, though she was just a year old. Hugging her, Mother smiled at me happily, but there was an inquiring look in her eyes. I kissed her warmly. In the last two years I had become a woman. I understood my mother, and my father, at last.

We drove up to Mother's house, where the first flowers were coming into bloom and the trees were putting forth green buds. It was the springtime of my childhood again. I took Marisa into the woods to see the small purple and white wild violets pushing up through the damp earth. There were fragrant lilies of the valley, narcissus and hyacinths. At the edge of the creek, now that the rushing waters from the melted snow had passed, I showed Marisa how to turn over stones to see baby crawfish and tadpoles. Mother often came with us. We didn't talk much about the past; we didn't need to, and we felt very close, sharing Marisa and the springtime.

A cable arrived from Gustav saying that there was a delay, and he would come a week later than expected. I was not surprised, but when another cable arrived a few days later — another delay — I began to feel anxious. I would have appreciated a letter with some explanation.

It was Mario who received a letter. He called me from New York. "Gustav asked me to help you find a house to rent for the summer; a house near here so that he and I can work. I don't know what he is thinking of doing with the business, but he said he would be coming soon."

"I don't understand. Why hasn't he written to me? Don't I get to say anything?" Another surprise — and this time I was angry. "I don't want to stay in a house down there; you're practically in the city! I thought we were going to Jamaica!"

"Renée, don't ask me. You know Gustav —"

"I am beginning to think I don't know him! All right, I'll come down to see you, but I can tell you now I won't stay in the city."

I left Marisa with Mother, and Mario met me at the train station. "Renée, there are small towns across the Washington Bridge in New Jersey. I'm sure we can find a house you will like. It isn't far for Gustav to drive over here." Mario was a kind man, large and rather clumsy, and very worried, waiting for Gustav to come and relieve him of the responsibility of the company. We drove across the Washington Bridge to Fort Lee, where we were fortunate to find a house with a large garden, which belonged to an artist who was away for the summer. I bought a car and drove up to Mahopac to get Marisa. Back at Fort Lee, we settled in, I started learning to cook and keep house by myself, and I waited for Gustav. Finally he cabled that he was coming. I drove to Idlewild airport to meet him.

As soon as Gustav came through the gate, I knew something was wrong. He was with another man, and when he saw me he stopped walking to finish their conversation. It was a lengthy farewell. Finally the two men shook hands and Gustav came over to me. He kissed me almost absent-mindedly, asking, "How are you feeling?"

"I'm fine now," I said, as by then I was carrying the baby as easily as I had Marisa. I asked quickly, "How is Etienne doing with the tours? Were people disappointed that you sold the business?"

"The hotel people were disappointed, but Etienne is so eager to do well that everyone likes him. He'll be fine."

"And you?" I searched for the parking ticket to avoid looking at him. "Did you mind very much leaving?"

"Yes," he said shortly.

I bit my tongue. He loved Haiti. Although he had decided to leave, of course he had some regrets. I said hurriedly, "You'll like the house we have. It is typically Victorian, built in 1870. There's a big garden."

"Good. How far is it from Mario's place?"

"It's in Fort Lee, just the other side of the Washington Bridge. Mario's house is a few miles this side. Do you want to stop there now? Mario is really impatient to see you. He asked us to come for dinner." When Mario had suggested it the day before, I said we had better wait and see, remembering how impatient Gustav was to be alone with me the last time. Now I wanted to go to Mario's house. Mario would immediately involve Gustav in all his problems.

"All right." We stopped at a phone booth to let Mario and Kati know we were coming, and drove into the suburb of New York. Mario was waiting impatiently.

"Come, come," he cried, as soon as he greeted us, and started to pull Gustav into the garage. Then, "No, let's go into the house; Kati is cooking dinner." But in a few minutes the men went down again. Kati, a calm, good-natured woman who had been married to Mario for thirty years, told me to sit and talk to her while she finished cooking the meal. It wasn't long before she said, "What's the matter, Renée? You don't seem to be as happy as you should be."

"I don't know, Kati. It isn't me; it's Gustav. He didn't seem glad to see me."

"Of course he was. He's just sorry about leaving Haiti. Don't worry, he'll be over it in a day or two. Mario will see to that."

"I doubt that business problems are going to be a pleasant distraction for Gustav," I said, laughing to take away any criticism of Mario.

"You may be surprised. You may not know Gustav as well as you think you do. We were in Haiti, you know, when Gustav first arrived there. He likes business."

"Yes, a business that is easy and fun like the boat tours and fishing trips. Any man would enjoy that."

"Did you know that Gustav had considered opening a restaurant with Mario? That was before your grandfather interested him in the hotel instead. Gustav has always been intrigued by business. He is sorry to leave

Haiti, yes, but I am sure he sold the business because there was no more challenge. He couldn't do any more in such a small place and the novelty had worn off. I wouldn't be surprised if he decided to stay here to see what he can do with Cressi products —"

"That's not likely," I exclaimed. "Can you imagine Gustav a business man in the United States?"

It was late when we left and drove across the bridge to Fort Lee. I was worried because I had promised the baby-sitter we would be home early. When I had suggested earlier that we should leave, Gustav said irritably, "You're paying her, aren't you?" On the way home Gustav talked only about the business, the mistakes Mario had made, and how problems should have been handled. I didn't pay much attention. The night was chilly and I drew my sweater tightly around me. Before, Gustav would have pulled me close against him.

When we arrived at the house I had to drive the sitter home. By the time I returned, Gustav had fallen asleep in an armchair by the fireplace. I woke him up, and with hardly a word he went upstairs. I put a screen in front of the fire and followed him, stopping to look in on Marisa. When I went into the bedroom, Gustav was asleep again. I lay down beside him but couldn't sleep for a long time, fearing what could have happened to turn Gustav away from me.

In the morning Marisa woke us up, demanding to see Gustav. He called to her, "Come on, come to Daddy," and she climbed onto the bed to play. I was relieved that Gustav was in a good humor.

I cooked breakfast while Marisa led Gustav on a tour of the garden. Shortly after breakfast Gustav left for Mario's. At noon he called and asked me to get the baby-sitter again. He wanted me to go with him into New York. I put on a flattering, new dress and was relieved when Gustav arrived and looked at me approvingly. "I want you to drive for me," he said, because he didn't know the city. I was glad that at least he needed me.

"Which is the best sporting goods store in New York?" Gustav asked.

"You mean for all sports?"

"Yes, the smartest store."

"Abercrombie and Fitch."

"Do you know where it is?" I nodded. "We'll go there first."

"Didn't Mario ever go there?"

"He may have, but Mario doesn't know how to present himself. He knows he has the best products but he acts almost apologetic, as though he expects to be turned down. You'll see, Aber—, whatever it is, will buy from me."

Gustav had taken a jacket and tie, but the afternoon was hot and he wore only his shirt, the sleeves rolled up, the buttons undone almost to the waist, the way he usually dressed in Haiti. When we parked in the city, he took a long sample case out of the car, but not his jacket. He didn't even button his shirt. I looked at him doubtfully but didn't say anything.

In the store we went to the sportfishing section and asked for the department manager. While the clerk went to get him, Gustav took a speargun out of the sample case and became intent upon fixing something on the handle. Dressed the way he was, with his sun-bleached hair, he looked as though he were on his boat in the islands instead of a department store in New York City. The manager came out and appeared startled when he saw him. Gustav looked up, saying pleasantly, "Hello. My name's Gustav. Do you do spearfishing?" In a few minutes the two men were deep in conversation with Gustav telling about the sharks off Tortuga Island. The manager was enthralled, but then Gustav stopped talking to introduce me. He was delighted to meet me, the manager said; in fact, he wanted to invite us out to his place on Long Island to do some spearfishing.

"The water is nothing like in the Caribbean but we manage to catch some good fish," he said. "We live quite far out; how about coming for the weekend?"

Gustav accepted the invitation with enthusiasm, and packed his speargun back into the sample case.

"What have you got there?" the manager asked.

"Oh, I'll bring this along with me. You can try out the gun and see how it works."

"All right. I'll be looking forward to it."

Leaving the store, we laughed at Gustav's success in catching the manager's interest. Gustav had greatly enjoyed it.

"Let's go see a show tonight, as long as we're in the city," he said, so that evening we went to dinner and a Broadway show. At the end of the performance Gustav wanted to leave quickly. Unfortunately, I didn't realize that my foot had gone to sleep. Gustav was already waiting in the aisle when I stood up, took a step and fell forward, catching myself awkwardly on the back of the seat. Gustav saw me but he abruptly turned and walked away. The people behind me stared at Gustav, then leaned over me with concern. By then I was able to walk and I thanked them, but could hardly hold back the tears. A moment later Gustav came back down the aisle. He put his arm around me, muttering, "I'm sorry. I don't know what comes over me, but I cannot bear to see anyone stumble." It didn't make sense but I was too upset to say anything. It wasn't until we were in the car and he said something of the sort again that I lost my temper.

"I don't understand what is the matter with you, Gustav, and you had better tell me! You may be sorry about leaving Haiti, but that is not my fault. Why are you treating me this way?"

Gustav was silent for a few minutes. Finally, he said, "After you left, I didn't have much to do, just waiting for the sale of the boats to be finalized. Etienne took over the tours, and the house was empty." He glanced at me. I waited, staring blindly out the window. "So, I went around to the American Bar and other places, and I ran into Barbara Lenz. Do you remember her?"

"Of course." I felt a sinking in my stomach as my eyes filled with tears again. Barbara was a striking blonde, divorced, who had a dress shop in Port au Prince, and a Haitian lover.

"Well, Barbara was no longer with Jacques and she was feeling down, so I took her out a couple of times."

"That's all?" I couldn't keep my voice from trembling. I cried, "I don't believe you! I don't believe you only took her out a couple of times and that is all there was to it!"

"She wanted me stay with her in Haiti, but —"

"Then go back!" I cried bitterly.

"Renée, I chose to come here to be with you." He reached for my hand, but I pulled away, unable to stop crying, hating him for letting this happen.

"It's over with Barbara, darling," Gustav said. "Listen —" But I didn't want to. He stopped the car. He pulled me into his arms and kissed the top of my head as

I continued to cry. "I love you, Renée. I missed you and I guess I was angry because you left."

"For just one month?" I sobbed.

"I know, I know, but, darling, it's over. It didn't mean anything. I don't want anyone but you." He lifted my face and kissed me, murmuring that he loved only me, and I must never leave him again.

That night he made love to me as gently and sweetly as the first time two years before. I made up my mind not to let what happened affect my feelings or our marriage. Gustav seemed to have forgotten it completely. The next day he was in fine humor, playing with Marisa, teasing and affectionate with me. We often went to the city and he was always impatient to return home. We never spoke of Barbara again.

Kati was right, the business was a challenge for Gustav and he knew how to make it succeed. Despite the competition, the stores wanted Gustav's spearfishing equipment. After a few weeks Gustav knew that he could do very well if he ran the business himself. He decided to stay in the States.

Again, he made his decision before talking to me about it, but of course I agreed, particularly when Gustav said we should buy a house. He said we didn't have to live as close as we were to New York, so I wanted to move back to the Connecticut countryside where I used to go horseback riding. I drove up to Greenwich to see a real estate agency. A pleasant woman greeted me, and we talked about the kind of house that Gustav and I wanted. Gustav's name intrigued her.

"That's an unusual name."

"Yes," I said, "particularly for an Italian." Her expression changed. Was she imagining a small, dark Neopolitan? I said nothing more but arranged a time when we would return to see some houses. I was amused when she saw tall, blond, bronzed and handsome Gustav. Before the afternoon ended she tried to persuade him to speak to her ladies' club.

We were fortunate to quickly find a property that we liked. "It is a stable but it has great possibilities," she told us and it was true. We drove to an estate on Round Hill Road and saw the English Tudor manor, built of stone and thick beams, with a slate roof. The garages and stables were the same. The buildings had cost more than a million dollars to build in 1928. We followed the drive through a meadow to the stables. There were eight large box stalls, and the tack room had a slate floor and walk-in fireplace. The loft overhead had wide doors that opened onto the view of the valley. Tall elms shaded the courtyard, and four acres of fields and orchards were included in the sale. We could live in the stableman's neat two-bedroom cottage at the far end while we converted the stables into a unique and beautiful house. We bought it.

For the rest of the summer we crossed the George Washington bridge every Friday at midnight, avoiding the traffic that poured out of New York for the weekend, and drove up to Greenwich to work on the house. We painted, put fresh wallpaper in the cottage, scrubbed the slate floor, polished the oak beams in the tack room, and moved up to stay there in September. The maple leaves turned to red and gold, the gnarled, old trees in the orchard still produced delicious Macintosh and cooking apples, and we adopted a calico cat that always followed us on our walks through the meadows in the crisp autumn air.

A month later, my pains started, and Gustav took me to the hospital. I was in hard labor all night. Before I fell into exhausted sleep I heard the nurse say the baby was a girl.

The doctor woke me up a little later. "Renée, can you hear me?" I nodded, and realized that Gustav was there also, holding my hand tightly. The doctor leaned over me, speaking very clearly. "Renée, I have to tell you that your baby isn't breathing properly. There is something wrong with her. We have her in an oxygen tent, and we're taking good care of her. Do you hear what I am saying?"

I nodded again. The doctor told me about the baby while I was still under sedation so that it would be less of a shock. I looked at Gustav; it was he who was in a state of shock. I put my arms his neck, drawing him close, and murmured, "I'm sure she'll be all right."

But she wasn't. I went to see her in the nursery, lying in an oxygen tent, not moving. She had dark hair and eyes like mine. The doctor said she was injured at birth, paralyzed by pressure on the brain. A specialist came from Boston. An operation could be attempted to relieve the pressure, he told us, but it might be fatal. We had to wait until she was a little stronger.

I tried not to think about the baby. The worst times were when Gustav came to the hospital. He was not only depressed about the baby but worrying about mounting costs. And he missed me. The doctor agreed to let me leave the hospital, and we drove home along the road bordered by fields of goldenrod and the glorious fall colors of the trees arching overhead. I felt much better.

Thank heaven Marisa was born first. I was so grateful to have my small daughter, now a vivacious, amusing little person. Gustav delighted in her, too, and she adored him. They would roll on the floor together, playing acrobats or wild animals, Marisa shrieking with excitement.

We seldom talked about the baby. We went to see her but she was always the same, not moving, not seeing. The doctors then told us there was no improvement and never would be.

Gustav must have told them to try the operation. One night when the baby was six weeks old, the doctor called us. The baby had died.

The leaves fell from the trees and the weather grew cold. I bundled up Marisa and we followed the trails of deer across the fields, occasionally seeing the shy, graceful animals. They came near the stables at night, leaving their hoof prints. We put grain out for them as winter came and snow covered the ground. We saw the deer in the moonlight approaching daintily to feed. We continued to work on the house, while Gustav designed plans for transforming the stables into a beautiful home.

8

California, Land of Enchantment

Gustav went to New York two or three times a week and contracted a packing and shipping company to replace Mario's garage operation. Mario and Kati went back to Italy. I started working for Gustav by typing his letters; the Cressi equipment was becoming known, and stores and distributors were asking for the products. A French manufacturer asked Gustav to represent his company as well, so the work and correspondence increased.

I wouldn't have minded packing and shipping to help Gustav but I hated writing those letters. I was no typist but I had plenty of time to put down what he dictated, then wait up to several minutes while he decided upon what he wanted to say next.

He often repeated himself with variations to be sure he was understood, as he was translating mentally from Italian to English or French. The letters seemed endless.

He refused to employ a secretary, saying he would never find one who could understand him in whichever language he used. I became bored and impatient so we often argued over the letters.

A large distributor of sports equipment located in Los Angeles was very interested in Cressi products, and wanted a contract for exclusive rights to distribute the products nationally. A company in New York was also interested, and I thought Gustav would certainly give the contract to this one. However, he went to Los Angeles to hear what the larger company had to offer. It was also an opportunity to see what the golden state of California was like, as we planned to make a trip across the country some time. He was to be gone for only a week. He was gone for more than two. Every time he called he was more enthusiastic about California. Something about his calls made me think there was more to it than the warm climate and the good business prospects.

I went to the airport to meet him when he returned, and I recognized the same manner he had had when he came from Haiti. On the way home he talked of nothing but the lucrative prospects of national distribution. Driving into the courtyard of our beautiful property, where the sun was filtering through the new green leaves of the elms and the spring flowers were in bloom again, Gustav rhapsodized about the tropical plants in the Beverly Hills hotel patio and the palm trees along the Santa Monica palisades.

He saw I was not impressed. He finally said in an aggrieved tone, "You don't seem very happy to see me."

"No, I don't suppose I am," I said flatly. He was surprised. Marisa was delighted to have her daddy back and she chattered happily through dinner. I washed the dishes and put Marisa to bed while Gustav wandered moodily around the house. I went back downstairs and found something to mend, ignoring him. He sat down beside me.

"You think something happened in Los Angeles."

"Yes, I do."

"Well, yes, but it wasn't important." I shrugged my shoulders. Instead of being annoyed he went on almost eagerly.

"I met a girl, a starlet—"

"How remarkable."

"No, but—" He was a little subdued. "It was really quite funny, what happened." I didn't say anything. He continued, "I didn't know she was married. I just took her out a couple of times — actually it was the president of the company who invited her to a small party he had arranged for me — that's how I met her. Anyway, we were going out to the parking lot after dinner in a restaurant, and her husband was there. Suzanne — that's her name — was very angry with him and they started to fight. He hardly looked at me although he had been following us. She told him to go away, leave her alone, but he begged her to go back to him. He was crying and I felt sorry for the guy. I told Suzanne to go back inside. I put my arm around the husband's shoulders, talking to him like a father. I told him how he should treat his wife if he wanted her to stay with him. He stopped crying and thanked me and said he was so grateful to me for telling him how to understand women."

I was determined to stay calm, to think before I spoke. What was the matter with Gustav? Was that normal behavior — do all men do this when they are away from their wives? He loved me and he was sure of my love for him, so it wasn't that. Did he have to prove to himself that he was still attractive to other women?

"Renée, it wasn't important. I certainly don't care about the girl. Forget about it. You know I love only you."

But he had wanted to tell me, the way he did the other time when he had provoked a way to tell me. Why did he want me to know? It wasn't to hurt me; it was to satisfy some need in him.

I swallowed my anger. I let him lead me to bed, and his love-making was as passionate as ever. Gustav again immediately forgot the adventure and I tried to also.

Gustav decided upon the company in Los Angeles. I was disappointed. I had a feeling that it would not be good for us to go to California, but maybe it was because I loved Connecticut and our stable house. Gustav did too, but he continued to rave about California. I began to believe that it must be a wonderful place to live.

Gustav thought that the drive to Los Angeles, stopping to see the country and attractions along the way, then finding a house, might take two or three months. His mother, in Italy, had always asked him to take Marisa and me to visit her. Gustav decided that this was the opportunity for us to go, and let Marisa stay with her grandmother until we were settled in California. Gustav could not leave at that time so Marisa and I flew to Milan where his sister, Maria Theresa, met us. We made the beautiful trip by train across northern Italy to Bassano del Grappa, at the foot of the Alps, near Venice. Gustav's mother was a lovely, quiet woman. She kept a perfect house, was president of the Red Cross, and went to church every morning. She spoke only Italian; it was Maria Theresa who spoke French, translating for her and the rest of the family of sisters, brother, in-laws, nieces, nephews and cousins. They were enchanted by Marisa. She looked so much like Gustav and her extrovert personality was just like his. She knew she was the center of attention and loved it. Two days later I went with my sisters-in-law to Venice for the day, and then to Cortina d'Ampezzo for two days. Each time we came back to Bassano we found Marisa and her grandmother absorbed in play, understanding each other perfectly. Marisa hardly noticed me when I arrived. I was surprised and a bit hurt. I had to remind myself to be pleased that she was happy there without me. She hugged me good-bye calmly when I left for New York the next day; of course she didn't realize how long we would be separated.

Our beautiful home was sold and our belongings were on the way to California when I arrived back in Connecticut. The maple trees and birches were beginning to turn to autumn reds and golden yellow as I regretfully saw the last of New England and we started across the United States. We didn't stop often along the way, after all, because Gustav was eager to get to Los Angeles. I was impressed by the space and stark beauty of New Mexico and Arizona; I even looked forward to seeing California. Gustav assured me, "You'll love it."

We approached Los Angeles about noon, October of 1955. Long before we reached the city we saw the yellow smog. We drove through towns that ran into each other and became industrial East Los Angeles. I was startled to see a policeman with eyes so red he seemed to be crying, and I realized that my eyes were burning. Gustav said (as I was to hear often about the smog in Los Angeles), "This is a particularly bad day." We drove through downtown while he explained that the beautiful part of the city was on the other side. We drove all the way to Santa Monica to see the ocean, but from the palisades directly above we saw only a thick blanket of dirty fog. We went down to the beach and contemplated the water that looked dark and cold. Gustav kept saying, "It wasn't like this before. The weather was beautiful. The sky was perfectly clear." I didn't say anything.

We drove back to Beverly Hills, to the hotel where Gustav had stayed before. Gustav went to the reception desk as I followed slowly, admiring the tropical decor. The desk clerk looked up and recognized Gustav.

"Hello, Sir. Nice to have you back." As Gustav filled out the registration form he continued genially, "Is your wife with you again?" Gustav did not reply and I pretended I had not heard. The episode of the starlet was long past.

The worst of the smog cleared away but there were never the clear skies I had imagined, except after a heavy rain. I was surprised then to see the mountains so close. We drove all over the western side of the city looking for a house to buy. We certainly did not want to live in the smog. We finally decided upon a lovely Spanish-style house built on the hillside above the Sunset Strip in Hollywood. The house was about fifty years old, needed repairs and paint but the price was only $28,000. The Spanish-colonial center of nearby Westwood was being torn down, so we bought huge beams, carved oak doors, wrought iron railings and old bricks. We went to the auctions of Hollywood estates to buy handsome oak furniture. Gustav refinished the doors and tables, and even installed the railings and the brick terrace. He didn't have much patience for measuring and fitting, though. One morning he called me out to see the gate he had just installed. "It's nice," I said, "but isn't it a bit crooked?"

"Crooked? Of course it is. I did it on purpose to give it the antique look."

I didn't like Los Angeles. It had none of the advantages of a city like New York, nor of country. We were living in suburbia. We never walked; we always drove. The only neighbor I knew was the crazy "Countess" whose husband was a Nazi or a character actor, I never understood which. We never saw the husband and suspected she had done away with him. A famous Latin-American bandleader lived on the other side but the house seemed to be occupied by a large number of unsavory ghetto types. Friends lived at least an hour away on the freeways. We made some trips north, as far as San Francisco; the countryside as well as San Francisco were beautiful, but all too far away from Los Angeles. We went to Catalina where Gustav went spearfishing, but the island and the ocean were nothing to compare to the Caribbean. Gustav had little time to travel anyway; he went to his office for long hours every business day.

We were anxious to have Marisa back home with us. She was gone much longer than we had intended, waiting for her aunt Maria Theresa to bring her. We had furnished a charming room for Marisa, with an old-fashioned canopied bed that I covered in chintz, with curtains and dressing table skirt to match. She was going to love her room and I waited eagerly for her to arrive.

When Marisa stepped out of the plane we were amazed to see how much she had changed. With her hair in pigtails, wearing a tailored plaid dress, she was no longer a baby but a little girl three years old. "Perhaps too much time has passed and she won't remember us," I said to Gustav, but Marisa ran directly to us. She flung her arms around me and didn't want to let go. She sat on my lap in the car and chattered, in Italian. She had forgotten all of her English. I had not yet learned Italian so Gustav and Maria Theresa had to translate for us. It didn't bother Marisa a bit. She followed me everywhere, knowing that I would be taking care of her again. When I couldn't understand what she wanted she ran to get Gustav or her aunt to come and explain.

Unfortunately Marisa was very spoiled in Italy. At dinner one evening, Marisa asked for something and Gustav said no. Marisa put her fists to her eyes, and Maria Theresa started to get up. Gustav sharply told Maria Theresa to sit down. Marisa began to cry. Gustav told her to stop but then she began to cry in earnest. Astonished, I demanded to know what was happening.

"I won't have her behaving like that!" Gustav said.

"But don't shout at her. She's only a little girl. What does she want?"

"What difference does it make? Did you see Maria Theresa jump up as though she were a servant? My mother obviously did the same thing and Marisa is completely spoiled. You are so wrapped up in her now you don't see it."

"Gustav, she just arrived. I have to pay a lot of attention to her or how will I understand what she's saying? I won't spoil her. I know she shouldn't cry like that —"

"She isn't crying; she's just pretending. Look at the way she is watching us right now." The little minx was peeping at us through her fingers, her eyes bright as they went from one to another. I thought it was funny but I turned to Maria Theresa and said, "Please explain to

Marisa that in this house she mustn't cry like that when she wants something. If it isn't good for her we must say no and she must accept that. Otherwise Daddy and I won't be pleased."

Marisa put her hands down and watched me solemnly as I spoke and Maria Theresa translated. She nodded obediently, smiled tentatively, then smiled happily as I smiled back. She never again pretended to cry.

Marisa was completely devoted to me from the day she arrived. There was no apparent reason for her preference for me over her father, unless it was that her grandmother had cared for her from the moment she was up until she went to bed, and so she thought that I would too. But I was not so attentive, nor was I a very affectionate parent. I hugged and kissed Marisa morning and night and when she left for school, but, like the rest of my family, I was not very demonstrative. Marisa didn't seem to need more affection; she was unusually independent. As soon as she woke up in the morning she started on one of her projects, as she called them. Imaginative and dexterous at the age of three she drew and painted, cut, glued and sewed; I had to call her two or three times to come to breakfast. She was cheerful and obedient, never protesting against bedtime at eight or objecting to any of my rules.

A week or so after she arrived Marisa saw some children being picked up by the school bus. She asked to go to school. We told her she wouldn't be able to understand anyone but she insisted until I agreed to give it a try. The first day I hung around the phone in case the school called to say Marisa wanted to come home. At noon I went to pick her up. She was jumping up and down with eagerness to tell me all about school. She brought picture books home and we described the pictures to each other in English and Italian. We were quickly independent of our translators.

A few weeks later, I started teaching Marisa to ride horseback, which she loved as much as I did. By her fourth birthday she managed her horse with ease, cantering along beside me through Griffith Park. When we discovered the roadbed of the eventual 5 Freeway, we raced our horses along it at a gallop. Gustav also enjoyed playing with Marisa, building castles in the sand at the beach. When we went camping the two of them proudly caught fish, cleaned and cooked them on the grill.

Illogically and disastrously, although Gustav and Marisa were so similar and enjoyed doing things together, there was an intense rivalry between the two of them. With unrelenting determination, they fought for my attention.

As Gustav was usually gone during the day, their habitual battleground was the dinner table. When I realized what was happening, I tried giving Marisa her dinner early, and, as our mother had done for my brothers and me, I read to her for half an hour before tucking her in to sleep. But it was impossible because I had to prepare dinner for Gustav. He never ate lunch, so he was as hungry and cross as a bear when he came home, and wanted to eat as soon as possible.

Marisa was a little chatterbox but she knew she must never interrupt when grown-ups were talking. She sat between us and watched, waiting for a pause in the conversation. Then she quickly spoke up. She always started with, "Mother?"

Gustav would say, "Marisa, we're talking."

"Oh. I thought you'd stopped."

"No, we haven't stopped. Renée, what was I saying?"

More conversation, and Marisa waited. Gustav knew she was waiting and would try again at the next pause. I knew she was waiting and that Gustav was always ready to stop her. I tried to pay attention to what Gustav was saying, hoping he wouldn't pause, hoping that for once Marisa would not try to speak when he paused. If Gustav did let Marisa speak, she usually had forgotten what she was going to say. Satisfied, Gustav accused her of simply wanting to interrupt.

I tried to explain to Marisa that she mustn't make her father angry because it made me unhappy. She said she was sorry and didn't want me to be unhappy, but she was a small child and unable to stop herself.

It should have been easier for Gustav to understand what was happening, but he too could not stop himself from playing this destructive game. They didn't realize what they were doing to me. I remember Gustav's astonishment the first time I lost my temper with him. I never would have admitted that I was angry with Marisa, too.

I was shocked by my anger. The tension had been building up inside of me, but I thought I had my feelings under control. One day, when they started bickering, I couldn't stand it a moment longer. My anger suddenly boiled up, the heat rushed to my face, and I felt a constriction around my head like a metal band about to break. I jumped up from the table, screaming at Gustav, "Leave her alone! Why can't you leave her alone!"

Gustav stared at me, astonished. "What is the matter with you? There is nothing to get so upset about."

"Then why do you get angry with her?" I demanded.

"Why do you shout at her? She's only a little girl. You're the one who needs to understand."

"I do understand," he said in a reasonable tone. "I'm only telling her she mustn't interrupt. She has a bad habit of interrupting—"

"Can't you just ignore her? I can manage her. If you just wouldn't shout — if you'd just leave her alone!"

Gustav and Marisa never did stop. I was not able to make either one understand what they were doing to me, and what they were doing to each other. We were a family and really loved each other. Why were we fighting?

Perhaps I was overly sensitive. One time I was so angry that I snatched up a knife, although I dropped it immediately. I had simply grabbed the first thing at hand, to throw it at the wall. I couldn't hit Gustav so I had to hit something else. It made Gustav laugh. He never took my anger seriously. In the back of my mind, I was afraid that Gustav was right. Perhaps I was overreacting. He charmed me out of my temper and tears. Circumstances permitting, he took me off to bed. I responded willingly, not wanting anything to become between us. But, as the scenes happened again and again, I could not forget so easily.

Our life developed a routine. Gustav went to his office almost every day, and Marisa went to school. I taught myself Italian and then photography, spending many hours in the darkroom. I worked in the garden and went horseback riding. But I was still restless. I wanted to take some classes at the university.

"What do you want to study?" Gustav asked.

"I'm not sure, but to start with, I want to attend some lectures on art."

"You're not going to learn about art by going to lectures. You have to do it the way I did, in Paris, getting right into the milieu, rubbing shoulders with real artists. You can't do that here in Los Angeles."

"I don't want to be an artist. I just want to understand art better. You say I have a lot to learn —"

"Then find something to do in the daytime. Why go to lectures in the evening when I am at home?"

"It's only one night a week, Gustav."

"I should think you wouldn't want to go out in the evening when I am here."

I didn't go to the lectures.

9

Back to Islands

There were times that were wonderful. We traveled every year, taking the American spearfishing team that Gustav sponsored to the international competitions held in different countries. We kept a car in Italy, and after leaving Marisa with her grandmother, we drove to France, Spain, and Jugoslavia, as well as to the Alps where Gustav was born, down to Rome, across to Portofino on the Italian Riviera, and wherever the spearfishing championships were held.

The year the championship was at Malta, the small, rocky island located in the middle of the Mediterranean Sea, I didn't go at the same time as Gustav. There would not have been much sightseeing for me to do during the two weeks of preliminaries; designating the competition area to be reserved, and giving time for the divers from other countries to become familiar with the underwater terrain and the local fishing. I stayed in Italy with Marisa and Gustav's mother until the day before the competition.

Flying over Malta, I was amazed by the size and complexity of Grand Harbor, which made Malta such a strategic naval base since the year 1800, when the English drove out the French and made Malta a British Protectorate. A representative of our host spearfishing organization met me. He explained that Gustav was attending a meeting, and asked if I would like to make a tour before going to the hotel. I was delighted to have a guide who showed me the Arab and Spanish architecture from previous occupations, the Tarxien Temples from a far earlier age, and the arid countryside checkered with stone walls to hold the soil on what used to be a richly agricultural island. That was long before the crusaders, of the Order of St. John, were driven out of Jerusalem and Rhodes. They came to the island in the Middle Ages, and established the order of the Knights of Malta, who each paid to the Crown the annual fee of one Maltese falcon.

My guide took me to the hotel and Gustav arrived shortly after. He was pleased with the program of the championship hosts, who had provided tours and entertainment for the visiting teams and their supporters. One or two times Gustav mentioned a young woman who particularly impressed him, but I didn't pay much attention to what he was saying. We dressed and went out to dinner with our U.S. team. It was always fun to travel with the Americans who had never left the States, and were thrilled by everything they saw in Europe. When we returned to the hotel, Gustav picked up an envelope of photographs at the reception desk. "These are pictures of the arrival dinner for the sponsors," he said, opening the envelope as soon as we reached our room. "Here! This is the girl I was telling you about. Can you imagine, she was even trying to sneak away from her husband to come and see me!"

I looked at the picture. A small, thin girl with a large bun on the top of her head, and eyelashes as long as antennae. She was clinging to Gustav's arm.

"She looks like an ant. Is she a Maltese ant?" I asked.

Gustav was not amused and nor was I, but I didn't care much anymore.

We returned to Los Angeles. I tried to be a good wife and mother but I was becoming short-tempered and rebellious. I knew I was the one who should keep calm and I felt guilty for provoking Gustav. Letter writing still exasperated me, I paid little attention to him when he came home from the office, and did not respond to his lovemaking the way I used to. I still loved him but no longer adored him. He had changed and so had I. I had become a woman with a mind of my own, with interests and opinions that were beginning to conflict with Gustav's. He did not want a wife who challenged him in any way. I resented him for not appreciating a more knowledgeable wife. I became depressed and tried to talk to Gustav.

"What do you mean, something is wrong?" he said. "You have everything a woman could want. You have a beautiful home, you travel, you have a husband who loves you. Look how many husbands drink, gamble, or go off with other women. You are really spoiled, you know."

"Yes, I know, Gustav, you give me everything I want—"

"Then what is the matter? You're just not being realistic, Renée."

I began to realize that Gustav also was not satisfied

with his life now as an American businessman. As much as he loved the challenge of business, it was not compatible with his favorite image of himself. He had changed his image from carefree bohemian to sophisticated, though casual, conservatism. He inspired respect now, rather than awe. He was no longer the extraordinary adventurer. He liked to say I made him leave Haiti. It was a long time before I realized that this was his excuse for having left the exotic West Indies and finding himself in Los Angeles suburbia.

Gustav met Gerard Blitz at the 1958 International Spearfishing Competition. Gerard was also a sponsor of the championships, as he was the man who originated the concept of Club Mediterranee, which had become a nucleus for divers from all over Europe. Gerard conceived his idea after World War II, when he was put in charge of an organization for the men who had returned from the war or concentration camps to find they had no homes left and their families were lost. The men waited in the centers for the government to help find their families. It was a difficult and depressing time. Gerard organized sports and activities to keep the men occupied. His program was so successful that, later, he put together a vacation package based on the same idea; that is, all sports and activities as well as room and board, included in one price. The accommodations would be simple, practically camping, but the food would be excellent. Water sports were the main feature of Club Med activities, available to every guest, who was called a G.M. — *gentil membre* (kind member) of the Club. The concept was an immediate success. Within a few years there were several villages, as the vacation centers were called, on the coasts of different countries around the Mediterranean. The theme of the villages was Tahitian. All construction was in open, Polynesian style, and everyone lived in bikinis and *pareus*, the Tahitian sarong.

There was a small organization of Club Med in Tahiti, where every G.M. dreamed of going some day. Gerard wanted to build a village there, but the problem for European travelers was that Tahiti, in French Polynesia, was on the other side of the world. The Messagerie Maritime ships left Marseille for Tahiti once a month, a trip which took a month, and then a month to return. Not many people could afford the time and money to do that. When Gerard Blitz met Gustav, he wanted to go after the much more accessible market — the United States. Gustav, the most famous man in the diving world then, would be the ideal man to promote the Club Med in Tahiti.

Gerard invited us to visit two of the Club Mediterranee villages. We went to Corfu, a peaceful little Greek island dotted with ancient, gnarled olive trees and a wild, rocky shoreline, in the deep, indigo waters of the Ionian Sea. The other village was on Sardinia, in the western part of the Mediterranean Sea, where many of the spearfishing teams congregated that year. After the day's activities and gargantuan meals, everyone still had the energy to dance under the stars half the night. Gustav was quite convinced that the Club Med concept and Tahiti, according to Gerard's lyrical description, would be tremendously popular with the Americans.

Gerard invited us to go to Tahiti, as Gustav was now interested in investing in the proposed new village. Most importantly for me was, if we could spend several months of the year in Tahiti, perhaps we could find a balance between business and the carefree life we had lived the first years of our marriage.

To get to Tahiti in 1959, we had to fly across the Pacific to Fiji, northwest of French Polynesia, to meet the French TAI flight that came from Paris by way of the Orient. Our plane arrived three days before TAI, so our introduction to the South Pacific islands was in Fiji.

We landed before five o'clock in the morning, while it was still dark. I could see nothing beyond the lights of the landing strip, no sign of an airport or any building. We were really in the middle of nowhere, I thought excitedly, and Gustav's grin showed that he was as pleased as I. We gathered our things together and waited for the door to open, then stepped out into the tropical night, into the soft, warm and perfumed air. It was like an embrace; I had never known that air could be so tangible, so subtly provocative; I fell immediately under the spell of the South Pacific. At the edge of the airstrip there was what appeared to be a picket fence topped with flowers; I had to go and see. It was a row of leafless branches that had been simply poked into the ground, yet were covered with pink and creamy yellow flowers. The scent in the air was their perfume — they were frangipani.

I ran to catch up with Gustav as he crossed the greensward at the end of the landing strip, walking towards the amber lights of a long thatched building just becoming visible in the early dawn. Because there was no airport, the airline office and customs control were in the Hotel Mocambo, which was straight out of Somerset Maugham's *Rain*. Bamboo walls set off brightly cushioned rattan chairs; the ceiling was a canopy of *tapa* cloth, with intricate designs painted on the wide strips of bark fabric. Lamps made of bamboo fish traps shed a soft, warm glow that dimmed as the morning light grew brighter.

"Come on, Renée!" Gustav called me impatiently. He wanted to leave quickly for Korolevu, a small village along the coast where the coral reefs were known to be particularly beautiful. The sun was rising by the time we were on our way through sugar plantations and small villages made up of a few thatched huts in the shade of large poinciana trees that spread their brilliant red flowering branches over the village and across the road. The Fijian families were up early; men waiting by the road for the local bus, children playing outside the huts. They stopped to watch us go by, frizzy hair framing round, black faces that lit up with huge smiles when I waved. The villages were very clean, with grass growing luxuriantly between the houses and flowering bushes along the road. We followed the coast for some distance, then rounded a sharp bend to find ourselves in dense tropical vegetation of towering banyons, palms and tree ferns. We had arrived at Korolevu.

The hotel was made up of a main building and cottages, called *bures*, that were scattered along the edge of the white sand beach. The temperature had risen considerably but the thatched *bure* was cool. I showered and slipped into bed to catch up on my sleep.

I woke up to see a pretty Fijian girl, a large, red hibiscus tucked behind her ear, tiptoeing into the room with fresh towels and a basket full of flowers. She grinned as she excused herself for waking me, and was obviously eager to talk. She chattered while she decorated the *bure* by putting flowers along the window sills, around the lamps, and even in the toothbrush holder.

Her name was Mila, and she told me about her village and the happy way Fijians live. She was to be married soon — perhaps I could stay for the wedding? I thanked her for the invitation, but we would be leaving soon, so I asked if I could visit her family the next day. Mila was delighted, saying she would come to take me to her home.

I changed into a bikini and went out to find Gustav. He had lost no time getting into the water and was far out by the reef. I explored the fringe coral in the shallow water along the shore, surprised to discover fish very similar to the ones in Haiti, but there were also varieties I had seen only in books. The most beautiful were electric blue, glowing like tiny neon lights. Sea worms, their feathery tentacles spread, looked like full-blown dandelions. Shells clung tenaciously to the rocks but I managed to collect a few. Gustav came back from the reef, exuberant over the rich coral banks and the crystal clear depths. He was so happy to be back in tropical waters that he was out from morning to dark.

The next day I went with Mila to her village. Mila's mother greeted me with courteous dignity, taking me into her one-room, sand-floored *bure*, meticulously clean and orderly. Children crowded in the doorway to watch while Mila served us tea and English biscuits. *Kava*, a mild narcotic, was served in a ceremony reserved only for men, fortunately, because I knew how the beverage was made. The women chewed the *Kava* roots to a pulp, then spit the murky grey juice into a bowl where it was left to ferment a little to enhance the flavor.

Although the English established excellent schools, and the Fijian Islands were a fine example of colonial development, the Fijian customs and community ties were still very strong. A village formed a cooperative unit; for example, when a man needed to build a new house, everyone participated. I was invited to inspect the community garden. When I admired the vegetables, the children promptly picked the ripest eggplant and cucumbers. Mila quickly wove a basket out of a green palm frond and they filled it with the vegetables, mangos and bananas for me to take back to the hotel.

I was sorry to leave Fiji. It seemed that no place could be more beautiful or have more charming people. But the legend is that Tahiti is the terrestrial paradise, and we would soon find out for ourselves.

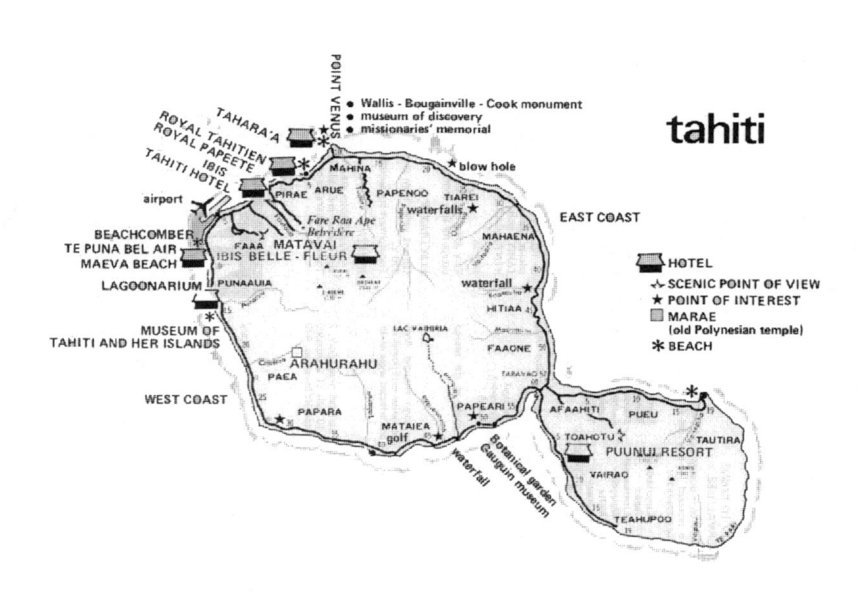

10

Tahiti

The old Bermuda sea plane landed in the lagoon, sliding in like a pelican, the spray curtaining the windows so that it seemed as though we had gone underwater. As the windows cleared, we saw that we were in a bay, originally the crater of a volcanic mountain, of which all of these islands were formed. There was no flat land for an airport near Papeete (Pah pay ay' tay), the only town; we had changed planes at Bora Bora, where there was the landing strip built by the Americans in the Second World War. We embarked in a small boat that took the passengers to shore. Coconut palms, frangipani, and brilliant red poinciana trees lined the coast.

"Look, there's Gerard and Claudine," Gustav said.

"Where? I can't see them. There are so many people!" All I could see was a patchwork of bright dresses and shirts, palm-woven hats with wide shell bands and flower crowns.

"*Ia Orana*!" we heard as we approached the quay. "That's hello, or welcome — pronounced Ee ah Oh rah' nah," I told Gustav, delighted to hear our first Polynesian word. I had read several books on Tahiti, and one had a list of words and phrases.

We were stepping on shore before I recognized anyone, and then Claudine and Gerard were placing flower leis around my neck and kissing me on both cheeks, the way the Tahitians always do.

"*Ia Orana* ! *Bienvenue* ! Welcome to Tahiti !" Other people were putting leis on our shoulders and kissing us, as Claudine introduced friends and Club Med G.M.s and G.O.s, the Club personnel. The arrival of the sea plane was the highlight of the week, and the "airport," a converted hangar, was crowded with people. Tahitians were playing guitars and singing at the bar, and we saw several pretty girls, pushed forward by their companions, dancing to the exotic rhythms of the Tahitian songs.

"They're bringing the baggage in now," Gerard said. This might have been the only place in the world where one could greet friends right off the plane, and then pass through immigration and customs. "It will take only a few minutes, then we'll go to Punaauia, where the Club Med is."

Gerard and Claudine, his attractive wife who had lived in Tahiti for several years and was responsible for the Tahitian concept of Club Med, drove us to the Hotel Rivnac.

"Rivnak?" Gustav asked. "That sounds like central Europe."

"Right," Gerard said. "Rivnak was a Czechoslovakian who lived here about 1890, on. He opened this little hotel, and had a *vahine*, a young girl, probably about fifteen, who lived with him. When he died, — Claudine, what is her real name, his *vahine*?"

"I don't know. She has always been called "La Matair, mother, as long as I have known her. She must be over sixty now. She has run the hotel ever since Rivnak died, with just two or three people to help."

"Do you know the story about the paintings?" Gerard asked.

"Rivnak was an artist?" I knew that Tahiti had always attracted painters.

"Not that I know of. I mean the Gauguin paintings."

"What! There are still some Gauguin paintings here?" Gustav exclaimed. "I thought that the last one was the glass window in a door that Somerset Maugham found at an old house — out in the country — that the man didn't want to sell until Maugham bought him a wood door, which he preferred."

"It was what everyone thought," Gerard said. "You know Gauguin paid some of his debts with paintings, or maybe he just gave these to his friend, Rivnak. La Matair had no idea of their value or perhaps she just didn't like them, and she had put them in a store room. The story is, a guest of the hotel happened to see them and immediately recognized they were Gauguins. He told La Matair he could sell them and get a lot of money for her. La Matair agreed to let him take the paintings to France — and it was the last she ever heard of them."

"That is typically Tahitian," Claudine said. "Always trusting, not having any idea of the value of anything, and not much caring. The family, a house, fishing, enough for everyone to eat, music and dancing; it is all that is necessary to live in Tahiti."

We arrived at the Hotel Rivnak, and La Matair, a

buxom, kindly woman, welcomed us. The hotel was at the edge of the lagoon, on a large property with tall mango trees, grapefruit, and papaya, with the fruit growing right out of the trunk of the tree. There were *bananiers*, heavy with ripe bananas, and flowering bushes everywhere. The main building was simply a vast, oval, thatched roof supported by coconut logs, with a low wall of woven bamboo that enclosed the dining area and lounge. The windows were sections of woven bamboo that were propped up or let down, depending upon the weather. There were no interior walls, no offices; the kitchen was separate, reached by a covered walkway. The openness gave a panoramic view of the ocean, beach, and gardens.

The Club Med organization in Tahiti occupied ten of the twenty cottages, called *fares* (pronounced far'ay) of the hotel. The guests' *fares* were built in the same style as the main building. I was enchanted by the simplicity. How wonderful it must be to live in a climate where the houses are always open to the cool breezes from the ocean and the perfume of tropical flowers! I changed into a bikini and *pareu*, putting my American clothes away for the duration of our stay. It was great not to have to think about what to wear; so easy to simply choose a favorite color of the brightly patterned Tahitian cottons.

It was Saturday, and Saturday night was famous for the fun and dancing in Papeete. After a dinner of delicious lagoon fish grilled on a bed of coral coals, with breadfruit that tasted like mild sweet potato, and a desert of fried bananas smothered in coconut cream, we drove to Papeete's favorite rendezvous for coffee and people watching. Vaima's Bar had tables scattered out across the sidewalk so that people passing through wove their way between the tables, stopping often to greet friends. Nobody was ever in a hurry. There were few cars, few offices, and few fancy shops in this town where life was so simple and carefree.

Gustav and Philippe, a spearfishing enthusiast we had met at the hotel and who joined us for the evening fun, were entranced by the Polynesian girls, and I couldn't blame them, as many of them were really beautiful; even the plainer ones had a natural grace and charm, enhanced by their shiny, wavy hair falling to their waists. They often wore a crown of the *tiare* (tea' ah ray) a star-shaped, perfumed, single gardenia, and even the men who, I noticed were also very attractive, always wore a partly opened *tiare* over one ear. The present generations of Tahitians are a mixture of three races, the original Polynesian mixing first with the English, French, and other European adventurers, whalers and merchants who came to the islands in the 18th century, and then the Chinese who were brought to work on the cotton plantation, when there was the embargo that prevented cotton from being shipped from the southern states to Europe during the American Civil War. This mixture created striking features; the large, soft Polynesian eyes enhanced by the Asian tilt, and the tendency of the Polynesians to become heavy was counteracted by the slimness of the Chinese. The young men were broad-shouldered and narrow-hipped. They wore crisply clean clothes and their thick, black hair was glossy and neatly combed.

Within an hour Philippe had a *vahine* (pronounced vah' he nay) or rather, she claimed him. She was a small girl, more Chinese than Tahitian, vivacious, shrewd and funny. Tina introduced us to several of her friends, who all went with us to Quinn's, the most famous night club in the South Pacific. It was simply a barn-sized room with a long bar, a large dance floor and a great dance band. Although a girl came out once in a while to do a solo, the main show was the customers, who jumped up and rushed onto the floor the moment the music started. They all seemed to be excellent dancers. The girls at our table, Gustav and I, were soon in demand, particularly for the *tamure* (tah'moo ray), the wild, gyrating dance that is almost impossible for a non-Tahitian man to do. There is always enthusiastic encouragement and laughter for good sports who try to keep up with their partners. I was astonished when a Tahitian caught my hand to go out and dance the *tamure* with him.

"I can't do that!"

"Yes, you can," Tina cried. "Come on, I'll show you. Look — stand with your knees a little bent. Now move one knee forward, and then the other, and let your hips move forward in turn. That's all there is to it!"

I could see that it was a lot more complicated but, all right, I could do that much, and I kept up with my partner at first. I gave up when the tempo accelerated, and the Tahitian girls' hips gyrated in provocative circles and breaks, timed perfectly to the staccato rhythm of the drums.

Tina was keeping a close watch on Philippe, but all the girls in our party wanted to dance with Gustav. "Do you think any of them are prostitutes?" Philippe asked Gustav in a low voice, in English, not wanting to offend any of the artless, laughing girls, all dressed in simple, brightly-colored dresses, wearing the wide crowns of fresh flowers on their clean-smelling, long black hair. The girls jumped up as eagerly to dance with their Tahitian friends as they did with the few white men. The complete lack of coquetry or guile must have been enchantingly intriguing for men. The girls were just as sincerely friendly with me, chattering with me as though I were one of them, although I heard later that it was great fun among Tahitian girls to entice a man away from his wife. It was risky for a marriage to go to Tahiti on a honeymoon. Gustav was enchanted by all the attention, and I was delighted to be dancing with so many handsome, admiring men. It was the last day, when Gustav left me at the airline counter while he parked the car, that one of the girls followed him and said, "If you should come back again, without your wife, I will stay with you." Gustav thanked her gravely and said he would remember.

Sunday morning was the big market day, so we were up early to go into town. Papeete was small, with narrow streets, and Chinese-owned grocery and hardware shops surrounded the marketplace, their family living quarters overhanging the sidewalks. The Chinese had left the cotton plantation when their indenture time was up, and opened up stores that made them the principal merchants in French Polynesia. The stores usually had two or three tables for serving early breakfast to the market vendors. We ate a delicious meal of marinated fish, crisp French bread, and local coffee flavored with vanilla, while watching the colorful local buses, converted trucks, called *le truck*, which had been arriving since 5:00 a.m. Some people left their homes at the far end of the Presqu'île, the smaller part of Tahiti, at 3:00 a.m., to bring produce, fish, chickens and pigs to market. *Le truck* was the transportation system for most of the population, and they arrived, now crowded with passengers who came to purchase food to be cooked in the *hima'a* (he' ma a) the ground oven used to prepare their Sunday meal. In the market were huge stalks of yellow bananas, red *fe'i*, the indigenous banana that must be cooked, large beet-like taro, breadfruit, and manioc roots. Sweet, green-skinned grapefruit, spiny red lychees, and other exotic fruits I had never seen, were heaped on counters down the center. There were neat packages of taro leaves containing *maoa*, the shellfish found on the reef, mixed with grated coconut meat. There were the fish; tremendous tuna and 60 pound bonito, and coral fish of such brilliant reds, blues and greens that it was hard to believe they were to be eaten. Such beautiful creatures should be in aquariums, it seemed, but they were very common fish here. There were also squid and octopus, delicious when prepared with onions in coconut cream. Another quarter of the market was for the Chinese farmers who grew lettuce, tomatoes and other vegetables introduced to the islands.

We took advantage of the trip to make Tahiti the theme of our new publicity photos of Cressi spearfishing equipment. We drove around the island, along the west shore of white coral sand beaches and blue-green lagoons, to the east coast, the windward side, with waves crashing high upon volcanic rocks jutting out into the ocean. The road was cut into the side of the cliffs as it went from one valley to the next. Around every curve there was another spectacular view of mountain peaks against the intense blue sky, waterfalls, and rivers flowing through fields of taro, papaya, and banana plantations. We came upon a black, volcanic sand beach, sparkling like mica in the sun. It would make a dramatic and unique background for the publicity shots of the Cressi spearfishing equipment.

We needed two or three people; the girls were easy to find but we wanted a particularly good-looking Tahitian man for the main subject. Claudine Blitz said, "Take Teva. He's probably the best-looking man on the island." He was the guide for the Club Med members, whom we had met the first day. Yes, he would be perfect, but he didn't seem pleased when Claudine asked him. I guess he agreed to do it only because he worked for the Club. We all went over to the black sand beach, and Teva turned out to be a good enough actor, coming up the beach with the Cressi gear and a large fish, while the girls posed gracefully in attitudes of delight over the big catch.

Gustav wanted pictures of Teva spearing fish out on the reef, a classic Polynesian scene, so the next morning the three of us went out across the lagoon in a *pirogue*, an outrigger canoe. Gustav was in a "director" mood and gave me specific instructions about how to take the photos. I was limited to the width of the reef, having difficulty keeping the camera squared, even though Gustav was holding me as the waves broke across the coral wall. I knew I had several good shots of Teva's muscular body at the moment he raised his arm, throwing the long spear into a crashing wave. But Gustav wasn't sure that I had gotten what he wanted, and I became exasperated and embarrassed by his pedantic manner in front of Teva. We finally got into the *pirogue* and were going back across the lagoon, when Gustav turned around awkwardly and the narrow canoe tipped over. I managed to hold the camera above the surface as I went into the water. Teva righted the canoe, emptying it by the simple method of pushing it back and forth, so that the water swept out over the sloped ends. The men retrieved the fishing gear and paddles. Gustav was furious with his own clumsiness, and Teva's gaze crossed mine. Neither of us said a word, but it was obvious he felt sorry for me. We didn't see him again before we left Tahiti.

Back in Los Angeles, I seemed to have become a person divided in two. Cooking and housework was performed by one part of me, while the other part was still in the islands. I redecorated one room of the house with Tahitian print curtains, shell leis, and woven bamboo fans. Having taken many pictures when we made the trip around the island I enlarged the best ones and covered the walls with scenes of Tahiti's green mountains and valleys. The garden became a jungle of tropical plants, bamboo, and ferns. I taught Marisa how to dance the *tamure*, and was content because we were to go back to Tahiti again in six months.

Gerard and Claudine came to Los Angeles in June. We were to leave for Tahiti in July, when the project to turn the Hotel Rivnak into a Club Mediterranee village would be finalized and under construction, to be opened as soon as possible. But then Gustav became reluctant about leaving his company in the middle of the summer season. Gerard was also uneasy about being away from the European Club Med villages at that time. Claudine and I protested.

"This will be the last really Tahitian Fourteenth of July! They are going to build the airport next year, and then there will be tourism. Everything will change!"

"Bastille Day celebrations haven't changed since July 14, 1789, and tourists won't keep the Tahitians from celebrating any more than they do in France —"

"It won't be the same. They'll start using plastic grass skirts, and fiberglass canoes, and florescent painted shells, and—"

"All right, all right!" Gerard and Gustav finally agreed to let us go for the Fourteenth of July festivities. They would join us later.

Marisa had already left for Italy to spend the summer as usual with her adoring grandmother. Claudine and I left for Tahiti.

11

Return to Tahiti

By then there was a direct flight from Honolulu to Bora Bora, and the flying boat carried us again to the perfumed shore of Tahiti. We drove to the house in Punaauia, purchased along with the property for the new Club Med village. It was a large Tahitian *fare* built of the coconut log supporting posts, woven bamboo walls and thatched roof. It was situated at the edge of the beach and shaded by large *purau* (poo'rau) trees that sprinkled the sand with their lovely yellow and mauve flowers. We were delirious with delight to be back in Tahiti, caressed by the soft, warm air and the gentle breezes from the ocean. We lay in the shallow water of the lagoon and absorbed the peacefulness, hearing only the murmur of the waves lapping the sand. Eventually we would drive around the island, when we felt like it, but that was as much as we planned. We were gathering strength, we said, to be able to spend the several days and nights of the Fourteenth of July celebration in Papeete.

The waterfront street was closed off to cars and most of the shops and offices were closed for four days. The nightclubs closed, to open marvelously decorated bars and dance floors in the *baraques* on the waterfront. All of the construction of these temporary buildings was hidden under walls of woven palm fronds, brightly colored appliquéd quilts that the Tahitian women had learned to make from the missionaries, and of course, flowers everywhere. Papeete was overflowing with people from the districts (the country) and the other islands. Schools and even the town hall became living quarters, each family occupying the space of one or two *peues* (pay'oo-ay, mats) where they slept comfortably with pillows and *tifefes* (tea'fay-fay, quilts). Before the competitions we watched the groups practice the dances in the school courtyards. We heard the *toeres* (toe-ay'ray, log drums), guitars and songs of the competing groups as we passed from one part of town to another. Everything was ready by the evening of the thirteenth of July when there was the Governor's Ball, the first event of the festivities. We were guests of the governor and arrived at eight o'clock, but at midnight the gates were opened to everyone. We danced until dawn with every man who asked us; even the governor's wife danced with her gardener, impeccably dressed in a crisp, white shirt and a *tiare* flower over his ear.

We went to see the parade of all the dance groups in their costumes of finely-combed *tapa* (stripped bark of the *purau* tree), the canoe racers, and the competitive fruit carriers, their shoulders bulging under the poles weighed down with stalks of bananas, breadfruit, pineapples and oranges. During the next four days and nights we attended all of the dance competitions, the spear throwing — at a coconut perched high on a long pole — and the copra competitions. Copra is the dried meat of the coconut and the main source of income for the islanders. The coconuts are gathered, chopped open with an ax, and the meat is scooped out with a curved knife. The dexterity and speed of the competitors was amazing. The *pirogue* races were the most important event of the celebration. There were even women teams and they churned the water white as they raced across the bay.

At the end of the festivities the *baraques* were taken down, the streets swept clean, and Papeete recovered its small town atmosphere. We met at Vaima's restaurant to drink rich Tahitian coffee and watch the bonito fishing fleet come in to the landing across the street. There was a wagon with barrels of ice keeping coconuts well chilled; we drank the sweet and slightly effervescent coconut "milk," which was really a clear, refreshing juice. A rooster posed elegantly on top of the wagon roof and crowed to other roosters in the town. There were few cars, no traffic lights, and people drove slowly, to wave to friends or stop to chat a moment.

One night Claudine and I went dancing. It was preferable to go without men because then we were sure to have many partners. We stopped first at Quinn's. To our surprise a policeman came up and told us we couldn't go in.

"Why not?" Claudine demanded.

"Excuse me, Madame," he said. "but it is because you are wearing pants."

Claudine's slacks were the latest, smartest style in Paris. At our amazed expressions he went on to explain. "It is the law now, because the government is trying to discourage the 'Quinn's girls'." Perhaps they were prostitutes but they were not at all like the one well-known French woman with her *baise en main*. If they were prostitutes, the Tahitians and the local *popaa*, non-Tahitians, were very tolerant and amused by them. The policeman said, "The 'Quinn's girls' have started wearing American jeans, so we have been told to keep out all women who are wearing pants."

We retreated gracefully and went to Bar Lea, a smaller place, also becoming famous for its excellent dance band. We were soon besieged with partners. One of my partners was Teva.

Nothing would have happened if Gustav had not written to say he could not come to Tahiti. Gustav was putting off the Club Med project. He wrote that it would probably be another year before he and Gerard would be free to work on the plans for the village. I realized it would be difficult for him to work with Gerard because they were too much alike. Each was president of his company and Club Med was Gerard's company. But I saw this as our only opportunity to return to our island way of life. Gustav had said he wanted to as much as I.

Or was I mistaken that Gustav really wanted to get away from the States? He enjoyed the challenge of business, and could not leave it for months at a time if he was going to attain his goal of being top in the market. But he still regretted leaving Haiti. Or did he really?

I begged him to come to Tahiti. He answered that he could not and that I was to return to Los Angeles. I told him then that I did not want to go back.

Gustav called Gerard in Paris and told him he wanted to break up the partnership. Gerard was disappointed and angry. When Gustav cabled me he had pulled out of the Club Med project, I knew if I went back I would never see Tahiti again. I decided then to ask for a divorce.

Gustav had said sometimes, after a fight, I would divorce him if I had the courage to live without him. It was true. I had never lived alone, never worked, and could not imagine living without a husband. But I realized that in Tahiti I could live without Gustav. I returned to Los Angeles to get Marisa, who was arriving from Italy. It didn't occur to me that Gustav would not let me have her.

Gustav said he would not try to keep me against my will. He was sure that I would change my mind, that I still loved him, and if he let me go now I would return to him. He decided to keep Marisa until "I found a house and knew what I wanted to do" — or, he did not say, until I returned. He was confident that if he kept Marisa I would return.

I went back to Tahiti alone. Claudine had left for Paris. Because Gustav had broken the partnership with Gerard the Club Med house was no longer available. I rented a *fare* at Fisherman's Point, and a dear old Tahitian woman came to live with me. I thought about what I was going to do to stay in Tahiti. I wanted to work, to be independent, to show Gustav I could take care of myself. I would not go back to him because I couldn't live alone.

One day I took a *pirogue* and paddled out across the lagoon. I was all right in the open water, avoiding the coral heads easily visible in the crystal clear water, but I didn't dare go near the reef for fear of tipping myself out of the canoe. My housekeeper said she would have someone take me out to the reef where the shells were. There were *ma'oa* and *benitiers*, like oysters, as well as beautiful and rare shells I wanted to collect. Somehow it did not surprise me that it was Teva who came to take me out to the reef. This time he did not mind.

Teva was adopted, a practice common in Tahiti, usually because the young mother does not want to take on the responsibility of a baby yet, and gives it to her parents, her cousin, or someone else to raise. Teva's adoption was unusual; he was promised to his adopting mother, Teriivahine, before he was conceived. She could not have children so when her cousin was pregnant with her second child, Teriivahine asked if she could have the child. Her cousin agreed, but when the child was born she changed her mind. She said, "Be patient a while longer and I shall make you another one." which she did.

Teva was adored by his adoptive parents. They eventually adopted him legally, which is unusual in Tahiti. It is not important because families are always very close, the child knows who his real parents are and he is shared by everyone. He even inherits along with the other children of his adoptive parents, as well as his real parents.

There were other children raised by Teriivahine and Teriitane but Teva was the *enfant gaté*, the spoiled one. In Tahiti the other children in a family usually spoil the favorite as much as the parents. It can be the oldest or youngest or the only boy or girl. There is rarely sibling rivalry in Tahiti.

Teva's father, Teriitane (Teri i tah'nay), was quite well off for a Tahitian. He had a *magasin*, a general store, and that was unusual because it was the shrewd Chinese who had almost a total monopoly on commerce throughout the islands. Teriivahine and her adopted daughters ran the *magasin* and Teriitane obtained a license to sell beer and wine. All the men in the surrounding district met after work to drink beer out under the *purau* trees. They met there weekends to play *petanques*, the European bowling game, and the young people played volleyball. The *magasin* was small but it prospered well enough. Teriitane was the first Tahitian in the district to have a car and Teva was the first child to have a bicycle. He went to the Protestant school in Papeete and played football (soccer), first on the district team and then on the Tahitian team competing with other countries in the South Pacific. He went to the Tuamotus, the coral islands, to dive for pearl shell, and learned to play the guitar in the unique way of the Tuamotu islanders.

His first real job was with the Institut de Recherches Medicales. He worked in the districts, taking blood tests and administrating medication for filariose, or elephantiasis. This is an illness that is now under control and everyone simply takes preventative medication every six months. The treatment to stop the initial progress of the illness was simple, but Teva had to verify that the patients actually swallowed the pills. If they were already infected the medication caused malaria-like chills and fever, which the Tahitians feared more than the imperceptibly developing filariose.

Teva was with Lillian then but they had separated long before I met him. Their two girls, Noella and Marceline, were being raised by Teva's family. The girls often stayed with Lillian, who lived near by, and Teva and she remained friends. Teva brought the girls to my house at the beach. They were seven and four years old, with dark Tahitian eyes and glossy black hair. They looked at me shyly and then ran down to the water to play. They made an aquarium by digging a hole in the sand and letting it fill with water. Then they found pieces of coral that were home to tiny, jewel-like fish. When the coral was taken out of the water, the fish were caught in the branches and carried to the pool where, back in their element, they swam out again. I prepared lunch for the children and we became good friends.

Teva was different from other Tahitians because he was a dreamer. Tahitians lived from day to day as in those days they did not need to concern themselves about the future. There were schools in every district and every island, medical care was free for children and anyone who went to the government clinics and hospital. If a man didn't want to work at a job he went fishing and planted taro and other foods, or collected copra. When his parents died he would inherit the land where he had probably already built his *fare*.

In the Tahitian language there is little vocabulary for intangible things. Tahitians seemed to have no philosophy of life; there was nothing to question. Ideas, dreams, were foolish, a waste of time to think about, even dangerous to talk about. Few Tahitians were like Teva, who wanted to make something of himself. He wanted to travel and discover what existed beyond the ocean. Teva had worked two years for Club Med and the Club members told him about the way they lived in other countries. He agreed that Tahiti was a beautiful place to live, but the Hotel Rivnak had closed in preparation for the construction of the Club Med village, and Teva's job was reduced to supervising the workers clearing the property.

Teva came to my *fare* every day at noon and we went swimming in the clear, cool lagoon. He came in the evening and we went dancing. His crisp, black wavy hair scented by the *tiare* flower over one ear, his slightly almond-shaped eyes, high cheek bones and dark satiny skin were irresistible.

I wrote to Gustav and asked him again for Marisa. He suggested that I return to California to spend the Christmas holidays with them. I was surprised to realize I wanted to go for Christmas, but a couple of days later another letter arrived. Gustav said he would send Marisa to me for the holidays. I had not yet written to say I would be glad to go back to California, and I was too proud then to say so.

Marisa came to Tahiti. She arrived at the airport that had been completed shortly before. The flight arrived earlier than expected and I didn't know it until I heard the plane. I raced the car to the airport, to find her waiting for me.

She was patiently sitting on a counter, her blond hair in pigtails and her bright blue eyes watching the colorful crowd. I hugged her and exclaimed, *"Tu n'avais pas peur?"*

Her eyes grew big and she wailed, "Mother! Don't you speak English any more?" I didn't realize I was speaking French, as I hadn't spoken English for several months, and only then was she afraid. I assured her I had not changed at all, and took her home to our house on the beach. Marisa was enchanted by Tahiti.

Christmas in Tahiti was not celebrated like in the States, so we decided to have a *tamaa'raa*, a typical Tahitian feast. It was the rainy season and it started to pour for days at a time, so Teva and his friends built a thatch roof from the *fare* to the garage, to protect the tables from the rain. I had no idea what a Tahitian feast entailed, and Teva said I had nothing to do, everything would be prepared on Christmas day. The night before, though, several women came and made flower crowns, that Marisa was delighted to learn how to make too. I thought they were making a lot but I had no idea how many people would be coming.

The next morning the men dug a pit in the garden, lined it with stones and built a roaring fire in it. When the stones were red hot a large, flat basket containing a pig, breadfruit, bananas and vegetables wrapped in leaves were placed in this *himaa*, the ground oven, which was then covered with *purau* leaves and earth to seal it. The food would take about three hours to cook. Fresh banana leaves were placed on the tables because there would be no plates. Halves of coconut shells served as bowls to hold the slightly acidulated and salted coconut cream in which to dip the food. Flowers were hung from the ridgepole and tucked into the palm fronds woven around every post. When all was prepared and the food was cooking in the ground, the Tahitians went to church.

When they returned, about fifty of them, the drinking and feasting began. The food was delicious, finishing with a desert called *po'e* (po'ay) that had nothing to do with poi, the Hawaiian gluey taro. *Po'e* is made of manioc, as is tapioca, mixed with fruit, and served with coconut cream. Marisa enjoyed it all as much as I did, and the music and dancing enchanted her. The party lasted far into the night; we crawled into bed long before it was over.

The next day the party continued as the tables and garden were cleared, the left-overs finished for lunch, and more beer and music ended the festivities. It was considered such a success, they decided to take advantage of the roof and repeat the program for New Year's. I didn't say no but I should have. I was so tired of Tahitian partying by the end of the holidays I hoped never to attend another feast.

Marisa and I talked about why I could not go back to Los Angeles, but how could I explain it to her? She insisted that Gustav had changed; he was kind, never cross with her now. I had to tell her that although he always loved her, if I returned he would become the same way he was before and make us unhappy again. It had happened when I tried to leave once before, staying in New York for several weeks. He would never change.

Marisa was eight years old, a very intelligent little girl with a lively imagination. Although Teva's oldest daughter was only a year younger, the girls found little to do together. Of course there was the language barrier, but it was more than the lack of communication. It was obvious that Marisa, even at that age, was too intelligent a child to be raised in such a limited environment. Her determined attachment to me was perhaps intensified because no one else interested her. I realized that she had to go to school in the States or Italy, but I could have her the best times, during vacations. Fortunately, Gustav already had a girlfriend, at first just a friend who had helped him take care of Marisa, and she and Marisa liked each other very much. I felt much better about her going back.

But when Marisa left I no longer wanted to stay in Tahiti. I had to get away for a while and decide what I should do. I missed good books, music, and going to the theater and movies. I could not return to California and be a lonely divorcée in that big country, even with Marisa. I had not been able to make any real friends when I was with Gustav and my family was scattered around the world. Perhaps Honolulu was a good compromise.

I left Tahiti as suddenly as I had left California.

I was pregnant. I had let it happen because I wanted another child. Did I "let it happen" so that I could feel I was not responsible? Or was it a way of burning bridges, to make sure that I wouldn't go back to Gustav? I did not love Teva as I had loved Gustav, but I loved living with him. Uncomplicated and calm, he was everything Gustav was not. I knew I still loved Gustav, but remembering the scenes and fighting over Marisa, the tension and screaming frustration that would well up in me, and being confined to the role of devoted wife and housekeeper, I knew I could never go back.

I was not in Hawaii for long before I missed Teva. He came as soon as he received my letter. He stepped off the plane wearing a finely woven Tahitian hat, the style of the true *tane* (tah'nay — man, or husband). He was very handsome.

Teva soon knew all of the Tahitians who lived in Honolulu. Most of them were entertainers, dancers and musicians in the several Polynesian shows. We often went to Queen's Surf, the beach where the Tahitians and their friends met every day. They played Hearts and Scrabble in English. I was astonished by how quickly the Tahitians learned languages. Teva studied with me seriously and learned correct English, not the Hawaiian "Hey, brudda" manner that was popular even with mainlanders who pretended they were Hawaiian born. Teva was asked to play the guitar and sing whenever we went to the Polynesian shows. The Tahitians were always asked to parties, where they sang and enjoyed themselves the way they did at home, and Teva became a favorite of the Hawaiians and *haoles* (Caucasians). I was not pleased because there was always too much drinking. People seemed to go crazy in Hawaii. Teva did not drink much, but in Hawaii he was staying out later and later, and I didn't want to object because he was often paid to sing. I didn't drink at all and the parties soon palled on me.

I went to Los Angeles and spent spring vacation with Marisa. She didn't realize I was pregnant; I didn't want her to think that the baby was taking her place and would take my love away from her. Marisa and I lived at the house while Gustav stayed somewhere else, and came to visit us often. It was true that he had changed, at least for that time. He was considerate and careful to avoid any conflict. He knew if he tried to talk to me about staying I would leave the house. He really was as charming and amusing as he had been when I first knew him. But his manner was forced now and I recognized the tactic. He charmed prospective clients and useful acquaintances in the same way. I knew, though, he still loved me and did not know what to do.

That year there was the great fire that burned a large part of Bel Aire, and it came close to Westwood and our house. Gustav and I climbed on the roof to watch, the wind bringing hot blasts of air. The planes dropped tons of water on the fires but it seemed to have little effect. We could see each time large trees or a house burst into flames. I stood near Gustav. He was wearing a tweed jacket that was very familiar to me. I wanted to put my hand on his sleeve, to slip my hand in his. I turned away. I didn't know what I wanted to tell him. I didn't know how he would react, or what it could change. Too much had happened; it was too late. I returned to Hawaii. Marisa would spend the summer with her grandmother in Italy as usual.

There was a lot of jealousy amongst the entertainers in Hawaii. Somebody reported to Immigration that Teva was living with a woman still married to another man. Such puritan regulations were no doubt on the books since missionary days and ignored if no one made an issue of it. An Immigration man came to our apartment. I told him my divorce was pending and that Teva and I intended to marry as soon as possible. The man was nice about it. He said that Teva could stay until the birth of the baby and then he would have to leave.

Four months later our baby was born, and it was another girl. Teva, with two daughters already, and I, who had had two daughters, were disappointed. Besides, she was a scrawny baby, not pretty at all. I felt so sorry for her that I gave her my name as well as the names of both grandmothers, who had insisted upon it if I had a girl. Mother had been given an Hawaiian name by an admirer when she went through the islands on her world expedition and she wanted Leilani to have the same. The tiny girl was baptized Renée Tetuanui Leilani O Ke Aloha Nui. She was an easy baby to care for and it didn't take long for her to capture our hearts.

Two weeks later Teva left for Tahiti and took Leilani with him. I had to go back to California to have her birth certificate corrected, to remove Gustav's name so that Teva's name could replace it. I flew to Los Angeles and Gustav again left the house to Marisa and me. Gustav was away from Los Angeles when I arrived and he did not come back for several days. When he came to the house the evening he returned, he was angry and aggressive. I knew he had been with his girlfriend; Marisa said she was practically living with them. After the company of an adoring woman, he must have found my conduct intolerable. But he wanted me to stay. He insisted I should stay a while for Marisa. Lord knows, I felt so guilty about Marisa that I was ready to try anything, to find a solution to have more time with her.

Gustav felt I should get used to living in California again, keeping house and taking care of Marisa. He came over often "to give us a chance to get to know each other again." He certainly chose the wrong tactic to win me back. He should have swept me off on a fantastic trip somewhere. By putting me back in the same role I had run away from, he was punishing me for leaving. If I stayed it was obvious that I had been wrong. With Gustav somebody always had to be wrong. I said nothing because at least Gustav was making the effort to be patient with Marisa, who unfortunately started her old habit of interrupting at meals. Gustav often came for dinner and he monopolized the conversation, as always. For Thanksgiving I prepared a traditional feast and we were at table longer than usual. The conflict between Gustav and Marisa finally erupted, just as though I had never left. The pressure in my head reached bursting point and I wanted to scream at Gustav just as I used to. Nothing had changed.

I left for Hawaii a few days later. Marisa came to spend Christmas with me. She did not ask any more why I could not stay with them. We knew we could have holidays together and we would write and stay close to each other.

On New Year's Eve we drove to the top of the mountain behind Honolulu and watched how the Chinese and

Japanese celebrated by setting off thousands of fireworks on the stroke of midnight. We saw the bright lights of homes all over the valley until the bells began to ring. Ten minutes later a great cloud of smoke, accompanied by a tremendous din, covered the valley completely, and we could see only the highest rockets bursting through the cloud.

12

A Tahitian Wedding

The divorce from Gustav was final at last, and I returned to Tahiti to marry Teva. We went to see the *tavana*, (tah'vah nah — chief, or mayor of the district) who would marry us. Because I had already stayed several months in Tahiti the residence requirement was waived. We wanted only the simplest ceremony; we had been living together for almost two years and had a child. In Tahiti it was of no importance. About half the population never married. They had children, changed partners once or twice, and if they did marry it was to please the church, or sometimes after many years when they were certain they wouldn't change partners again. But I was not Tahitian and I wanted Leilani's birth certificate to be legitimate. Teva's parents asked us to have a religious ceremony as well as the civil one. They wanted to have a party and we agreed to let them invite a few close members of the family.

The civil ceremony was to be at the school, which doubled as the town hall. Two friends were to be witnesses and we were all to meet at the school at five o'clock. The wedding had to be late because Teva, captain of the football team, played in a match that afternoon. I dressed, turned Leilani over to her grandparents, and waited for Teva. He was to come home by four-thirty, shower and dress, and we would go to the school. He came at five-thirty. The *tavana* was unable to wait so he had delegated his adjoint to replace him. It was Saturday and the adjoint had already been drinking beer with friends in the coconut grove. He had been given the tricolor belt; the presiding personage must wear the tricolors of France when performing any civil ceremony. When Teva arrived to pick him up he forgot the belt that he had left on the ground. They returned to get it but his friends had gone and he couldn't tell which coconut tree they had been sitting under. It was raining heavily and got quite dark so they couldn't find it. Well, we had to be married without it, and Teva left the adjoint at the school while he came to get me.

"Are the witnesses still there?" I asked.

"Sure. They don't mind waiting," Teva said. "They bought some beer and they're enjoying the delay."

Teva showered and changed, so it was almost six by the time we arrived at the school. We were greeted with great cheer and the ceremony commenced. It was in Tahitian because the adjoint didn't speak French. I couldn't understand anything so Teva nudged me every time I was to reply *oui* for "I do." Afterwards we all went across the road to the church but the pastor had given up waiting for us and had left. We would have to return the next day. We drove back home.

While we were gone, the family had arrived to set up tables decorated with garlands of flowers and lanterns, and a feast for the members of the immediate family. There were sixty of them. The celebration was already well under way; toasts continued through the meal and long after. Everyone, including Teva, became quite drunk. I was getting very tired. I told Teva that I wanted to go to bed; he said fine, he would join me soon. I went into the house and lay down in our room, leaving the door open for him. I woke up in the middle of the night, remembering this was my wedding night, and reached for my husband. I was surprised to find it wasn't Teva beside me; it was someone else who had also succumbed early and found a quiet place to sleep. Teva was in the living room, stretched comfortably across two chairs. The house was still full of people and they were sleeping on benches, mats on the floor, everywhere. I returned to my side of the bed to go back to sleep. I woke up again quite early and went out to the kitchen, where some women were already preparing breakfast.

"*Bonjour!*" they greeted me cheerily. "Have some coffee."

"*Bonjour*. Thank you. Can you tell me why everyone is still here?"

"Oh, yes. They had such a good time that no one wanted to leave. They decided we should all go around the island today!" It was the traditional thing to do. Buses, the converted trucks, were hired, decorated with flowers and furnished with cases of beer and guitars. No one bothered to look out the windows; they drank, played the guitars and sang the whole way around the island. Teva was no more enthusiastic than I was to make the trip but of course we had to. We were exhausted when we returned home. It wasn't until the third day that we finally got to the church and finished getting married.

(When our son, Tane, was fifteen and my American friend, Judy, asked him about Tahitian marriages, Tane said he would never marry.

"Oh, you probably will," Judy said. "You want to have children, don't you?"

"Sure, but I don't have to marry because I have children."

"People usually do," Judy insisted.

"Why should I?" Tane replied. "My mother didn't."

"Of course she did. She told me about the marriage."

"No, she didn't. I ought to know."

Judy told me about the conversation. Surprised, I asked Tane, "Why did you tell Judy that Papa and I didn't marry?"

"But you didn't, did you?"

"Of course we did."

"You did? How come you never told me?")

In 1962 there were not many good jobs for Tahitians. The airport was finished but tourism developed slowly. Gerard Blitz stopped all work on the Rivnak property; he had decided it wasn't far enough from civilization. All Club Med villages were always well away from towns and tourist centers, so he was considering the island of Moorea (Mo o ray'a) instead. Teva was without work. He wanted to go back to Hawaii. He had been offered a job singing in a Polynesian club with Emma Mariterangi, who was the most popular Tahitian singer in Hawaii. Teva insisted that he would not be tempted by the partying, when he would be working, entertaining and seeing friends every evening. I would have to work in Hawaii too, so there would be the problem of child care while I was working, but I finally agreed to go back to try it for nine months. I was pregnant again.

I agreed to leave Leilani in Tahiti with her grandparents, as I would not have been able to take care of her as much as she needed. She still woke up several times a night because of the pain that would take several months to go away.

While I was still in Hawaii, six months earlier, I had received a cable from Teva. It said simply "Don't come to Tahiti yet. Sending Leilani to you." It made no sense because Teva knew I intended to leave for Tahiti as soon as Marisa returned to California.

I rented a crib and baby bathtub, asked my friend, Laura, to come with me to hold Leilani while we drove home, and eagerly waited for the arrival of the plane from Tahiti. Among the crowd of people getting off was a Tahitian woman carrying a baby. There was no other baby so it had to be Leilani. The woman saw me and nodded, smiling when I pointed to the baby. Soon they came through Immigration. It was a hot day but the baby was wrapped up in a blanket and bonnet, which the Tahitians thought was necessary no matter what the weather. A missionary probably taught them that. I took my baby in my arms, thanking the kind person who brought her, and was delighted that Leilani smiled at me with beautiful dark brown eyes. She was lovely. I unwrapped the blanket, took off her bonnet, and then I saw it.

I could hardly believe it. But it was there, hidden by the bonnet that framed her beautiful little face. One side of her head was covered with a deep red, swollen mass of thick, wrinkled flesh. There were pustules in the mass, and the rest of her head was covered with cradlecap. One ear was forced forward and the red flesh was advancing towards her cheek. I was white with shock.

"What's the matter?" Laura cried.

I turned so she could see Leilani's head. Laura gasped but recovered quickly and took Leilani from me. "Get in the car and drive. We'll take her home first and call a doctor. Oh God, it's Sunday. All right, tomorrow. Just drive." Laura talked, holding Leilani carefully because she flinched if anything touched her head.

We prepared a warm bath with a support so that little more than Lani's face was out of the water. We played with her and gently sponged her head so it was soaked by the warm water. She was cheerful and good and let us finally remove a lot of the cradlecap and scabs. I fed her and rocked her to sleep, but I did not sleep that night.

The next morning I was waiting at the doctor's door when the office opened. Seeing my anguished expression, the nurse took us straight in to see him. The doctor looked at Leilani quickly and said first thing, "Was she born with it?"

"'Why — no," I answered, surprised.

"Then let me assure you immediately," he said, "it will go away. Don't worry any more. It will go away by the time she is seven years old."

I stared, afraid to believe him. I burst into tears. He found a book behind his desk, told me to sit down, and took Leilani in his arms. Handing tissues to me to mop up the tears, he said,

"These skin maladies have been studied a long time. There are basically two kinds. When a person is born with it, it is permanent and only plastic surgery can repair some cases. When a baby is not born with it, it is a growth that will spread for a certain time, stop, slowly diminish and disappear." He showed me pictures of children who had been far more disfigured than Leilani, and completely recovered.

"Now we'll stop it from spreading any more, very simply. One X-ray treatment will do the trick."

Within a week of patient washing and medication, Leilani's head was free of infection. The flesh was very sensitive to any touch, but she was going to be all right.

Tane was conceived about six months after we were married. It was not convenient because I had to work, but fortunately I felt fine. My job was tour director and guide for French Canadian visitors to Hawaii. The French Canadians were enthusiastic tourists who wanted to see and do everything. After the sightseeing and shows of Oahu and Waikiki, we flew to Kauai, Maui, and "The Big Island" of Hawaii. The Hawaiian drivers were as knowledgeable as the best guides, always helpful and amusing, and they taught me more than I had learned in the training program. Wearing the full muumuu dresses that have been the costume of Hawaii since they were imposed on the women by the missionaries, I worked until almost my ninth month.

We had a boy. More than eight pounds and in such a hurry that I barely made it to the hospital. I woke up, still in the delivery room, and couldn't believe it when the doctor told me, "This time you have a boy!" He was in a crib beside me, I could hold his hand, and tried to pull him over so I could see if he really was a boy. Teva and his parents, promptly informed by cable, were ecstatic. We named him Tane (Tah' nay), which means man, but it was originally the name of the Polynesian god of the open sky, artisans, and all that is beautiful in the world. I was impatient to return to Tahiti and have all of our family together.

13

Teva

Teva did not want to go back to Tahiti and he didn't want to stay in Hawaii. There was too much jealousy and trouble amongst the entertainers. His partner at the club, Emma Mariterangi, left for a highly paid tour to Lebanon; Teva quit the club for what he thought was a better job, and ended with nothing. He wanted to go to the mainland.

"Renée, of course I want to go back to Tahiti eventually, but what can I do there now?"

"There'll be work soon when tourism really gets going, and you'll get a good job because you speak English now."

"Do you think I want to be a guide? Or work in a hotel? Do you see me working in an office? Anyway, those are dead-end jobs."

"Your father wants you—"

"My father! He says he wants me to take over the store. I told him that I could turn it into a real store, not just the Tahitian equivalent of a convenience store as it is now."

"That place supports your parents, your children, your sister and her children ."

"That's just it. When I tried to take over the book-keeping and make some improvements my father always objected. He says it's a good business now, why change anything? That's the way the old people think; they don't know things are going to change in Tahiti and we will have to change too."

"Teva, we must go back. Living is cheap there; we don't pay rent, and you can catch fish every evening the way you used to. Breadfruit, bananas, papayas and even coffee is free — we don't need much more."

"We have to have a car. There's going to be electricity soon. We can't live the way we did before, practically camping on the beach."

"You're exaggerating. You're spoiled by Hawaii. Anyway, we'll be able to buy a car as soon as we start working—"

"As soon as you start working. You can get a job right away. They need people like you who know what tourists want. How do you think I'll feel when my wife is working and I'm not?"

"Then what do you want to do?" I was exasperated.

"Renée, I think I have a good idea. Tourism will create a lot more in Tahiti than new hotels. Think of the taxis and buses, the cars that everyone will need to get to work on time, and all of the services like deliveries for the hotels."

"Yes. So?"

"Remember the American, Bob, I told you about, who stayed so long in Tahiti that time when you went back to California? He lives in San Diego and has a big car repair garage there. He wants to set up a garage in Tahiti. As he can't do it, not being French, he would make me his equal partner. He said that if I go to California he will give me a job and teach me mechanics."

"Go to California? What about me and the children?"

"It wouldn't be for long. He said I can live with him and his wife so I won't have to spend much money."

"Since when have you been interested in mechanics? Do you think you'd like the business? Do you think you would learn everything in a few weeks?"

I talked him out of the idea and we returned to Tahiti. He was right; I got a job immediately but he did not. There were many singers in Tahiti so entertainment did not pay well. Teva considered commercial fishing but of course he needed a boat, and we didn't have money to invest. He talked constantly of the opportunities that would open up for him if he went to the States. He could work in the troupe of Tahitian entertainers in San Diego at night and make some money besides learning the garage business. He wrote to a friend in the troupe, told him about his job in Hawaii, and was assured that he would be hired immediately.

"That's what your friend tells you. I think the entertainers on the mainland are no different than the ones on Hawaii," I said. "How are you going to get around in San Diego? It's much bigger than Waikiki. You can't walk, or count on buses, when you are working day and night. You'd have to buy a car, and there goes your money."

He said that I would always have a better job than he

did, if he couldn't start a business of his own. I could not convince him of the problems and uncertainty, and we argued constantly.

I finally agreed to let him try.

Teva's problems started as soon as he arrived in San Diego. Bob met him at the bus station and they went to his house. Bob's wife was not enthusiastic. She had not been to Tahiti and she was shocked to see a man as dark-skinned as Teva. Although Bob talked animatedly about his garage, he was obviously embarrassed by his wife's determined silence. She spoke up, though, when they were alone; she was not going to have a colored man living in her house. Poor Bob took Teva to see the garage, but Teva had to find a place to live. He called his friend in the entertainment troupe, and was welcomed to spend a few nights sleeping on the couch in his apartment, but there was already a full house. The job with the troupe was not available yet; the person he would have replaced had decided not to leave. Teva got a job as guard on a large sailing boat, which gave him a place to sleep but little money. Bob seemed to have changed his mind about having a garage in Tahiti.

Teva had a rough time but he was determined not to go back to Tahiti a failure. He finally joined another troupe of Tahitians and went to Las Vegas.

I didn't know until after Teva left that his parents were very angry with me. They thought I wanted him to go. They couldn't understand why I had let him go if I didn't want him to. I was to learn that as far as families were concerned, Tahiti was a matriarchal society and the women must keep a firm hand on the men, who generally have much less sense of responsibility towards their families. Teva's parents, though, were devoted to each other, and it was his father who was the head of the family. He was a tall, handsome man, pure Tahitian, serious and quiet-mannered. All of the family respected, loved and obeyed him. He never had to say anything twice. His wife was quite a bit shorter and stout by the time I knew her. Although she was far from vain and never wore jewelry, she had had a prominent gold tooth since her youth, when gold teeth were admired. When they married they were given new names, Teriitane (ta nay, man, husband) and Teriivahine (va hee nay, woman, wife), as was the custom. I knew their former names were Opura and Tetuanui but they were never called by them again. Teriitane and Teriivahine were devoutly Protestant and Teriivahine lead the singing in church. They went to the *fare putuputuraa* every Sunday and Wednesday evening. This meeting house is still a single, large room without chairs or other furniture. The people bring mats and cushions, put the children down to sleep, and spend the evening praying, discussing the Bible, and singing. The traditional hymns taught them by the missionaries are sung in Tahitian; variations are composed and become integrated in the culture of each district. Neither Teriitane or Teriivahine drank, but Teriitane was not above paying a workman with beer instead of money. Otherwise he was a perfect citizen and a leader in the community. He would have been *Tavana*, chief or mayor, if he had not owned a business.

Teriivahine was a loud and bossy woman but she always deferred to her husband and they never disagreed. Unfortunately she squabbled constantly with the rest of the family and neighbors. She was not liked very much, however she was treated with courtesy out of respect for her husband.

Teva, being the spoiled child, didn't work in the store or do any of the chores that the other adopted children did. He went to school to the mandatory age of fourteen; at that age young boys usually helped their fathers collect copra, cultivate the taro fields, or fish. Girls learned domestic skills and the weaving of mats, baskets, and other household articles that eventually acquired greater value as souvenirs for the developing tourism industry. Later the girls started working outside the home, in private houses, then in hotels and restaurants. Tourism developed slowly, the population was small so there was work for everyone in the construction of the airport and new buildings, and widening the single circle island road. This was at the time I returned in 1962. Until the airport was finished there were few tourists besides those who arrived on the Matson Line and other ships passing through the Pacific. They rarely stayed in port for more than twelve hours. A few taxi drivers who spoke limited English took the adventurous on the tour of the island.

Teva had worked for the Medical Institute but for some reason his father disapproved and made him leave. He went to the Tuamotus to dive for pearls but he didn't have the stamina of divers native to the coral islands. He decided to be a taxi driver and persuaded his father to buy a car for him, but he didn't speak English and soon became bored driving from the market to the inter-island boats, carrying the islanders and their purchases. Teriitane was loosing all patience with Teva. He said he was irresponsible and couldn't do anything right.

"That was a pretty quick judgment," I said, when

Teva told me. "At least you wanted to work and were trying."

"Actually there is another reason he has no confidence in me. When I was seventeen and went to the Tuamotus, I met some guys who had a boat and we smuggled alcohol. It was prohibited in the coral islands and we were caught. I didn't go to jail; my father went to jail for four days in my place. I can never make that up to him."

"You were just a kid. You finally got the job at Club Med and stuck with it. It wasn't your fault that it closed down. Maybe when it opens in Moorea you can work for the Club again."

"Yes, but for how much money? When you are a G.O. you live at the Club and you are paid pennies. I was paid $60 a month. Do you want me to go and work in Moorea for a salary like that?"

It was a long time before I understood Teva and his parents. I learned that although Teva had the will and ability to work, he resented me as he resented his father, for helping him. I had thought he would realize he couldn't manage alone in the States and he would return. I gambled and lost.

When Teva and I had returned to Tahiti from Hawaii, after Tane was born, I had expected to have my family, with Leilani and Teva's daughters, together at last. Leilani was a year and a half old and her soft dark hair had grown until it almost hid the "birthmark," which no longer bothered her. She didn't like me. From the day we arrived she would have nothing to do with me. When I tried to keep her at our house she cried and clung onto her sisters. "What is the matter with her? Why won't she play with you here? She just stares at me and cries — why is she so afraid of me?" I asked, bewildered, but not for long. I could hardly believe what the girls said.

"Because Grandmère always tells her when she's bad that you'll spank her."

"What! When she's bad? and I'm not even there?"

"Yes. She always says it. That you'll come and spank her." I could not understand why Teriivahine did that, because I knew she liked me. When I had told Teva he was angry but not surprised.

"It has nothing to do with you, Renée. It's because she wants to keep Leilani. It's often that way here; the grandparents want to keep the children. My mother doesn't realize she is hurting you and Lani; she just doesn't think. Children are the most important thing in Tahitians' lives, and when their own children are grown they want to keep the grandchildren or adopt others."

"Why didn't you warn me that this could happen! I would have found some way to manage in Hawaii if I had known!"

"You wanted me to bring her here when she was born."

"That was different. You had to leave Hawaii and I had to go to California. I wasn't giving her to your mother."

"Perhaps not then, but when you agreed to leave her here the second time, and for almost a year —"

"We weren't giving her to your mother! Why didn't you warn me this could happen!"

Teva couldn't warn me. He couldn't say anything against his parents. It would have been disloyal. They were raising his other children; why should he have objected to their caring for this one?

I protested to his mother. She looked a little embarrassed and promised she would no longer threaten Leilani that way, but the damage was done. It was weeks before Leilani trusted me and stayed to play at our house. It was years before she would stay the night. When it became dark she always said, "Mommy, I want to go home now." Home was with her grandparents.

We lived in the big family house because Teva's parents had built living quarters behind the *magasin*, the store, half a kilometer away and they preferred to stay there. Our house — which they said was ours — was not exclusively so, as it was taken over for family occasions, and even a *barraque* was built in the front yard one Fourteenth of July, with a dance floor, orchestra and bar. There was a wide lawn in front, with two towering breadfruit trees, grapefruit trees and a cashew tree. I was astonished when I saw the fruit of the cashew. It was exactly like an apple, with the cashew nut growing out of the bottom. In back of the house was a field where there were more breadfruit, banana, papaya and guavas trees. The house had a corrugated iron roof — when breadfruit dropped on it the noise was as shattering as a bomb. Typically, the large kitchen was apart and reached by a cool, covered walkway.

The grandparents, Grandpère and Grandmère, brought all the cousins over once a week to rake up the breadfruit leaves that were a foot and a half wide and two feet long. Sometimes we went into the valley to pick coffee. The bushes had long been neglected and were twelve feet tall, loaded with plump red beans. Nobody could afford pickers as there was not enough population in Tahiti to supply cheap hand labor. We would pick several sacks of beans, and then, hot and tired, we went to the river to swim in the clear cold water. Fresh-water shrimp hid under the rocks; they were caught at night by torchlight, using pronged spears. Fried, steamed, or *o'a* (raw, slightly marinated in lime juice), they were delicious. The coffee beans were pounded in an *umete* (oo' may tay), a large wooden bowl, or were taken to one of the small machines along the roadside, which looked like coffee grinders but stirred and rubbed the beans to slip off the skin. We spread the beans out in the sun to dry thoroughly, then roasted them in a heavy skillet, stirring and shaking the beans until they were an even dark brown. Tahitian coffee is one of the best in the world. Some Tahitians always put a vanilla bean in the coffee jar. Our neighbor had a small vanilla "plantation," as such gardens were called even if they were not large. The vines grow on slender, feathery-leafed bushes that shade them. The flower is an orchid; the plant came originally from Central America. As there is no insect in Tahiti that can pollinate the flowers they must be pollinated by hand, each flower treated separately, with great care. The seed pods, just like string beans, turn brown on the vine and the scent is so strong that it can be smelled a hundred feet away.

Fish was, and still is, sold from pick-up trucks; fat, glossy bonito and yellowfin tuna, or strings of silvery mackerel, delicious raw as in sashimi. The fishermen blow a conch shell horn as they cruise slowly by the houses, and also yell "*E'a* ! *E'a* !" which sounds like a paper boy shouting "Extra! Extra!" Reef fish and mahi mahi, famous in Hawaii too, is sold from racks set up along the road side as soon as the fishermen come in.

Bonito are caught with special pearl shell hooks, and the best are made from scarce yellow pearl shells from the Philippines, so they are precious to the Tahitians. The *bonitiers*, fishing boats, follow the sea birds that swarm over the schools of bonito. When they get into the middle of the school the fishermen, using twelve foot bamboo poles with expert dexterity, haul in the fish as fast as they throw the lines out. They jerk up the line and receive the bonito, weighing up to sixty pounds, flat on the chest with a resounding thump. It takes a powerful man to be a bonito fisherman.

When we went fishing in the lagoon, we asked the people who lived nearby if certain fish were poisonous at that time, because the fish who feed off the coral when it is growing too fast become toxic. The reactions to the poison vary. One is the sensation of cold water being too hot to touch. Sometimes the reaction is far more serious, and yet Tahitians, with their passion for fish, will take the chance. They tell me not to eat certain fish but they eat it. Parrot fish is one of the best to eat, but when they become huge and are called the napoleon, they are mortally poisonous.

Every morning Ah Son came by in his jeep from *chez le Chinois* (at the Chinese place), as all the grocery stores were called because they were almost always owned by the Chinese. He brought a supply of freshly baked French bread, a few canned goods, soap, kerosene for our lanterns, and a selection of tissue paper for making kites. The children spent hours making kites, using bamboo strips for the frames and painting designs on the paper wings. If they ran out of glue they used the sap from the breadfruit tree.

I had little contact with French people, and there were no Americans who lived nearby. There were less than two hundred Americans in all of French Polynesia, over one hundred islands. I had no social life. I had the children and was working on a glass bottom boat, just as in Haiti. I had a Deux-Chevaux (two horsepower), one of those ridiculous-looking French cars that are so simple and practical that I soon learned to repair it as easily as a bicycle. Two nights a week, Juliette, the shy Chinese-Tahitian twelve-year-old who was my baby-sitter, and I took sleepy Tane and a pillow and went to the movies. The movie house was just down the road and was no more than a wide thatched roof, with benches for seats and, in front, several large *peue* mats to put the babies down to sleep. Dogs often wandered into the building to settle under the benches. The films were old American, usually westerns, the soundtrack English dubbed in French. There was also a Tahitian translator who interpreted freely, adding his own comments on the scenario, which evoked hilarious responses from the house. The audience paid little attention to the story; it was the action and excitement that delighted the Tahitians. They chattered during any lengthy dialogue, screamed during the chases and fights, and hooted when lovers kissed.

I studied Tahitian although Teva had said it was a waste of time since everyone on the island of Tahiti spoke French well enough.

"But you're always speaking in Tahitian."

"Sure, amongst ourselves. We had to learn French because that's what is taught in school, and it is necessary to talk with the French and people like you."

"Wouldn't it be nicer if I could speak Tahitian to your family?"

"No. My parents never speak Tahitian to the children because if they don't speak French before they start school, they have a hard time catching up with the

French and Chinese children. The Chinese always teach French to their children."

"I'm talking about the adults in your family, as well as friends."

"I'll tell you what Tahitians talk about — children, the family lands, and religion. Do you think you'll be wanting to discuss those things?"

I learned some friendly and courteous phrases, and picked up a lot of household and common words, some so precisely Tahitian that they cannot be translated into French or English.

The trouble started over Leilani when I went to the *magasin* and found her with a bag of candy. "You mustn't give Lani candy like that," I said to Grandmère Teriivahine.

"I didn't give it to her. She is so smart, she climbs up and gets what she wants."

"What? You let her eat whatever she wants? Does she eat a lot of candy?"

Grandmère shrugged her shoulders. I knew that the stores generally gave candy instead of giving small change for purchases, which apparently pleased the Tahitians.

"She is going to have cavities and trouble with her teeth."

"Why? Noella and Marceline don't have cavities."

"Because their mother's family has strong teeth and they were lucky to inherit their teeth from her instead of Teva. His teeth aren't strong and nor are mine." A year later Lani had fifteen small cavities. I bought fluoride pills for the children. Tane had white spots on his teeth by the time he was ten, but they went away and the children had no more trouble with their teeth. It was unfortunate that the Tahitians had no idea that the change of their diet to the canned foods and soft drinks they loved ruined their teeth. Many wore at least partial dentures by the time they were eighteen.

I was sure that Grandmère was still managing to keep Lani from liking me, and of course she spoiled Lani so much I didn't like her either. I had to get her away from the grandparents. I went to the gendarmes. They said I should simply take her, and if the grandparents wouldn't give her up, they would go and take her away for me.

I had tried keeping Lani when I managed to have her at the house with Tane. After a while she would start to cry to go home, and it wasn't long before Grandpère came to get her. I insisted that he should leave her with me even though she cried. She would get used to staying with me — it was like getting a child used to a baby-sitter or starting school. There was no way to reason with the grandparents. We were yelling at each other, to be heard over Lani's screaming in my arms. I would give in because I couldn't stand the fighting, while Lani struggled desperately to go to her grandparents.

I sent for the gendarmes and we went to the *magasin* to get Lani. Grandpère was white with anger when he saw us, not only because I had called the gendarmes to help me, but all the family and neighbors knew what was going on. It was a blow to his prestige in the community. He kept his dignity but Grandmère grabbed a broom, trying to hit me. Grandpère stopped her, and the gendarmes asked me if I wanted them to take out a warrant against her for attempted physical violence. I was horrified to find myself in a brawl — and with Teva's parents.

"No, no, please just let's go!" I begged. I couldn't even hold Lani, she was fighting so hard, so one of the gendarmes took her, kicking and screaming. It was awful. We went to my house, where Tane and the neighborhood children were playing, but Lani would have nothing to do with them. The gendarmes left and I tried every way to calm her, but she curled up in a ball and wouldn't stop crying. She wouldn't eat and finally fell asleep from exhaustion, as I did too. In the morning the crying started all over again. There was no way to calm her. About six o'clock Grandpère's car came into the yard. I saw Grandmère, her face puffed up, her eyes red with crying. Before I could stop him, Grandpère marched into the house and grabbed Lani.

What could I do? Go back to the gendarmes again, have more scandal, and still not persuade Lani to stay and love me?

I reacted the way I unfortunately always did — running away from the problem. I was so angry that I would no longer go over to the *magasin*. I could only hope that when Leilani was older she would see that Tane loved me, that I was not a mean person, and she would eventually accept me as her mother.

Grandpère wrote to Teva about the shame I had caused his parents, not only because of the gendarmes but that I would have nothing to do with them.

Teva wrote to me that he could never forgive me. I could hardly believe that he took their side. I wrote back that I might never forgive him. Besides, if he had come back it would never have happened. He had been gone six months. He didn't answer and I didn't write him again. A friend brought her auntie, who had just arrived from the Marquesas Islands, to live with us and take care of Tane when I went to work. She didn't speak a word of French but we managed. I missed the other children.

I knew that for the Tahitians it was very important never to carry a grudge or trouble into the new year, so on New Years Day, several months after the fight with the grandparents, I went to the *magasin* and made up with Grandpère and Grandmère.

Teva finally wrote to me, from Las Vegas. He was still with the Tahitian entertainers, making a fair salary, but he gambled and had no money. He wanted to come home but was afraid he could not get work in Tahiti. "I am ashamed of myself," Teva said in his letter. "You are a good mother to our children and I need never worry about them. As for my parents; they are good people, but my father will never have confidence in me. He will never let me lead my own life. I owe him too much. You know now what he is like." Oh, Teva, I did, but how hard it was for me.

A year later Teva wrote to his father. Teriitane was fiercely proud. "There, you see? I knew he would come back. Listen —" Teva knew his father would rush to tell me.

Teva had had enough of the United States and wanted to come back to Tahiti. He had a child, born the year before, and wanted to bring him home. The mother was Japanese-Samoan, a dancer in the troupe. She had agreed to let him take the boy, whom he had also named Tane.

Grandpère said, "I am going to Papeete to buy the tickets for him and the baby, to come home immediately."

"But, Grandpère, he hasn't asked you for the tickets."

"What do you mean?"

"I don't think you should send them without asking."

"If he doesn't come now he might change his mind!"

"I think he wants to tell us about his plans to come home, and to find out how we feel about the baby. He is probably saving for the tickets himself, and will come when he is ready."

Grandpère refused to listen. "I am going to send the tickets. After all, he is my son so I can do what I want."

He went to the airline company and sent the tickets. "We should have an answer tomorrow," Loma, my friend at the office, told him. The next day I went to the office, hoping against hope. Loma looked at me with sympathy. "Renée, he refused the tickets."

I waited another year and then applied for divorce. I thought I was in love with another man; I had to get on with my life. It took a year for the divorce to become final; it was another six months before Teva, traveling with the troupe that year, learned about it. He happened to come back shortly after, on his way across the Pacific, for three days. The first thing he asked was, "Why did you divorce me?" I was too surprised to answer, and then realized that if he didn't know already, there was nothing to say.

14

Club Med, G.O.s, and Island Girls

It was in fact Claudine Blitz and I who first went to Moorea to look for a property for the Club. We went in a *bonitier* fishing boat to Papetoai, a tiny village in the bay of Opunohu. A friend of Claudine, a very intelligent and charming Tahitienne, was delighted to see us. Her *fare* was built several feet off the ground and not only the walls but the floor was made of woven bamboo. It was a bit springy but firm, and the house was very cool.

Papetoai is the village where the English missionaries built the first church and school in French Polynesia. They were welcomed there by the Pomare family, the reigning family of Tahiti, which occasionally had to retire to Moorea to regroup its forces, then return to battle with other chiefs of Tahiti. The last Queen, Pomare IV, resided there with her entourage in colonial-style houses like the English had built, all several feet off the ground because of the seasonal floods. Some of these houses remain; at least one should be preserved as a museum, but families still live in them.

There were no beaches in the bay of Opunohu so we drove along the crushed coral-surfaced, circle island road

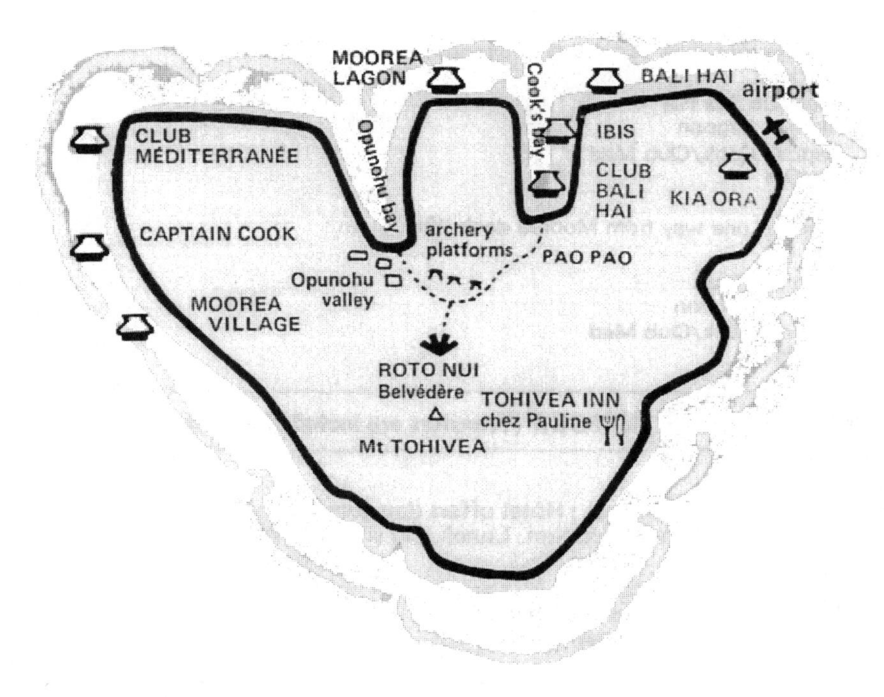

to the coconut groves on the west side. We walked under the canopy of palm fronds that shaded the path through to the white sand beaches protected by the lagoon and coral reef. There were two *motus* (mo' toos) coral islands, on the reef about five hundred yards out, that enhanced the beautiful site. It was a perfect location for the Club Med village and Gerard Blitz was able to lease the property. Clearing and construction started shortly after, and a few months later the Club Med Moorea opened.

Tane stayed with his grandparents when I first went to Moorea although I had never left him with them for more than a few days; I would not risk losing my son as well as my daughter. In Moorea I found Nini and Viri, a couple whose house was open to all the children of the district. They had two grown sons and two adopted daughters. Several of Nini's seven sisters' children stayed with Nini while their parents worked. Nini and Viri's *fare* was perched at the top of the beach which sloped to the clear, shallow water of the lagoon. Viri spoke some French and I explained why I had come. He translated to his wife but Nini hesitated to accept because she had asthma and wasn't sure that she could take care of a year-old baby any more. They said to bring Tane and they would find someone else to take care of him if Nini couldn't.

When Nini saw Tane she wanted him, and Viri said he would help to take care of him. Since that year I have shared Tane with them, his *faaa'mu* (fah ah ah' moo — adoptive) parents, as they liked to say, although there was no doubt that I was his irreplaceable mother. But Mama Nini was his best friend and fishing companion. From the age of two Tane went out with Nini, day or night, in an outrigger canoe, to fish.

At the end of the first Club season when I went to get Tane and return to Tahiti, Nini cried and didn't want to give him to me. "You have to give him to Renée," Viri told her. "He's her baby too." When I returned for the second season and asked if they would take Tane again, they were overjoyed. "But," Viri said, "not for the same amount of money." I would have gladly paid double for his good care.

"No," Nini insisted that Viri tell me. "He walks now

so we don't have to carry him any more and he's less work. Instead of 4000 francs you pay us 2000." I paid them 5000 francs, about $50.00 U.S. a month.

My first year was the second year of Club Med Moorea. The village was small; there were only thirty-seven *fares*, double occupancy. Several of us G.O.s had our own *fares*, so there were rarely more than sixty G.M.s. They were still mostly French people who came for a month or two. One of my responsibilities was to appoint the *fares*, with single G.M.s shared occupancy. As French names are sometimes the same for men and women, occasionally I had to do some fast switching at the last moment. There were some G.M.s who were willing to try the original arrangement.

The French were often surprised to see an American hostess in a French organization (There are many American G.O.s now). After several days, some would say, "But you are not like other Americans."

"Have you ever been to the United States?" I asked. Generally they had not. "Then how do you know?" I always added, "I am a Connecticut Yankee," and the tag stuck.

The year before, one of the few Americans who found out about the Club in Moorea was a handsome man who came at the time that Maeva's G.O. boyfriend had gone home to France for several weeks. Maeva was a beautiful Tahitian girl whom I happened to know because her family lived next door to our house in Punaauia. She had had a child by an Italian who stayed several months in Tahiti but had left before the child was born. Maeva was only sixteen then so she gave her baby to her parents to raise. She liked her G.O. boyfriend well

André Roosevelt, our grandfather and world adventurer, with Balinese dancers.

Scenes from the moving picture, *Goona Goona (Love Potion).* Filmed in Bali, 1928 by André Roosevelt and Armand Denis, the photo above is from a glass plate negative.

Our parents, Leila Roosevelt and Armand Denis, after the Africa-Asian expeditions, with the pet cheetahs they brought back to Connecticut. ⇨

Our very conservative, Belgian paternal grandparents.

Leila Roosevelt with her cousin, Eleanor, at the White House in 1932; after Leila's trip around the world in a second-hand Ford truck.

Mother and Father with the King and Queen of Belgium, sponsors of the first Denis-Roosevelt expedition: London to Capetown, South Africa, and back.

David, Renée, and Armand, with our nurses in Bali.

David, Renée, and Armand, leaving boarding school in England.

The author
Tahiti, 1975.

Armand, Renée, and David
California, 1995.

Seminole Indians who lived in the village down by the canal at the Anthropoid Ape Foundation, Florida.

Gus, Joe, Fifi, Midge, and Magnolia.

The author with a friend.

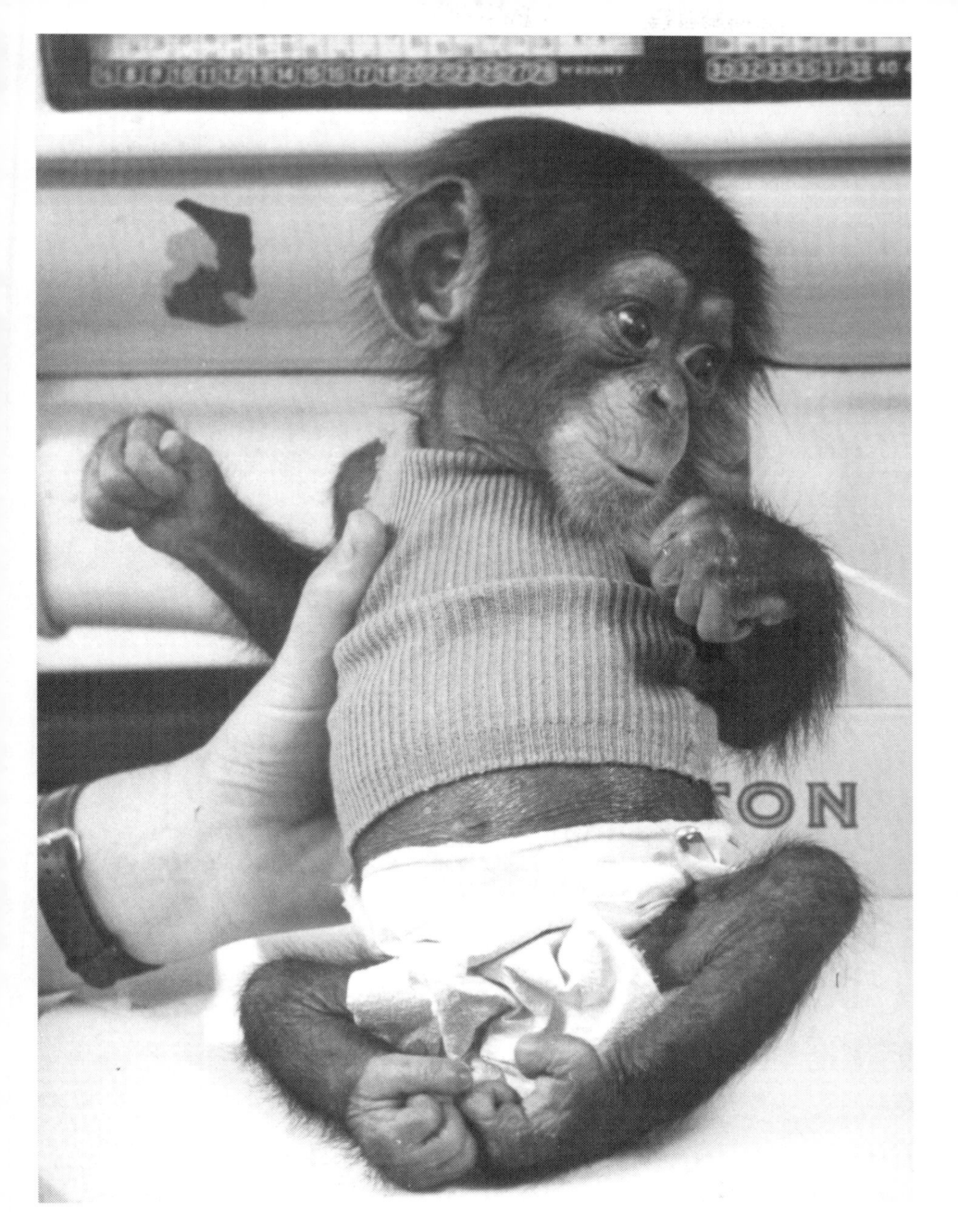

Roho, one of several baby chimps born at the Anthropoid Ape Foundation.

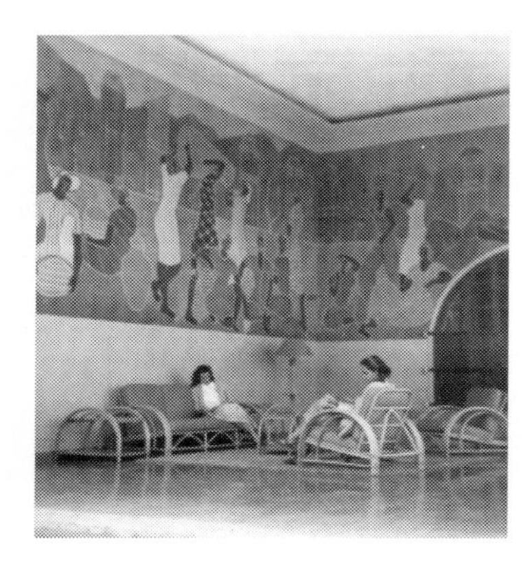

Lobby of the Hotel Ibo Lele. *(We are considering the possibility of reopening the hotel.)*

Scenes of Haiti.
Photos by André Roosevelt.

Gustav Dalla Valle, spearfisherman and Scubapro manufacturer, in his element. *Photo, Los Angeles Times.*

Back to Islands. *Photo of Bora Bora, Erwin Christian.*

Club Mediterranée: Moorea dancers and the author as public relations hostess. *Skin Diver Magazine, 1965.*

Club Med G.O.s, Fariua and Annie.

Tahitian Show, Moorea Village.

Cheyenne Brando, a delightful demie, at Tetiaroa.

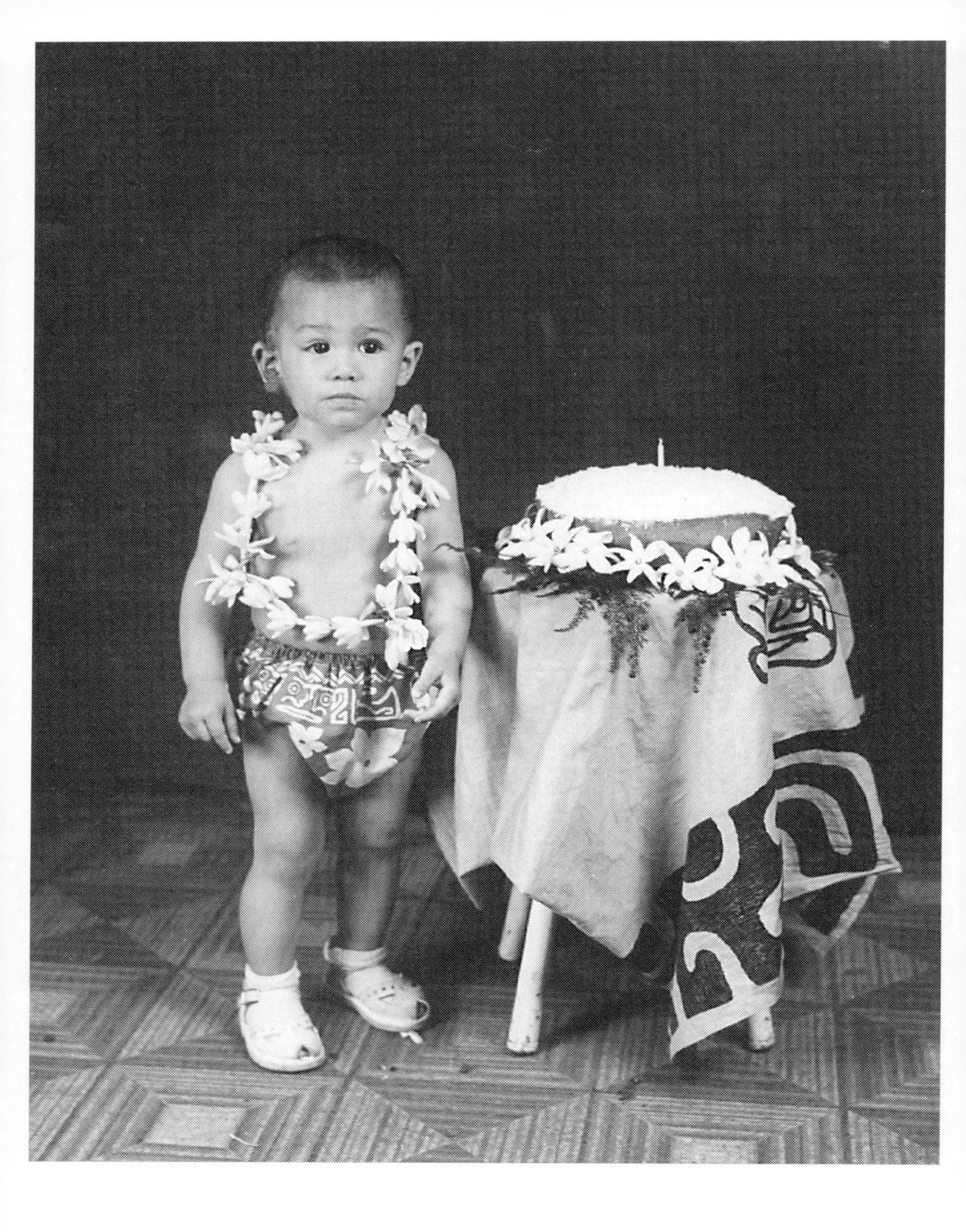

Leilani Tetuanui O Ke Aloha Nui, daughter of author's second marriage.

Tane with Tehotu & Cheyenne Brando at the Schoolhouse on Tetiaroa.

My son Tane.

My daughter Leilani.

General Charles de Gaulle with Grandpère, Teva's father, at Punaauia reception.

Teva, my second husband.

Terii, Tane's son and author's grandson.

The lagoon in front of the Rupé Rupé Ranch, my riding stable on Moorea.

My *fare* (house) at the ranch.

Scenes of the cobra god Naga and the peasant priestess, in Burma. Photos of the Denis-Roosevelt Asian expedition, *Life Magazine,* March 4, 1940.

enough but she fell totally in love with the American. It wasn't until after he left that she realized she was pregnant again. Of course, upon his return, the G.O. was very angry about it but he loved Maeva and took her back. Maeva was to go to Tahiti to have her baby but she waited too long. One evening, the first week of the Club season, Maeva's pains started. Fortunately one of the G.M.s was a doctor and delivered her little daughter without any trouble. Maeva had already promised the baby to a friend who took her away immediately. A few days later she brought the lovely baby girl back to the Club to show her to us. She had dark Tahitian eyes and blond hair. Maeva looked at her, smiled and said, "I can't imagine this baby was mine." The little girl was raised in the village near the Club, came over often, and grew up to be a strikingly beautiful *demie* (half Tahitian, half another race). Her father learned about her, returned to Tahiti and took her back with him to the States. She didn't stay there long, though, as she preferred her Tahitian home and way of life.

The Americans and other English-speaking G.M.s were my particular responsibility, getting them adjusted to the unfamiliar rustic accommodations, and acquainted with some of the French. Many French spoke English but often they would not speak to the Americans. Tables in the dining room were for eight, and G.M.s were supposed to sit at the next available places as they came in so that everyone got to know each other. It didn't bother the French to rudely pass up a table with some Americans, or to ignore them at the same table. I warned the Americans not to be offended, it was the French chauvinist complex, their problem, and if they would not be friendly, to ignore them. The Americans appreciated the warning and never let the French interfere with their enjoyment of the Club activities.

We had other nationalities besides French, and sometimes three or four languages were spoken at one table. I took a group of G.M.s on a tour of the other islands. There were four Americans, one German, a Chinese couple, a French butcher with his wife, and a royal princess of Spain on her honeymoon with her Italian husband. At Club Med, everyone wore bikinis and pareus most of the time; jewelry and stylish clothes were considered pretentious. There were no social barriers. Even the Tahitian personnel came to The Bar in the evening to participate in the games. Taina was a terror at broom hockey on the team against the G.M.s and they loved her. The Club Med formula really worked.

The second year I was at the Club was the best because, due to economic restrictions, we were not many G.O.s so we had to work hard to make the season a success. Most of us were recruited in Tahiti instead of France. There was always some contempt for the "G.O.s de Paris," as we called them, because they did not pull their share of the load. It was every French G.O.'s dream to come to Tahiti, and having arrived, they would not let work interfere with their pleasure. They formed cliques and chose a few companions among the French G.M.s, paying minimum attention to the others. It was hardest on me as I was responsible for the dining room and housekeeping personnel, arrivals and housing, as well as being the only hostess. The other local G.O.s helped me.

There was Fariua, who replaced Teva as "The Tahitian" of Club Med for all the G.M.s who came, sometimes several times, to French Polynesia. His wild good looks, extremely powerful body, and friendly savage charm attracted all the women. The men admired him too for his athletic prowess and generous nature. He was always ready to help, whether it was to participate in the games and shows in the evening, or cook the octopus he had caught that day. A pearl diver when at home in the Tuamotu coral islands, he was the spearfishing monitor at Club Med. The water was clear in Moorea, but he could dive so deep that he almost disappeared from sight. He would spear a fish, attach it to his belt, spear a second fish, then slowly return to the surface, already looking for another fish for the next dive.

We had great picnics on the *motu*, the coral island across the lagoon, grilling fish on a coral fire. Fariua grilled them whole, the blackened skin sealing in the juices. We broke open the skin to eat the tender flesh, crunching the crisp fins and heads. Some fish we ate raw, pulling off the tough skin, dipping the flesh in the sea water to salt it, and sprinkling it with lime.

Such things the G.M.s learned at Club Med as well as scuba diving, other water sports and horseback riding. Of course I rode the horses whenever I could. We had docile island horses at first but then had horses shipped from New Zealand. A couple of them were so high-strung that few G.M.s could ride them. A big roan called Tonga became my exclusive horse. She was enchanting when I wanted her to run; she always started with a very collected canter, hardly advancing until she suddenly stretched her legs and took off. We rode through the coconut groves, on the beach or in the water. Fallen coconut trees the horses simply leaped over.

The weeks before the Club Med seasons opened were especially pleasant. There were about eight of us

G.O.s to start with, living in *fares* scattered over the spacious grounds. I often woke up very early to walk along the beach when the sun was just rising, and the waves, touched with silver, were curling softly along the shore. The lagoon was azure blue and still, disturbed only by the flutter of tiny fish as a school of them leaped out of the water, to fall back again like the patter of raindrops. Brilliant yellow and blue demoiselles drifted lazily around the coral under the pier where sea anemones swayed gently with the current. It was the best time to take a *pirogue* and paddle out to the reef. The tide was so low one could walk on the reef and find shells, or fish off the far side. It was very quiet and the air was still cool from the night. When the rising sun touched the top of the palm trees on the *motu* it was time go back.

Everyone gathered in the kitchen for breakfast of strong café au lait and French bread, the baguettes, still hot and crispy from the brick oven. The workers came in from the neighboring villages and we organized our crews for the day. Several housemaids cleaned the *fares* with a couple of men who did repairs. The maids were not very efficient at first but I developed an excellent system with them. I inspected a few *fares* each day and made notes, one for each *fare*, listing anything done carelessly. When the women gathered for lunch I gave each the lists of her *fares*. The others could not see what was on the lists but they could see how much writing there was on them. There was a lot of laughing and teasing and the lists were shorter every day.

I needed more waitresses, though the four who were with me for the three years I was there, were the most efficient and good-natured girls I have ever worked with. Their hours were from seven to ten, eleven to three, and seven to nine. If the village was full they would not take their days off, sometimes for several weeks. The more people and work, the more they enjoyed it.

A few days before the opening one year, one of the French G.O.s returned from the Tuamotu Islands where he had spent the several months while the Club was closed. He had a lovely girl with him, sixteen years old, who's name was Vaiana. Xavier introduced the shy but smiling girl to us, saying she wants to be a waitress. She was a sturdy girl but too young, I thought, and said so to Xavier. He answered, "You said you were looking for waitresses so I brought you one."

"Well, tell me more about her. Where is she from?"

"She comes from Rangiroa." It was an atoll about an hour's flight away.

"Is her family agreeable to her working here?"

"Sure."

"Is she your *vahine*?"

Xavier shrugged his shoulders. "Not really."

"Not really? Well, does she sleep with you?"

"Not yet. She can sleep in my *fare*, if that's what's bothering you."

I was nonplused. I watched the girl. She insisted upon helping at table, it being time for lunch. She was obviously delighted though still shy, never looking at anyone except the cook or me when she was not sure what she should do.

I watched the men as they arrived. They all stared at Vaiana. She was such a lovely, fresh, true island girl. Her skin was a warm brown, her eyes almond-shaped and shining, her straight black hair long and thick. She still had a baby chubbiness but it was charming with her smooth round arms and legs and her high, perky breasts. Her face was not childish but it had not yet become the face of a woman. She glowed with young animal health and happiness. It was easy to see what all the men were thinking of. I knew I had a problem.

After lunch I called Vaiana aside to talk to her. Though she had never left her island she spoke French quite well. Of course she had gone to school, taught in French, until she was fourteen. I got to the point right away, knowing how much she wanted to stay. "Vaiana, are you going to live with Xavier?"

Vaiana stared at the ground. "I don't know," she said finally.

"I must tell you, Vaiana," I said, "although I think you might be a very good waitress, I don't like to have a girl so young working in the village. I will want a letter from your parents saying that they are willing for you to stay and work here, but it still depends upon your situation with Xavier."

"I will sleep with him."

"What do you mean? Sleep in his *fare* or really sleep with him?"

"I don't know. I haven't really slept with him yet but he is good to me."

I felt helpless. There was nothing to be done today; she had already been with Xavier the last couple of days so she could continue to stay with him until we heard from her parents.

The next day Vaiana came to me asked if she could sleep in my *fare*.

"What is the matter, Vaiana?"

"Nothing. But I don't want to stay with any man."

"That is exactly the problem, Vaiana. That is why I

think you are too young to stay here. You must have seen the way all the men are looking at you. They won't leave you alone."

"Even if I am staying in your *fare* and you tell them to leave me alone?"

I sighed. "They won't listen. I can't watch you all the time. I think you should go home."

"No!" She all but stamped her foot and there were tears in her eyes. "Please let me stay! I'll work hard and then I'll go and lock myself in the *fare*. I won't give you any trouble."

Opening day was in two days. I certainly couldn't leave the village to take her to Papeete to put her on a flight to Rangiroa. I took her over to Viri and Nini's *fare*. Viri worked in the kitchen so Vaiana could go home with him when the restaurant was closed. She would be safe there at night.

The generators were started up that day so we had electricity. In the evening all of the G.O.s gathered at The Bar to hear the new records. By this time there were about fifteen G.O.s and we danced until midnight, when I went back to my *fare*. I opened the door, to find Vaiana asleep on the bed. She woke up, frightened, when I entered. I saw her face was streaked with tears and she started to cry again. I took her in my arms while she sobbed desperately. Finally she calmed down enough to tell me what had happened.

"I went with Viri to their house the way you told me to," she hiccuped, "but right after, Ian (another French G.O.) came and said I shouldn't have left, that you needed me so he came to get me. I got in the car with him and he brought me to the village, but instead of bringing me to you he took me to his *fare*. I tried to stop him but he made love to me. I didn't want him to but he wouldn't stop. I cried and cried until he finally let me go. I came here to wait for you."

She continued to cry while I held her and stared out the window, seeing nothing, raging against all the sons of bitches who could not leave a young girl alone. I had to get Vaiana away. I didn't say anything more then and she finally fell asleep again.

The next morning Ian was waiting for me. He had some money in his hand. He said, "Renée, I want Vaiana to stay with me. Please give her this and tell her I'm sorry I forced her, but I want her to be my *vahine*."

I pushed his hand away. "I am not going to speak for you. Vaiana doesn't like you. The money won't make any difference. You can give it to her if you want but she is leaving just as soon as I can get someone to take her to Papeete."

I couldn't spare anyone to take her that day. Vaiana didn't work in the dining room. She stayed with Cecile who had the little souvenir shop, and took care of Cecile's baby. When I saw her she was radiant again. She loved taking care of the baby, she said happily, so she could stay with Cecile and Robin. I pointed out that there wasn't enough room for her to sleep in their *fare*. Robin said, "Cecile needs a nurse for the baby and Vaiana was very good with him. We'll hire her and you don't have to worry about her any more."

"So where is she to sleep?"

"Renée, you are making too much of an issue of this. Why should she be any different from other Tahitian girls? They all have a *tane* by the time they're sixteen. Vaiana just has to decide which one she'll stay with and there'll be no more problem."

I was disgusted. I knew Robin also wouldn't hesitate to grab her if he could get her alone. "I tell you she doesn't want any *tane*!"

"What makes you think that if she goes home she won't be with a *tane* in a few weeks? Do you think she'll stay at home? She left with Xavier, didn't she? She might as well choose a *tane* where she wants to stay." Robin was right, as much as I hated to accept the fact.

She might have lived with Viri and Nini but they knew she would soon want to stay at the village in the evening to see the entertainment and to dance, as any young girl would. They didn't want to be responsible for her. I told Vaiana what Robin had said.

Vaiana listened, nodded and said, "I know. I have already decided. I'll stay with Maurice."

I could hardly believe her matter-of-fact acceptance. She looked a little embarrassed. "Maurice comes to see me when I take the baby down to the beach. He's nice. He hasn't tried anything but he wants me to be his *vahine*. So I'll tell him yes."

Maurice was very good to Vaiana and she was happy. She took care of Cecile's baby until the end of the season, then she and Maurice went to Papeete where he opened a restaurant. It was an immediate success. I stopped by occasionally to see Vaiana who always hugged me fiercely in her affectionate, childish way. I went away for a while and it was some time before I went to the restaurant again. Maurice was there but Vaiana was no longer with him. Maurice shrugged his shoulders when I asked him why. "She is eighteen now and makes her own decisions." She had made decisions ever since she left home, I remembered.

Shortly after, I was walking in Papeete and heard someone calling me. It was Vaiana. She was pregnant.

She hugged me as enthusiastically as ever and bubblingly told me that she had a new *tane* and they were getting married. He was a French militaire; they were going to live in France. I kissed her, wishing her a very happy marriage, knowing that her husband would always love this delightful island girl.

There was another young Tahitienne, Maia, who wanted to work at the Club. We G.O.s had arrived at Moorea as usual before the season started. Ian had a new girlfriend. She was a *demie*, Tahitian-Chinese, very pretty, and though she was also sixteen years old she was sophisticated in the way the Tuamotu island girl, Vaiana, was not. She said she wanted to be a waitress but I doubted it because she was distracted and careless.

After a morning of dusty work supervising the new housemaids I was cooling off in the shallow water of the lagoon. Maia came down the beach and joined me. She hardly glanced around before taking off the top of her bikini. Her skin was already a copper gold. Maia's figure was lovely and I complimented her. She was pleased, looking down at herself with satisfaction.

Maia told me she had not been with Ian for very long. She had been with another *popaa* (po' pa a, non-Tahitian) before. She had a closet full of pretty dresses he had bought for her, she said proudly. He also bought her her teeth. That is, he had ordered them, she had gone to have them fitted, but then they quarreled. Probably Maia had met Ian and did not hide her enchantment with this new admirer. The *popaa* left without paying for the teeth. Ian did the gallant thing and paid for them.

Now Maia could practice her perfected charms and she didn't waste much time working. She often went out on the fishing boat and sunbathed bare to the hips. The men thought she was cute but they were not very interested. Except Michel, another French G.O.

Ian soon found out his *vahine* was flirting with Michel but he did not vent his anger on her. It was Michel who got the black eye. Ian then told Maia she had to choose which man she wanted to stay with, and she chose Michel.

At the end of the season Maia and Michel went to France and were married. A year later they came back. Maia was pregnant and she soon gave birth to a perfect doll of a baby. It was a pretty picture to see Maia with her exquisite little girl who was a miniature of herself. Eventually the novelty of young motherhood cooled and her mother took over the care of the baby. Maia had her perfect figure back and could be admired again by all the men of the village.

The Club season ended so I didn't return to Moorea for several months. Ian had started a boat rental business. I went over to his place to reserve a boat for some G.M.s and was surprised to see Maia there, wearing as usual the briefest of bikinis, a round belly not at all detracting from her prettiness. I guessed what had probably happened.

"Maia! How nice to see you. Is Ian here?"

"He's gone off for something, but I can rent you a boat," Maia said.

"Oh, then— ?"

"Didn't you know?" She smiled sweetly. I divorced Michel and I'm with Ian again. I'm going to have his baby."

I didn't go back to Moorea again for a year or so. Somebody mentioned Ian so I asked if Maia had a girl or a boy.

"She had a boy this time, who looks just like Ian. Ian is keeping him. Didn't you know that Maia left?"

An American tourist told me later about a girl who enchanted him. It was Maia. She was working in a hotel in Tahiti. The American said she works hard and goes over to Moorea every month to see her children. He had tried to date her but she would have nothing to do with him. She told him she hates men.

15

On My Own

The rain, having kindly let the new G.M.s arrive in a pleasant tropical mist, began to fall with a vengeance. Buckets of water fell non-stop on the *fares*, the dining hall, The Bar, and the dirt paths between. It didn't stop the next day, and hardly stopped for a week. It was not cold of course; the G.M.s went swimming, scuba diving and water skiing, and the ambiance in The Bar was not inhibited by the rain. But the new plumbing was. The village had tripled in size and the new *fares* were built with adjoining bathrooms. Sand had drifted into the pipes during the time it took to install them over several acres. The pipes began to block up as the water level of the island rose. Shower stalls had water standing in them permanently and toilets were backing up. We waded along the muddy paths, through ponds and streams running everywhere.

Finally the rain stopped and the G.M.s had perfect sunny weather to make up for their discomfort. But the plumbing continued to be a problem, along with other facilities that were faulty. As they didn't break down completely the French did not see the importance of repairing them right away. The Americans were on my back all day long and although I begged and pleaded, I could get little cooperation from the French. The new chief (manager) of the village was a buffoon. For some reason the Paris office promoted entertainers to chiefs and expected the accountants to handle business. Nobody felt responsible.

Gerard Blitz decided to send another village chief, his daughter Hélène. That sounded fine to me — surely she must be like her father, capable and dynamic. We soon learned she was an alcoholic, foisted off onto one village after another, chief only because of her father.

I was dismayed that Gerard would do that to his favorite village, and was quickly disgusted with his daughter. Nothing changed. The demands of the G.M.s and trying to get repairs done exhausted me. I should have been more patient. Eventually the problems were taken care of, after I left.

Gerard said, "The chief of a village is the same as the captain of a ship. I am very sorry that Hélène and Renée cannot get along together. Renée will have to leave." I am sure he arranged for me to overhear the discussion in the office. I went to my *fare*, packed, and got up early the next morning to take the boat. The waitresses and housemaids heard I was leaving. They crowded around me, crying, and I cried too. I went to get Tane but I couldn't tell Nini that I was leaving the Club. I promised I would bring him back when I could.

We went home to the grandparents, who were delighted we would be staying again in Tahiti. I cleaned our house, left empty and waiting for us as usual, as I thought about what to do, and on the third day I went to the Club Med office.

Gerard was there with the director of the Club Med villages in French Polynesia and they asked me to come in right away. Gerard told me how difficult it was to let me go.

"Renée, you can go to any other village of Club Med. You can take Tane with you and work in any Club in the world." He took my hands and then hugged me. I thanked him, smiling, and said I would think about it. He was being so kind but I only wanted to leave before I started to cry again.

I returned a couple of days later. Gerard had left Tahiti but the director repeated what Gerard had proposed. I had a different idea. First of all, I didn't want to leave Leilani, and the convenience of making the trips to spend vacations with Marisa; nor was I enthusiastic about taking Tane from one country to another. It sounded exciting to live in many countries but life in Club Med villages is pretty much the same anywhere, whereas schools in different countries are very different.

I wanted to do tours of the island of Tahiti for the G.M.s of Club Med. They could make reservations at the Club as they did for other excursions and I would meet them at the pier when they arrived from Moorea. The director thought it was a fine idea and he promised me all the help I might need to get started.

I bought a Volkswagen minibus for $3500, that year of 1965. I applied for the business license and learned that my taxes would be $45 a year. The salary for a chauffeur for the first six months was $200 a month. I was required to have a chauffeur even for a minibus, because the syndicate protected the taxi drivers from the travel agencies that were importing buses for transfers and tours. To drive the minibus myself I had to apply for a taxi driver's license, which I did, as the minibus carried only eight passengers and was considered a taxi. I was warned the taxi drivers would give me a lot of trouble; they might even slash my tires. The second week after I started thirty people signed up for the tour. The trip became a caravan, with four or five taxis following the minibus. I paid the drivers their regular fee; they had little more to do than drive because I grouped the people at each stop to tell them the facts, history and stories of Tahitian life. All the taxi drivers wanted to work for me.

Tourism developed steadily after the airport was finished. The hotels filled as soon as they were built. Prices were very reasonable — the price for my tour was $9, including the fee for the Paul Gauguin Museum and the Botanical Gardens. Everyone had work and the development of the island's economy, based upon the island's only product, tourism (with copra, little else), was progressing smoothly. All that changed after Autonomy, when the control was turned over to the local government. But I was not concerned with progress or politics; I lived out in the district and it was a long time before changes affected the daily lives of families more than ten miles from Papeete.

Some questions people ask are not always covered in the guide books, nor are they answered satisfactorily by Tahitian guides. What is strange to visitors may be so ordinary to the Tahitians it isn't worth mentioning. For instance, the boxes and what appears to be newspaper delivery tubes beside every driveway around the island, are not mail boxes. They are for bread, the French baguettes, that are delivered, still hot and crusty, twice a day by the Chinese bakers. Going to the market, visitors would see neat rows of *peue* mats and bundles of articles tied up in pareus along the sidewalk, with no one nearby.

"Don't people steal?" I was asked.

"No. The owners probably came last night to be at the market early, and they slept on the mats spread out on the sidewalk. They may be gone elsewhere all day today but will be sure to find their possessions untouched when they return."

In the market I told my tour people not to try and bargain — acting hesitant about the price or turning away may bring forth a voluntary better offer. If one buys several articles, the vendor always adds another as a gift.

The Tahitians have a different idea of big business. An American wanted to order a large number of a particular shell necklace, as much as a hundred. The Tahitian was surprised he wanted so many the same. The American asked how much discount he would get for ordering so many.

"Discount? No, no discount. I think I'll have to charge $1.10."

"What? The price of this one is $1.00."

"Yes, but I'll have a hard time getting the women to make one hundred because it will be so boring."

At the Paul Gauguin museum there are no originals, not only because of the value of the paintings but because of the humidity that requires special conditions to preserve them. There are several of his carvings, copies and prints of paintings and drawings, but the museum is mainly a collection of articles, his carving tools and household furniture as well as photographs showing how he lived in France, Tahiti, and the Marquesas Islands where he died in 1903.

The Society Islands were named by Captain Cook, after the British Royal Society under whose auspices he made his trip to Tahiti and Moorea in 1769, to make the scientific recording of the transit of Venus between the sun and the earth. The island was discovered by the English navigator, Samuel Wallis, in 1767, and also by the French explorer, Bougainville, by chance, nine months later, each thinking he was the first, but actually the Spanish came upon the island several years earlier. They were not interested enough to claim any of the several islands they touched upon, so the claim of Captain Wallis in the name of King George III, was maintained. The French came later, wresting control of the islands from the English, to have a base for their commercial and colonial enterprises in the South Pacific. Many family names, as well as many words in the Tahitian language for articles that did not exist before in Polynesia, and the days of weeks, etc., came from the English missionaries, seamen and merchants of the early days. Commerce was established, exporting pigs to New Zealand, and importing cloth, called the *pareu* like the original *tapa* bark cloth. Unfortunately, firearms and alcohol were also imported. The Tahitians of two hundred years ago were not so peaceful as they are now; the district chiefs and warriors were always raiding each others' territories to acquire more land and slaves, or for a question of honor.

People asked about the colorful chickens that are everywhere, even on hotel grounds. They do not belong to anyone. If you want to have chickens and eggs, you call and feed them twice a day, then they will lay eggs under your bushes. The chickens are not indigenous; there are no animals native to French Polynesia. Pigs, chickens, and dogs were brought by the Polynesians in their outrigger canoes when they crossed the Pacific from the west and established themselves on these islands. Cows and cats were brought by Captain Cook, horses were introduced by the Spaniards. The Tahitians named the cows after their pigs — *pua'a toro* ; they called the horse *pua'a horo fenua*, the pig that runs over the land. The cat was called *uri pi'i fare*, the dog that sits on the roof.

I built a house at the Pointe des Pecheurs — Fishermen's Point — where Teva and I had first lived in Punaauia. The property belonged to Teva; he had inherited it from his real parents. He and his two brothers had inherited several small properties in Punaauia and they divided them equally as all Tahitian families do, although Teva would receive properties of his adoptive family as well. All the land was inherited by the Tahitians, except for what they had sold to the government, the French, and a few foreigners who, in principal, were not permitted to own land in French Polynesia.

There were only half a dozen houses at the Pointe, hidden from each other by the breadfruit and mango trees, *bananiers*, and hibiscus. We had the most dramatic view of Moorea's mountain peaks nine miles away across the Sea of the Moon. There is a wide pass between the reef and the shore at the Pointe, so at one side there is a white sand beach, bordering the lagoon protected by the reef. The other side is the open ocean and the shore is covered with volcanic stones, rubbed round and smooth by the constant waves. The current is very strong and boats are easily swept out through the pass. The Tahitians who live along the shore always keep an eye on *pi-*

*rogue*s if they come too close to the point. These men are expert at swiftly racing out the pass to guide struggling canoers to shore.

It was the children who amazed me. Only a few meters from the pass, where the beach curves in from the point, the children played in the shallow water, infants of three or four as well as older ones. There is some current even where the water is shallow, but I never heard of a child being carried away or drowned. The bigger children take their responsibility for younger children very seriously and no mother worries as long as there is a ten-year-old on the beach. They seem to know just how far out they can go; if they do go too far they ride with the current and make their way back to shore farther along.

My *fare* was built quickly despite the four day week; the workers did not work on Mondays because it was the day to recover from the weekend. We had electricity in the district by then. Before, if one didn't have a generator, and few people did, we had kerosene lamps, refrigerators were kerosene operated, and irons that were a real fire hazard as the flame was apt to shoot a couple of feet in the air. Electricity was cheap at first. We didn't pay for water — in fact, when a house was built the town hall gave about fifty feet of pipe to be installed from the main source, free.

Someone who interviewed me asked, "How is your life similar to that of a housewife in the States?" I thought, he should be more interested in how different it was. Small differences perhaps, but with the wonderful climate, lack of traffic, and having a large family eager to help with the children, being a housewife and mother was very easy in Tahiti.

Grandpère took all the children to school every morning. School was from 7:30 to 11:00, which is noon, lunch time, in Tahiti, and from 12:30 to 3:30. I usually picked up Leilani and Tane after school, though sometimes Tane came back along the shore, stopping to fish along the way. The school was about two miles away; the name was, and still is, "2+2=4." The French government maintains the schools in every district and on every island of French Polynesia. The curriculum is basically the same as in France, which is fine for French children who come from or go back to France, but it isn't very practical for the Tahitians. We used to pay about 25¢ for each child's lunch, which was a better meal than they would have had at home, because Tahitians rely too much on the cheap and delicious French bread. The bread is baked two or three times a day; it is common to see children going home from the Chinese store with a dozen baguettes, two feet long, twice a day. There used to be little choice of imported foods and supplies, but what there was came from countries as diverse as France, the United States and China. An essential still is Chinese punk coils — "Fish Mosquito Destroyer" manufactured by the "Blood Protection Co." There are far more mosquitoes, though, in Florida or any northern state in the summer than there are in Tahiti. French Polynesia has few bugs and no snakes. The only disagreeable insect is the centipede, and although the bite is painful it is not more harmful than a bee sting. One rarely sees them; the chickens assure that.

The greatest danger in Polynesia is in the water. The *nohu*, or stone fish, is an ugly beast that conceals itself in gravelly sand and broken coral. It lies in wait for small fish swimming by, then suddenly opens its cavernous mouth, sucking in the hapless prey. The danger for children and any person strolling along in shallow water, was the row of spines along its back, which can be mortally poisonous for small children or an adult with a heart condition. The *nohu* does not inhabit clean sand beaches and has been practically eliminated from the shores of the larger islands. The fish is good to eat, so the Tahitians gladly keep the shores clear of the *nohu*.

We had no television at first; television arrived in Tahiti about 1965 and the programs, mostly American films dubbed in French, were from 6:00 p.m. to 10:00. One day Tane came home from the neighbor's house where he was watching television, and asked me, "Do you know what is Chicago?" Of course I said no. "Chicago," he informed me, "is a country where they make war." "The Uncorruptibles" was very popular with the Tahitians.

They had little idea of American geography. When I returned from a trip to the States, someone asked me if I had seen Teva. "No, I went only to California, and he is in Florida."

"Why didn't you go to Florida?

"Because it's too far."

"Too far? Isn't it on the same island?"

I said one time I went home to Connecticut, and explained it is near New York. "You know where New York is, don't you?"

"Yes," one Tahitian said, but another one asked him, "What part of France?"

When Tane was about six, he came in from play one day and said, "*On m'a dit que je suis americain*! One of the kids said I am an American! I'm Tahitian!"

"You are both, Tane," I explained, realizing my children knew nothing about their heritage, and they didn't speak English.

"Papa is Tahitian so you are Tahitian, but I am American so you are American, too. You were even born in America." He considered this a moment and then ran back out. I heard him say proudly, "I am an American!"

Tane became very interested in America. He asked me the name of the capital. I told him, pronouncing the word slowly, Wash-ing-ton. He repeated it slowly, then asked, "Why does the capital of the United States have a Chinese name?"

The Tahitians kept their radios turned on all day. The programs were half in French, half in Tahitian, and of course there was a lot of music, French and American songs, but mostly Tahitian. I was surprised to recognize some Tahitian tunes — one was clearly "Red River Valley" and another was "Pistol Packin' Mama." The radio was very important for the Tahitians as it was used to send messages to let families know when someone was arriving from another island, or informing family scattered throughout the islands that someone had died and the funeral would be the next day at four o'clock.

Our house was close to the cemetery, which was at the edge of the beach. Many of the graves were covered with roofs and surrounded with low whitewashed walls, so the cemetery looked like a miniature village. The frangipani trees shading the graves were always laden with their perfumed, creamy white and yellow flowers. Young girls climbed the trees to pick the blossoms to make leis. When there was a burial all the family and neighbors came, dressed in white, and their singing at the grave side was sad but beautiful.

There is the family Bible, its fine print in old-style,

elegant Tahitian. Tane went to Bible class, the family being very religious, and I went sometimes to hear the singing. In the Catholic church the Gregorian chants are in Tahitian too. The whole congregation sings; I was startled the first time when a big woman near me began the hymn without warning, no organ or accompaniment. The other women joined in immediately, while the men sang the bass in beautiful harmony. I overheard Tane discussing religion with a French boy, saying that Christ had used magic to make the blind man see.

"Not magic," the other boy said. "It was a miracle."

"Miracle?" Tane repeated thoughtfully. "Anyway, he had a lot of *mana* (the absolute power of the old Polynesian chiefs)."

Leilani's "birthmark" disappeared by the time she was seven years old, just as the doctor in Honolulu had promised. Her face is oval like mine, she has my nose and mouth, but her eyes are pure Tahitian, her skin a smooth golden-brown. Her hair is dark brown and wavy. She always speaks very quietly, and she giggles like every young girl. Tane's features are more like Teva's, but both children are typical of *demis*, half Tahitian, half Caucasian, with the beautiful Tahitian eyes. *Demis* are proud of their mixed blood, but they are Tahitians first and always. They want to be as dark-skinned as possible, and no matter how educated they are or how much they may have traveled, they remain simply, proudly and happily, Tahitian. They are modest, considerate, and as friendly as their reputation.

The tour was a lonely business even though I met many interesting people. I had to get up very early to leave the house at five or six o'clock, depending upon what time the UTA flight arrived. I helped welcome the G.M.s and get them on their way to the pier in Papeete. Before they left on the boat I gave them my card with the description of the tour. I was busy at the same time welcoming the G.M.s who were arriving on the same boat to take the tour. Some were on their way back to the States or Europe and would spend the night in Papeete, so I directed them to hotels and pointed out to others the marketplace or best places to shop until nine. I hired my taxis, picked everyone up and we departed on the tour. I had already talked a lot but now talked with hardly a break for six hours. Returning the people to Papeete, I picked up the children at school or the grandparents' house and finally returned home. I washed the minibus and fixed dinner. By nine o'clock at night I was asleep.

Marisa came one year for her summer vacation. She was not getting along with her father. They seemed to go through successive spells of understanding and disagreement. Even Marilyn, Gustav's next wife, a wonderful surrogate mother to Marisa, was tired of trying to keep the peace between them. She had a baby and although Marisa loved the baby, she was sick of the arguments about how to raise it. Marilyn didn't want to hear about how I had raised Marisa (I was astonished to hear of Gustav's approval of me) and placidly went about it in her own way. Marisa wanted to come to Tahiti; Gustav was willing to let her stay with me at last.

Marisa was delighted to be back in Tahiti again. She was sixteen, as tall and slim as I, but there the resemblance ended. She could be recognized as a California golden girl but was already becoming an example of the striking northern Italian blondes. Marisa insisted upon helping me with the arrivals, getting up before dawn to go to the airport and to the pier.

We had all arrived at the pier one morning before the boat from Moorea came. Some Italians wanted to go quickly to the marketplace before leaving; Marisa, translating for the responsible G.O., told them they had time. One woman didn't want to take her hand luggage to the marketplace so again Marisa checked and told her it would be safe to leave it. When the woman returned and did not immediately see her bag, which had been put on board with the rest of the luggage, she screamed at Marisa for losing it. Marisa explained very calmly where the bag was but I saw she was trembling. She walked away from the pier, then exploded with rage, to keep from crying. I was very proud of my daughter. After Marisa had come on the tour with me two or three times, she did the tour by herself when there was a small group. But generally, after I left on the tour, she took *le truck* to return home.

Marisa was thrilled when I had written to her that the Club had given me a colt that they didn't want to bother to raise. He was a Marquisan, of the Marquesas, the north-eastern islands of French Polynesia. The Spaniards had taken Arabian horses to the Marquesas islands, where they multiplied, grew wild, and evolved into the smaller, sturdier horses that can be found on many of the islands now. I had raised Akahiki (Chief, in the Marquisan language), and broken him in to bit and saddle when he was two and a half. It was not difficult because I had let him roam freely in the yard where he followed me around like a puppy. He was used to the bit in a couple of days and made no objection when I got on his back. He wouldn't let anyone else ride him, but surprisingly, he seemed to sense that Marisa was the same as I, and let her ride.

One day Akahiki was ill. He had eaten a plant that stops horses from defecating. I called the veterinarian who came and gave Akahiki a shot while Marisa and I held him quiet. The vet told me that if there were not the desired results in a couple of hours we should give the horse an enema. I stared at him. We should give the horse an enema — how? Akahiki was gentle to ride but he was a stallion and quite nervous. The vet said, "You simply insert a funnel and use the hose."

I nodded, speechless. Marisa, seeing my face, could hardly wait for the vet to leave so she could ask me what he said. I started to translate and burst out laughing. Marisa shrieked as I told her the vet's instructions. We suddenly heard a noise from Akahiki, we turned to see him straining and, to our great relief, he shat. No enema. We howled and clutched each other, exhausted with laughter.

We were invited to the Club Med in Moorea (Hélène was no longer there). I was an honorary G.O. but I couldn't stop the tours to go, so Marisa went without me. I wrote a note to my friend, Fariua, knowing he would take good care of her. Marisa reported he acted just like a father, watching to see whom she danced with and lecturing her if he imagined she might do anything he disapproved of. It was great fun for her, but I felt as I suppose my mother felt about me when I first asked to go out on a date. I knew how easy it was to succumb to the Club's romantic atmosphere. Marisa was sixteen, not yet interested in boys, but I was relieved when she came back.

If Marisa was to go to school in Tahiti she had to

learn French. I arranged for lessons but after a few lessons she wouldn't go any more. I didn't blame her for not liking the teacher but Marisa discouraged me from finding another. I took her to the Lycee Gauguin, the equivalent of an American high school. We talked to the secretary about her studies and in which class she would be able to start. Marisa suddenly got up and left the office. I hurriedly excused myself and followed her.

"It's too late for me now. I'll never catch up," Marisa said. I had to admit I agreed with her. French is a very difficult language; French literature, mathematics and sciences would be totally beyond her. "I had better go back and finish high school," she reluctantly decided.

I wrote to Gustav, explaining the situation. He and Marilyn missed Marisa and were pleased she was willing to return to them.

I again went into a depression after Marisa left. I continued with the tours for another few months but it was too lonely without her. Of course I had Tane whom I adored, but he was just a little boy. I needed someone to talk to and laugh with.

16

Taaroa

Marisa and I had gone to the Tahiti Village Hotel every Sunday afternoon to dance. Several hotels had an orchestra for dancing Sunday afternoons; the Tahiti Village was the nicest, besides being less than a kilometer from the house. It was a Tahitian style hotel, and the open terrace where we danced overlooked the beach. Sometimes we went down early to swim; after the music started, we showered, slipped into mini-dresses or pareus like the Tahitiennes, and went up to the terrace. We took a table near the dance floor and ordered a couple of lemonades. There was rarely time for them to arrive before we were asked to dance. Sometimes we arrived too early, there were few dancers on the floor, and the Tahitians were a bit shy because they knew they would be teased for being in a hurry. But as soon as one cavalier came to the table the rest no longer hesitated. We preferred to dance with the Tahitians because they were always good dancers and did not talk a lot. The French usually wanted to talk; we went there to dance, not to meet men. The Tahitians waited until the second or third dance, choosing a slow one, to start talking. They asked questions that were invariably the same, one right after another.

"What is your name?"

"Are you married?"

"Do you have a *tane* (fiancé, boyfriend)."

"Where do you live?"

"Can I take you home?"

Depending upon how charmingly he asked the questions, and how well he danced, we answered more or less truthfully; of course the answer to the last question was always no.

The music stopped at six-thirty and everyone had to clear out. A few who had attached themselves to the bar rather than dance had to be persuaded to leave, but there was never any trouble. I imagine if they went on to Quinn's Bar or another nightclub they may have ended up in a brawl. Sometimes one or two Tahitians, more persistent than the rest, would escort us to our car, where we left them with promises of returning next Sunday.

I noticed one tall, powerful-looking Tahitian who didn't dance often but watched from where he stood near the bar. The second Sunday I saw him he asked me to dance. That was indicated in the way the Tahitians often do when they have made eye contact with you a couple of times, even while dancing with someone else. If you wanted to dance with him you let the eye contact happen at the start of the music, he raises an eyebrow questioningly, you nod slightly, and he comes over with a satisfied smile. The big Tahitian and I danced several times together but he didn't say a word except to thank me politely each time. The next Sunday he asked me the expected questions, to which I replied evasively as usual. He cheerfully escorted us to the car.

Every Sunday he danced with me more often and became more persistent with his questions. I teased him the way it was customary, insisting I had a *tane*, but it is pretty obvious that if one goes dancing every Sunday afternoon there is no *tane* around. Sometimes Leilani and Tane came from the grandparents' house, up the beach a ways, and joined several other children sitting on the edge of the terrace, to watch us. I called them to the table for lemonade. They sat and drank solemnly, glancing at the other children who giggled but were envious. The dancers always asked Marisa and me, "Who is your friend?" meaning each other, and it amused us to tell them. It is not unusual for young Tahitiennes, married or not, to have children ranging from babies to fifteen and older, so my family did not discourage my admirers. They too had come to dance; the rest was *baratin*, a bit of flirting that was fun for both partners, and then, perhaps something might come of it.

At the end of the holidays Marisa went back to California. I didn't like to go to the Tahiti Village alone.

I was leaving the bank one day when a car stopped beside me. It was an American car so I was surprised to see the driver was the Tahitian with whom I had danced so often. He got out of the car and we talked a while. His name was Taaroa. He asked, "Why don't you come to the Tahiti Village anymore?"

"My daughter left and I don't like to go alone."

"You can go with me," he said, to my surprise, because it was not usual for a man to go anywhere with a woman unless she was his *vahine.*

"No. I don't want to go there with a man. The others won't dance with me." Actually they would, but everyone would have immediately presumed I was his *vahine.*

"But this way you're not dancing at all," he said reasonably.

"That's true," I laughed. "Well, if I can some Sunday I will go."

"I'll be there." He kissed me good-bye on both cheeks, Tahitian style.

A couple of Sundays later I took Leilani and Tane and went to the Tahiti Village. Taaroa was there and we danced several times. Seeing Taaroa's interest, the other Tahitians were even more persistent in asking me to dance, to tease us both. The children became restless and ran off to the beach. A couple of hours later they wanted to go home. I had to get up at four o'clock the next morning so after supper I took them to the grandparents' house to sleep. When I returned home and was about to unlock the door, a voice startled me. I wasn't frightened because it was not unusual for an admirer to come politely to the door. After a few minutes conversation they generally asked if they could spend the night.

"No," I'd say. "My *tane* is coming — in fact, you'd better leave quickly or he'll be angry." This way his honor was saved, and so was I.

"Oh, excuse me," he'd reply. "I didn't know." And sometimes added, "Perhaps another time?"

Taaroa came into the light of the porch. "*Bonsoir*, Renée," he said, with a rather foolish but confident smile.

"*Bonsoir*, Taaroa," I laughed, "and how did you find out where I live?" Of course it was simple enough; he had only to ask anyone in the neighborhood. The Tahitians are very romantic and they would have told him immediately. It would not have occurred to them it might be dangerous for a woman. A Tahitian attracted to a woman would not harm her, although if he had reason to be jealous of his *vahine* the same Tahitian could be very violent. Some women were not displeased by the black eye or bruises they received because they thought it showed their *tane* really loved them. But a Tahitian falling in love is very gentle.

"I followed you when you left the hotel," Taaroa said. "I was waiting until you put the children to bed. Where did you take them?"

"To the grandparents. I have to get up very early in the morning."

"Then you stay here alone?"

"Now don't get any ideas, Taaroa," I laughed. It was pointless to pretend that anyone else was there. I knew that Tahitians would watch a house for hours, even days, when they were interested in a woman.

"No, I just want to talk to you," he protested.

"All right," I said, pocketing my key. We sat on the steps and I asked him what kind of work he did. I was curious to know how he owned an American car. They were expensive and so was the gas; usually only the taxis were American. Taaroa explained he had inherited a lot of land on Bora Bora from his mother. He said proudly,

"I sold some land and started a business. Haven't you seen my truck?"

"What kind of truck?" I asked, and then remembered seeing a large, red truck like a big military weapons carrier, and had thought the driver looked like Taaroa.

"Yes, that's my truck. I cut trees."

"What! You're the one who is cutting down those magnificent old trees in town?"

"It has to be done. The streets have to be widened. I plant trees too," he said quickly. " We'll be planting many flame trees and other shade trees to replace the old ones."

"That will be beautiful, I must admit. I hate to think of the town changing, though. It was fun to sit at Vaima's waiting for the Moorea boat departure across the street. Now there is the new pier and that big parking lot —"

"I hate to see the changes too. Do you know the district of Pirae?"

"I go through it of course on the way around the island, but I don't know it."

"I bought a valley there, and land to the top of a hill so I can see the ocean. I have put a road through and planted hundreds of flowers, and every kind of plant that grows in Tahiti. I can make anything grow. Now I'm building a house."

His enthusiasm was catching. He made everything sound exciting. He talked in a very intense way but sometimes laughed at himself. We talked until it was quite late for me. "I have to go in," I said.

As I expected, he asked, "May I go in with you?"

"No. You must leave now." I got up and he did too, standing close to me. I suddenly felt hot. I was totally attracted by his powerful body, his almond-shaped Tahitian eyes and crisp black hair. He always wore a *tiare* over one ear; maybe it was the flower, cologne or soap, but he always smelled clean and good.

"You'll come to the Tahiti Village next Sunday," he said.

"Yes."

When he kissed me slowly, the strength of his arms

around me and the warmth of his body soon made me want him very much. I pushed him away gently and he let me go. He was probably quite satisfied with my reaction, sure I would give in to him eventually. He was right.

After Marisa left I lost all interest in doing the tours. I was working too hard, getting up at four most mornings, not getting home until four in the afternoon, having to take the children to the grandparents after only two or three hours with them, and falling into bed at eight. The final installment on the minibus was paid a couple of months later so I decided to stop doing the tours. Club Med picked up the program but they never had any where near the success I had. I doubt that someone paid to be a guide put his heart into it the way I did.

Finding a job was easy, every hotel and travel bureau needed English-speaking personnel, but I wasn't sure I wanted to work in a hotel, — certainly not in an office. A friend needed someone to replace his personnel manager whom he had just fired. He asked me to go and try out the position. He would give me a cottage for Tane and myself, meals and expenses paid, plus a good salary. I agreed; the hotel was small enough so that I wouldn't be stuck in an office all day. I would be training and supervising, as I had done at the Club. The former personnel manager never left the office, so the maids and gardeners had become careless. At first they were sullen with me. I used the same methods as I had done at the Club and soon got the personnel organized. I was pleased when the hotel manager told me, "I asked the housemaids how they liked their new boss. They said you are strict but good."

A few days after we moved to the hotel one of the bellboys told me that someone had come to see me. I knew it was Taaroa. I had told him before leaving home that I would let him know when we were settled in at the hotel and I had my work organized, but of course he wouldn't wait. He insisted upon seeing our *fare* and pulled me into the bedroom, but I protested we both had to go back to work. I arranged for one of the maids to stay with Tane that evening and we went out to dinner. When we were driving back from the restaurant, Taaroa suddenly drove the car off into a side street, watching closely in the rear view mirror.

"What is the matter?" I demanded, but he wouldn't say. After a few minutes he turned the car around to drive back to the hotel. We went into the cottage and I paid the maid. Tane was sleeping soundly. When Taaroa started to kiss me I pushed him away.

"Taaroa, who are you hiding from?"

He sighed and said, "My wife, Mareta. We were divorced several months ago. She applied for the divorce and then changed her mind. By then I had met someone else and I wanted the divorce. So it was granted, but she does not want to accept it. She is very jealous and keeps following me."

"What about the other woman you were with?"

"Cristelle. She isn't here any more. I know that my ex-wife thinks you are her — she is tall and has long, dark hair like you. Mareta hates her because she thinks I divorced her because of Cristelle.

"You don't love Cristelle any more?"

"No!" I was surprised by his vehemence. "She is a witch. I was in love but then I became sick of her. She was always complaining and she was jealous too. When I went to see my children she was sure I was still sleeping with my wife. There was trouble all the time. She wanted me to go to France with her but I wouldn't go."

So she was gone from Tahiti. That left a jealous ex-wife, but perhaps when she realized I wasn't Cristelle and he wasn't going to return to her, she'd leave us alone.

Taaroa came every evening and it was a good time for us. Taaroa loved his work and when he and his workers stopped at four o'clock, he went to his valley. One afternoon he came to take me to see what he was doing. I arranged with Tane's favorite housemaid to meet Tane at school and let him swim in the hotel pool, and went with Taaroa. When we reached Pirae he turned the car onto a dirt road, drove quite a ways and stopped at a stream. A truck could have crossed but not a car. We walked down to the edge, then Taaroa simply picked me up to carry me across.

His valley was absolutely beautiful, just as he said. Sprays of bougainvillea of every color bordered the road. Frangipani trees were already in bloom. The *fare* on the hillside was almost hidden by bushes and flowers. We climbed to the top of the hill from where we could see the ocean and Moorea in the distance. I was enchanted and Taaroa was very pleased. He held my hand tightly, leading me back down to his *fare*.

Then for several days Taaroa didn't come to see me. I became worried and depressed, but I was sure he loved me and would come when he could. I worked hard in the hotel so the time would pass quickly.

I was standing on a stepladder, tacking down shelf paper, when one of Taaroa's workmen came into the room. I climbed down and he told me, "Taaroa couldn't come. He wants you to meet him tonight at eight o'clock, at the bridge before you reach the road that goes into his valley."

I arrived there at eight. Taaroa was on the bridge,

staring down at the river when my lights caught him. I stopped the car and he came over quickly. "Slide over," he said, taking the wheel. He drove into the valley and off the road to where the car was hidden in the trees.

Taaroa was depressed. "I've been staying with the children. Mareta says she won't fight the divorce any more, she knows I won't go back to her. Now she wants to leave Tahiti. She won't take the children with her; she doesn't want to take care of them anymore. I have taken them to my cousin's house in Taravao."

"That's way out. You've been driving out there every day?"

"Yes, but they'll be fine there. My cousin has only one boy and she's really happy to have my two."

"That's good. When is Mareta leaving?"

"The next flight for Noumea — Sunday. Have you started your divorce yet?"

"What? Oh, my divorce. I wanted to write to Teva but I don't have any address."

"You don't have to find him. Go to the court, you can get it for desertion, and even if Teva can't be located the divorce will be granted. Then you can marry me."

Really marry again? It seemed good enough to live together the Tahitian way, but if he wanted to, well —

Two nights later I was awakened by knocking on the door. It was three o'clock. Taaroa was there with a woman whom I guessed to be his ex-wife, Chinese-Tahitian, looking very sullen.

"I had to bring her. She wouldn't believe you were not Cristelle. I told her you are a good woman and you will take care of the children." He turned to Mareta, "Now you see her. This is the woman I am going to marry."

I sat down in the doorway, still groggy with sleep. I didn't know what to say. Mareta stared at me. Then suddenly she flared, "Take him then! Take my children too! He has lots of land so it'll be a good thing for you!" She started to cry and turned away. I was sorry for her but I couldn't say anything. Taaroa bent down and kissed the top of my head.

"Go back to sleep. Everything will be all right." He led Mareta away.

Again Taaroa didn't come back for two days. He was surely waiting for Mareta to leave so I didn't worry. Then I learned that if it was why he didn't return, it wasn't the only reason. Papeete was a small town and news traveled fast. Taaroa was seen in a restaurant with the "other woman." She had returned from France.

I realized then who she was. "Cristelle" is such a French name I had presumed she was French, but she was a *demie*, French-Tahitian, whom I had known slightly at the Club. She had inherited a lot of land or money, had a beautiful home and often traveled to France.

Taaroa came to the hotel the next day in the afternoon. He demanded to see me so I left my work and went to the *fare* with him. As soon as the door was closed he began to kiss me, almost violently. I was so glad to see him I didn't protest and I let him make love to me. I pretended I didn't know about Cristelle.

But it was again three or four days before he came back. This time I stopped him before he could kiss me. "Taaroa, are you seeing Cristelle again?" I asked.

Taaroa let me go and was silent for a few minutes. I waited. Finally he said, "How did you know about her?"

"Toti saw you with her in a restaurant a few days ago."

He sighed and fell back on the couch, staring at the roof. "Yes, it was Cristelle. She has been back a couple of weeks. She came to the valley to find me. I told her I didn't want to see her any more. She cried and begged me to stay with her again. That's not all. Mareta didn't leave."

Good God, there was still the two of them. I was beginning to suspect he was enjoying it, but he was so depressed since the night in the valley it was hard to believe that. I said calmly, "What do you want to do?"

"I don't know, Renée. Everything has changed. I have changed. I wish it were the way it was before when it was just you and me here. I liked my work, and now there won't be another job for a while. The truck is only half paid for — even the valley isn't paid for yet. I don't even go there any more. I don't know why."

"Perhaps you really love Cristelle."

He exclaimed again the way he had the first time he told me about her. "No! She is a witch, a *sorcière*! She put some kind of spell on me!"

"Well, whatever kind of spell it was," I said sadly, "it seems to be working. You had better decide what you want to do."

Perhaps he could not make up his mind but my decision was made for me. The next morning the night watchman came to see me.

"Last night a woman came looking for you. Renée, it was that crazy wife of Taaroa. She couldn't find your *fare* so she asked me to show her the way. Of course I didn't. You had better leave here before she comes back again. Renée, I saw she had a knife, and I know she is crazy enough to attack you."

That was enough. I went to the manager and told him my mother was in an accident; I had to leave immediately. He was annoyed but what could he do? I packed, went to get Tane at school, pick up his transfer papers, and left for Punaauia. The next day we took the boat for Moorea.

The first week was bad for me but, surprisingly, I recovered quickly. I knew I was very, but only, infatuated with Taaroa. He may have sincerely wanted to marry me but everything had changed. Apparently the slightest difficulty confused and discouraged him, as it did so often with Tahitians. I made myself face the fact that if nothing else, Cristelle's money made her attractive to him.

After a month in Moorea I became restless. The school in Papetoai was poor, and I couldn't find a job I liked. I decided to return to Tahiti, so we took the boat and returned to Punaauia.

The day after we arrived there was a cocktail party at one of the hotels and the hostess there, a good friend, asked me to help her. I was talking with a group of travel agents when a Tahitian came and politely asked me to go to speak to someone else. I went with him, asking who it was. It was Taaroa. He took my arm and pulled me aside.

"What do you want?" I demanded, wondering why the gods were playing with my peace of mind, to let Taaroa find me again so quickly.

"Renée, please, I have to talk to you," he started, and went on to tell me how unhappy he was, that Cristelle was a witch.... Not all that again, I told myself and closed my mind to what he was saying.

"Taaroa, I don't care any more. I don't know why you want to tell me this. Just leave me alone now." I felt sorry for him, but then I wasn't sure he really needed my sympathy. "What do you do with Cristelle all the time if you can't talk to her?"

He stared at me a moment, then laughed. He said quickly, "Come with me now —" and when I shook my head, smiling, "I promise I won't try to make love!"

"No, Taaroa. I like you still but no longer enough to risk any more trouble. Please don't come around any more." I kissed him on both cheeks, to show I wasn't angry. He sighed and let me go back into the hotel.

I saw him occasionally during the next couple of years. Sometimes he came to the house late at night and asked to come in. I went out instead, sat on the steps and let him talk a while. It was always the same story. He was fed up with the way he lived, he was no longer with Cristelle (sometimes she was gone, but I heard also about the fights they had), and he wanted to go away. I heard he did go to Europe finally, and when he returned he started another business. It also was a fiasco which put him in debt again so he had to sell more land.

I woke up one night aware that someone was in the room with me. "Who is there?" I demanded, trying to turn on the light.

"Shh — don't — you'll wake up Tane." It was Taaroa. He was already beside my bed and started to kiss me. I was pinned under the sheet and could only protest, which did wake Tane. He wasn't really awake but he asked for a glass of water.

"Get him some water," Taaroa whispered, letting me get up and following me out to the front room. I went into the kitchen for the water and whispered to Taaroa to wait for me. He waited while I closed the bedroom door and gave Tane the glass of water. Talking to Tane as though to calm him, I quietly went back to the door, turned the lock and went back to bed.

After a moment Taaroa called me softly. I didn't answer. He tried the door, called again but I still didn't answer. Shortly after he came to the window. I couldn't keep from laughing. "Why did you do that?" he asked reproachfully, but then he laughed too. "Give me your hand," he said. I put my hand out through the louvered window. He kissed it, said good night softly and left.

I didn't see him again, and heard he had gone back to Bora Bora.

17

New Caledonia

I was having lunch with a friend at the Vaima Restaurant when a tall, aristocratic-looking Frenchman stopped by our table to say hello. My friend, Yves, introduced us.

"Renée, may I present Gilbert Dumont? Gilbert, this is Renée, an American friend. Will you join us for coffee?"

"Delighted," answered Gilbert, but he was looking at me carefully because he did not expect me to be American. I was usually taken for French or even a *demie*, part Tahitian. I could tell right away he had come to the table to meet me, and was disappointed to hear I was an American. I smiled at him charmingly and knew I would not like him.

"Gilbert, you are in luck," Yves said. "You are looking for a Tahitian girl who speaks English, to be hostess over the holidays at your hotel. Renée is not Tahitian but she is about as close to one as an Anglo-Saxon could be. She was hostess and assistant manager at the Club Mediterranee, and had her own tour business, but right now she is looking for a job for a couple of months."

I looked inquiringly at Yves. "Gilbert operates a hotel on the Isle of Pines, one of the small islands that is a part of the territory of New Caledonia." That was a French territory not far from Australia, north of New Zealand, and I knew that the Isle of Pines was considered one of the most beautiful islands in the world. I was interested. I let Yves speak for me.

"I know you want a girl to dance as well. Renée does Tahitian dances and has great success teaching tourists; they love it no matter how awkward they are. You'll get the kind of atmosphere she creates at the Club, and have a supervisor who will get your personnel in shape."

"I have been here only a few days," Dumont said. "I will let you know if I do not find a Tahitienne." I was sure he wanted a girl who would accommodate him in bed too. I didn't expect to hear from him.

Few Tahitians, or Tahitiennes, spoke English. He could easily get a girl for his personal wants, but he needed someone like me for the hotel. He called me a few days later and I decided to go. It would be for two months, over the Christmas holidays. Dumont purchased my ticket and departed quickly with a very sweet Tahitienne. I left a few days later.

The plane flew over Fiji on the way to New Caledonia, and I saw with fond remembrance the brilliant green vegetation and sparkling rivers, hoping that New Caledonia was the same.

It was entirely different. I knew it was a large, or rather, a long island of about 8,500 square miles. Its mountains were brown and barren, and smoke was rising from circular areas on the mountain slopes. I asked the steward if they were volcanoes. He replied they were brush fires, that it was an exceptionally dry season. He insisted the island was green and beautiful the rest of the year, but it was hard to imagine. The desolate terrain looked no better as we approached and landed at Tontuta, the airport that was about fifty kilometers from Noumea, the capitol. As I made my way through customs someone hailed me. I didn't expect Dumont to meet me. It was not he; it was an ex-G.O. who now worked for a diving outfit in Noumea. I was pleased to see a friend and on the way to Noumea he told me about the country.

"New Caledonia is unique because it was formed entirely differently from any other island. At the time when the world was developing its land masses, this area was at the bottom of the ocean. During the upheavals, the forces in the center of the earth squeezed the ridge of the ocean floor at this location, thrusting it up out of the water, creating the island of New Caledonia. Therefore it is made of a different substance than volcanic or coral islands throughout the Pacific. It is rich in minerals and nickel is its greatest wealth."

"Why is it called New Caledonia? It has no resemblance to Scotland."

"Captain Cook must have discovered it during the rainy season."

I went to the hotel where Gilbert Dumont said he would meet me. He was not there but he telephoned. "I'll be coming over in a couple of days. Your room at the hotel is reserved. Look around Noumea until I arrive."

I was delighted. The hotel was cottage style and very pleasant. When I went into the dining room I saw the crew of the UTA airline I had flown over on. They were celebrating a birthday. I knew some of them who had come to the Club. "Come and join us!" they hailed me. It was a hilarious evening. After dinner we went to a couple of nightclubs, both with Tahitian entertainers, as all the entertainers were then. We danced far into the night.

The next day I set out to explore Noumea, which had been the second largest military base in the Pacific, after Honolulu, during the Second World War. Little evidence of that occupation remained. From the bus terminal one could take the jitneys serving each of the valleys that made up Noumea and see the whole town. The center was the Place des Cocotiers but there were few coconut trees. Instead there were rows of glorious royal poinciana trees, their scarlet flowers in full bloom. Around the square were a few fairly modern stores, but the rest of the town was like an American western film set. Instead of cowboys there were *indigènes*, the Melanesians. They were not tall like the Fijians; they were short and generally squat, with broad, full-lipped features. But when they spoke their language was melodious; a group of women sounded like birds twittering. The Melanesians were only part of the population. There were the Caledonians, many of whom are descendants of the settlers, and convicts sent from France from 1864 to 1894. I never learned if they were proud of their colorful ancestry, the way the old families of Australia are. They were very provincial in Noumea. There were Indonesians, Polynesians, Tonkinese, and other races and nationalities of the Pacific and Orient. The Vietnamese and Javanese women were often dressed in sarongs and tight little shirts as they do in their countries. The Melanesian women wore bright flower-sprinkled printed mother hubbards. I watched a softball game, of which the women are passionately fond, and the players, wearing the voluminous dresses, hiked the skirts up around their thighs as they raced around the bases.

I had a brother-in-law in Noumea, Teva's oldest brother, Titema. He and his wife had lived there for several years; he was a supervisor at one of the nickel mines. I did not know them and was rather shy about visiting because Teva and I had been separated for four years. But when I called they were delighted, and invited me to go to their house for dinner. There was Titema, his wife, and a little boy, their adopted daughter's child whom they were raising. They welcomed me warmly. I was fascinated by Titema's resemblance to Teva. He was nine years older but he had the same features and build, the same expressions and manner of walking. He spoke as quietly as Teva did. It made my heart ache. He wanted to hear all about the family whom he had not seen in ten years. I asked how it was to live in Noumea. Titema said the people were jealous of Tahitians, who were the best workers in the nickel mines and were the entertainers. They had their own community and lived as though they were in Tahiti. The pay was good in Noumea but they all wanted to go back home eventually.

When Gilbert Dumont came over to meet me he had his Tahitienne with him, and said he would be staying in Noumea a couple of days. I was to go to the Isle of Pines and wait until he arrived.

"There is going to be a shake-up in the personnel. If they are not correct with you, remember you are the one to give the orders."

He was making my introduction to the hotel very

difficult but he seemed not to realize it. I wondered what kind of a reception I'd get.

I took the flight across to the Isle of Pines. There were many small islands and from a distance they looked like sailing ships with many masts. The water of the ocean had receded over centuries, leaving fringe coral shelves a couple of feet or more above the surface, the undersides washed away by the waves. Norfolk pines with short branches stood tall and straight, like masts. The pines were dark green, the ocean a sparkling deep blue, and the sand was pure white, so powdery fine that it actually floated a moment when thrown on the water.

I was met at the hotel by Sacha, much to my surprise. He was a young Frenchman who had lived a short while in Tahiti and had a reputation of being sly and manipulative. I wondered how he got along with the authoritative Gilbert Dumont. He seemed to be genuinely pleased to see me; I hoped he had acquired his reputation unfairly. He introduced me to the guests, emphasizing my qualities of hostess and my travel experiences so I was soon involved in conversation and felt very welcome. There were several French people but most of the guests were New Zealanders and Australians. They were quite thrilled when — because I knew how popular he was everywhere in the United Kingdom — I told them I was the daughter of Armand Denis.

The personnel, mainly Melanesians, did not welcome me. They were rather sullen and suspicious, so I simply smiled when I was presented to them, then returned to the guests. Sacha seemed to have everything running smoothly enough. When Dumont arrived he called the personnel together, told them I was now their chief and they must cooperate with me. He was very blunt and tactless; I learned he was always that way, even with the guests. He was not popular.

The Melanesians watched me and I was very careful not to antagonize them in any way. Fortunately the cottages were cleaned very well and only occasionally lacked a towel or soap. I had a harder time getting the confidence of the boys in the dining room, but when they saw I would help, pouring coffee and making toast when there were a lot of people, they began to accept me.

I had arrived at the hotel about three weeks before Christmas. I accompanied the guests on explorations of the small island. There were the caves, the entrance to them a fairyland of giant ferns under towering trees, where huge, red bats lived. They fed only on fruit, the mangos, papayas and other fruits that grew here as in all the Pacific islands. The Melanesians ate the bats and said they were delicious, but they were not served to the guests in the hotel. We visited the village where the faithful attended church services dressed in white as the Tahitians did, wearing woven and lacy white hats. We visited some homes, rude huts made of branches coated with mud or clay. They were clean enough but very primitive, without decoration. Only the meeting house had carved and tinted panels by the entrance. The Melanesians collected shells, of which there were many and varied, including the extraordinary, ancient convoluted nautilus.

Diving in the waters around New Caledonia was fantastic. The coral was the most beautiful I had ever seen, and the variety and size of the fish sent spearfishermen into raptures. We saw snakes curled up on stones at the edge of the water and they slipped slowly into the water when I approached. They were easy to catch, making no attempt to bite, fortunately, because when I told Sacha I had picked them up he said they were poisonous.

We worked hard the day before Christmas to prepare the feast for Christmas Eve. The party started about seven with cocktails and canapés, and the atmosphere was very festive. About nine-thirty we went into the dining room where two long tables were prepared for all the guests. Sacha had gone home so there was only Dumont and me as hosts. Dumont grouped the French people around him at the end of one table while I seated the English guests. We were not more than fifty people but after seating everyone I saw there must have been a miscount because there was no place for me. It would have been simple to add another place but Dumont coldly dismissed me and returned to his table. The guests noticed but I said gaily that I would be right back. I went out to the kitchen where the Melanesians had prepared a table for themselves. They were surprised to see me, but jumped up delightedly to give me a seat and treat me like an honored family member. When I returned to the hotel guests after dinner I had a hard time brushing off questions about where I gone.

The morning of New Year's Eve I caught a cold. I had started to help with the preparations but by afternoon I had red eyes and a runny nose so I went to my room to rest. Pretending to be quite ill, I sent word that I could not participate in the festivities that night. I was not going to risk another embarrassment like the Christmas dinner. Two giggling housemaids came to tell me that Dumont was furious but I didn't care.

When I showed up the next day all the guests greeted me with pleasure and inquired solicitously how I was feeling. I felt fine, the way one does surprisingly after those twenty-four hour colds or allergies that stop as suddenly as they start. Everyone was very gay. All the personnel, well aware of Dumont's anger the night before, grinned at me and were particularly efficient and courteous. That evening Gilbert went into his office, dashed off a letter and handed it to me. It was my dismissal.

I was glad to leave. Although Dumont refused to pay me I had my return ticket. My departure was rather obstreperous. The little transfer bus was waiting outside. I had given my Tahitian shell leis to the personnel. They all brought me flowers and shells and crowded around the bus with the guests. Even a friend of Dumont climbed on top of the bus, clowning. Dumont glared from inside the building. I left in a shower of best wishes and fond farewell.

18

Social Director

The lobby was impressive, with Italian marble floors and columns, and showcases in the center of the large hall. The far side, glass the whole way across, promised a magnificent view of the ocean, swimming pool and gardens, if you dared to open the glass doors to go out on the terrace. There was air conditioning, so the whole hotel was sealed against the sun, tropical breezes and perfumed night air. There were elevators, and one walked along marble corridors to a wall-to-wall carpeted room beautifully furnished with plasticized fake bamboo furniture and Thai silk curtains. "Maeva" meant welcome. I was sure I would not want to work in that cold, luxurious hotel.

I had stopped by a couple of days after the hotel had opened to make a phone call, and Yvonne Laurent was there. I knew she was the social director and perfect for the job. She was very attractive, had great poise, and knew everyone in Tahiti. I would have been willing to know her better but she had an irritating habit of breezily saying hello, then turning away as though she was too busy to say anything more. If she did stop to talk a moment it was in a condescending, amused manner that made me think she must be quite shallow.

When Yvonne saw me that day she exclaimed, "Renée! I want to see you. Can you come into my office as soon as you are free?" That was a change.

"All right. I have only a phone call to make." She waited, and then took me into her office as though we were old friends. A young woman was working at the desk, which was a large packing case. Yvonne explained, "I'm holding out for the furniture I want, and the manager has promised I'll get it because obviously I can't be receiving guests before a packing case, can I?" The woman shot her an angry look for some reason. "My assistant, Nicole."

Yvonne was organizing entertainment and activities for the clients. "Of course there is always the tour of the island but I want to offer other tours as well. Picnics, hiking, horseback riding. I see you riding along the beach in Punaauia and I know you did a great tour for Club Med. Would you be interested in putting together a horseback riding and picnic excursion for the hotel?"

I thought a moment. "It would not be easy. It is difficult to get enough riders together at the same time to offer it at a reasonable price. It has to be worth while for me to do it instead of the island tour. I would like to give it a try. There is a perfect place for a picnic by a river deep enough for swimming, and although it is a beautiful ride it takes an hour and a half to get there from the stables. Not many people will want to ride for three hours."

"It sounds wonderful! Can you take me there to see it tomorrow? But I won't have the time to ride — can we go by jeep?"

That was typical of modern adventurers. "I don't know. Part of the way, yes."

A bellboy came into the office and put a list of names on the desk. Yvonne glanced at it and exclaimed, "Oh, these arrivals at five o'clock! Nicole, can you do the arrival tomorrow morning?"

Nicole jumped up from her chair. "No! That's enough! It isn't my job to meet planes."

Yvonne shrugged her shoulders. "It isn't my job either but somebody has to go. I'm here until eleven at night; I can't be up at five in the morning. Somebody else has to go. Renée! If I can persuade the manager to hire another person, would you accept the job of meeting the planes?"

It was the best way to make contacts for the tours, meeting and assisting at the airport. "Yes," I said. "I could do it."

"Wait here." Yvonne went immediately to talk to the manager. She came back in five minutes. "He wants to meet you but doesn't have the time now. Can you come this evening? Come at six-thirty. He can see you then and we'll have dinner afterwards."

"Don't you go home?" She was married and had three girls. "Of course, when I can, but there is too much to do here," she said lightly. Yvonne and her husband, Jean Jacques, had worked for a travel agency when they first came to Tahiti; then Yvonne had switched to UTA airlines and Jean Jacques opened a shop of South Pacific artifacts. He traveled often to New Guinea, New Hebrides and other islands. The hotel had on display many of the artifacts of Michoutouchkine, who had the largest collection of South Pacific artifacts, but Jean Jacques also had many extraordinary pieces.

I returned to the hotel at six-thirty; Yvonne was waiting and took me to the manager's office. The manager was Ivan Falesitch, an attractive Swiss, about 40 years old. He had an abrupt manner of speaking, and looked at me keenly. "Yvonne tells me you are the best guide on the island and she wants you to organize some tours for us." We discussed tours for a few minutes; he nodded, satisfied. "So you'll be able to do the arrivals too? Can you start tomorrow morning?"

I could and did. Later Yvonne and I rented a jeep and drove a ways along the trail I proposed for the horseback riding tour. Yvonne thought it was great, but as it turned out I was never able to get two couples together the same day for the tour. Also, Nicole quit abruptly, so I worked with Yvonne to organize activities.

There were problems in the hotel. The whole first floor was inundated when a water pipe unaccountably broke. The large glass panels of the entrance simply shattered one day. The water in the swimming pool became murky and impossible to clean, no matter how much it was treated or changed.

The Tahitians said the problems were caused by *les esprits*, the spirits. The hotel was built upon land where there was once a *marae*, the site of a Polynesian temple, and the spirit of the god, to whom it had been dedicated, was angry. They said that a *tahua*, sorcerer or healer, could tell what should be done to placate the spirit. Ivan Falesitch was desperate, and asked for a *tahua* to come to the hotel.

An old man arrived with his assistant or co-sorcerer;

Ivan welcomed them with respect despite his doubts, and started to explain what was happening. But the *tahua* raised his hand to stop him — he barely understood French anyway.

"I know already what is the trouble," he said. The assistant translated. They walked directly to the display cases of artifacts, all of us following and watching the *tahua*'s face. He went to every case, sometimes stopping and frowning, before he returned to the first one he had gone to. Still, he did a tour of the hotel, the halls and corridors where there were more artifacts. We all followed silently. The two men talked together in low tones, particularly before the statues of gods or spirits, that they called *tikis*. They returned to the first display case where there was a particularly malevolent looking *tiki*. The *tahua* spoke and the assistant translated.

"This is the one. He does not like to be in this box. He wants to get out." That made sense. "And he wants a *vahine*."

"Aah," we all sighed with relief, as it was only natural in Tahiti for a god as well as a man that he can be happy only if he has a woman.

The *tahua* knew what he was talking about. The *tiki* was removed from the hotel. Fortunately Jean Jacques had a female *tiki* who must have pleased him because there were no more accidents in the hotel. Jean Jacques benefited too, as his business picked up and thrived.

A few days later Yvonne didn't come to work as expected. I had delayed starting my tours until everything was running smoothly. When she didn't come the next day I called her and she asked me to go to her house.

"I guess you know that Jean Jacques and I have not

been getting along for some time," she said, "I suppose you've heard the rumors in the hotel."

"Rumors? No. I don't work in the hotel," I reminded her.

"Well, you know Tahiti. They say I am staying at the hotel all the time because of Ivan Falesitch. To make a long story short, I am going to leave."

"You can't leave! Yvonne, there is no one who can do the job you're doing. Who would take your place?"

"You."

"Oh, no," I laughed. "Out of the question. I cannot be a social director. I'm not social."

"Look, Renée," she insisted, "it is not as 'social' as you think. How much of the time am I actually being a social hostess? The big dinners for the local and Paris VIP's are just about over now that the hotel is opened and running; besides, there is Ivan's wife to hostess those. What is more important is the organization of activities and entertainment. You are better at that than I. I told Ivan you are the only one to replace me. You'll get a good salary. Please go and see Ivan."

I liked the fact that Ivan let us call him by his first name. With the advent of more Frenchmen coming to Tahiti to establish larger businesses as well as modern hotels, the stuffiness of director versus personnel relations was moving in.

I went to see Ivan and he assured me that I would never have to hostess dinners or do anything that would make me feel uncomfortable. "The title of social director is used on all the information and publications so we can't change it, but forget the meaning. You know what is important for the clients, you work out the hours you need to be here, but I really want you to continue meeting the planes. That first impression is so important, particularly for the Americans, and makes it easier for the rest of the personnel during the guests' stay." I agreed to become social director, continue meeting tour groups and VIP's, and organize the entertainment programs.

The Camera Show was my greatest success and, after a couple of years when I had worked a season in each of the other big hotels (big for Tahiti was about 200 rooms), the Beachcomber and the Taharaa, it became a tradition in the hotels to have the show every Sunday noon on the dining room terrace. As many local people came as the tourists.

The program highlighted the Tahitian dances, of course, but we also demonstrated different ways to tie the *pareu*, and showed how flower crowns, hats and baskets were made, the girls sitting attractively in a group for photographers. We demonstrated how coconuts were husked and the delicious cream extracted, and how Tahitians shinny up a coconut tree. As there were no trees directly in front of the terrace, I had to take the people out into the garden. The first Sunday we were not prepared. Unthinkingly I said to the dining audience, "Follow me!" which they all promptly did, leaving the bills for their meals on the tables unpaid for.

19

French Bureaucracy Foiled

"Renée! Renée! Wake up!"

I closed my eyes tightly and hoped he would give up and go away. I hadn't had a day off in three weeks. The hotel arrival list had no VIP's or tour groups due today, and no special events scheduled. A rare day of nothing happening. So why was Manunui pounding on my door?

I sat up and looked out the window at my garden. Pink and yellow frangipani had fallen on the grass still wet with dew, and sprays of purple bougainvillea spilled over the hedge of hibiscus and ferns. They needed to be cut back. Everything grew so fast in Tahiti one hardly had time to keep up with it. I looked forward to working in the garden. It was so peaceful there at Fisherman's Point, with only the sound of the ocean waves and the mynah birds in the mango trees.

"Renée!" Manunui wasn't going to leave.

"Yes, all right!" I answered, "I'm coming." I reached for my *pareu*, wrapped it around me and went to the door.

"*Bonjour*, Manunui. What is the matter?"

The chief porter at the Hotel Maeva Beach, Manunui was a big, dark-skinned Tahitian with curly black hair held back with a strip of red fabric, his round, earnest face already perspiring despite the early hour. He blurted,

"At the hotel, Renée — a tourist died!"

"Died! Oh, Lord. Who is it?"

"An American, so they sent me to get you." He pulled a piece of paper from his pocket and gave it to me. "Mr. Dudley, member of tour group," was scrawled on it. "He died during the night — a heart attack," Manunui said. "His wife didn't even know. He looked like he was still asleep in the other bed but she couldn't wake him up. I'd just come to work and they sent me for a priest — they're Catholics. The doctor came and we laid out the body."

"His wife — how's she taking it?"

"She was crying a lot but she's better now. The tour conductor asked for you."

"Yes, of course." My job as public relations manager (Social Director) covered many duties; also, I was the only American in the hotel. "Thank you, Manunui. You go on back. I'll be there as quickly as possible."

I hurriedly showered and dressed. In a hotel, I thought, it is bound to happen some time. In fact, it was not the first death at the Hotel Maeva. A week before the hotel opened, an old Tahitian man had been hired to scrub the kitchen floors late at night. Celebrating his new job that evening before going to work, he and a couple of pals had drunk a case of beer down on the beach.

The swimming pool was located between the beach and the hotel and the pool lights were not yet turned on. The Tahitian, too drunk to follow the path through the garden, staggered into the pool and drowned. His family came to get him, the wife in a terrible temper. His sons picked him up, their shoulders cowered in embarrassment as she cried over and over, "*Ill l'a merité!* He deserved it!" But she finally broke down, sobbing, as he was carried away. He was buried by four o'clock the next day, as was necessary because of the heat.

The manager was waiting for me at the entrance of the hotel. I was surprised that he did not appear disturbed.

"*Bonjour*, Renée," he greeted me. "Thank you for coming. Mrs. Dudley is quite calm now and she will tell you what she wants done. Yes —" He was distracted by the porter who had just arrived with a golf bag; then he turned to me again. "Gaston will help you, Renée. I'll be back, uh, about noon."

"Monsieur!" I exclaimed. "Don't you think —?" But he was gone.

Gaston, the Front Desk Manager, was busy with the early morning arrivals. He stepped back from the counter when he saw me. "Renée, thank God you're here. Mrs. Dudley is in your office with the tour conductor. They're waiting for you. It is not going to be easy," he said sympathetically. "Mrs. Dudley says she wants the body sent to the States."

"Sent to the States?" That had not occurred to me, but of course she would. I looked at Gaston doubtfully. "Has it been done before?" I did not know of any tourists dying in Tahiti.

"First things first," Gaston said practically. A debonair Frenchman with years of experience in European hotels, he was not flustered by the death of a guest. "I sent a driver with the death certificate to get the *Certificat d'Etat Civil* . He'll be back soon, and then we'll find out about the shipping procedure. You go and see Mrs. Dudley now." He turned back to the reception desk, smiling courteously.

As I crossed the hall to go to my office, a Tahitian woman rushed up and grabbed me by the arm. "Mademoiselle," she cried, "I heard what happened!" I paused politely but I didn't recognize the woman. Middle-aged, her graying hair pulled typically into a knot at the back of her head, the only remarkable thing about her was the most lugubrious face I had ever seen. "You don't know me," she said sadly.

"I don't believe so," I murmured, reaching for the doorknob. "Excuse me, but —"

"You will, you will," she said, clutching my arm more tightly. "I came to see the bereaved."

"Oh? Do you know her?"

"No, no. But I have come to give her compassion. I," she paused for emphasis, "am Madame Berthaut." I still looked blank. "I am the undertaker."

"What!" I exclaimed, and then remembered having heard that the undertaker in Tahiti was a woman. I said, "Thank you, Madame, for your concern. It is very kind. We'll let you know if there is anything —" I slipped through the door, closing it firmly behind me.

Mrs. Dudley, an attractive woman of about fifty-five, was sitting in the armchair by my desk. She appeared calm but her hands were clasped together tightly. Her eyes were still red from crying but her hair was neatly combed and she had even put on a bit of rouge.

Linda, the young and pretty tour conductor, sitting on the edge of her chair, was even more distraught. "Oh, Renée," she gasped when she saw me. Mrs. Dudley looked up with relief.

"Mrs. Dudley, I am terribly sorry," I said, taking the hand she gave me. She smiled gently.

"Thank you, dear. Actually, it was not unexpected. Our doctor warned him that he should slow down. But he said he would rather take the chance, lead a full life, rather than live like an old man before his time. The doctor, this morning, assured me that he didn't suffer at all."

I patted her hand, admiring her self-control. She told me that the priest was arranging a special mass for her husband. Then Linda said, "Renée, you know the tour group has to leave tomorrow morning. Do you think it would be possible for — Mr. Dudley to go on the same flight?"

Mrs. Dudley looked at me anxiously. In twenty-four hours? I had not the faintest idea how bodies were shipped, but I well knew the ponderous wheels of French bureaucracy.

"I really don't know if it can be done so quickly, but I promise you, Mrs. Dudley, we will do everything possible to get him on that plane."

She clasped my hand gratefully. I kissed her check and said, "If you will excuse me now, I must start immediately to make the arrangements."

"Certainly," Linda replied. "I'll stay here with Mrs. Dudley."

Out at the reception desk, Gaston had finished with the arrivals and the guests had gone to their rooms. I asked, "Where is Mr. Dudley now?"

"In his bed," Gaston said. "We can't carry him

through the halls while guests are coming out of their rooms for breakfast. Besides, where else could we put him?" Of course. There was no such thing as a funeral parlor in Tahiti.

"Gaston, Mrs. Dudley wants to have his body shipped to the States on the same plane they're taking tomorrow morning."

"What did you tell her?"

"That we'd do our best."

"That's right. But you know it is impossible."

"Yes. Regulations, documents, permits." I thought a moment. "But really, it shouldn't be all that difficult. Did you get the *Certificat d'Etat Civil*?"

"Not yet. The driver came back and said the *tavana* has gone fishing." The *tavana* was responsible for such duties as legalizing documents. The *Certificat d'Etat Civil* was the official record of a death certificate.

"Well, he should be in his office soon."

Gaston looked at his watch. "I can stop by there on the way to Papeete to apply for the shipping permit. Where is the *tavana's* office?"

"You know that old green grocery store that's at the sharp curve, mountain side, about kilometer ten? It's behind there."

"Oh, is there a road back there?"

"No. The office is in the back room of the store. You go down the right side, past the beer cases."

"*Mon Dieu*. All right, I'll find it."

I turned to the problem of getting a coffin. Where had I seen coffins in town? The telephone directory — that'd tell me. What was the word in French? Cercuil. I flipped through the C's. *Cadastre*, *carrelage*, *charpentier* — no *cercuil*. Wait — *charpentier*, that's carpenter. The furniture store where they made me the kitchen cabinet; that's where they were. Hop Chin Lee was the name of the store. Going into Gaston's office so that no one would hear me, I found the number and dialed.

"*Allo!*" The voice was Chinese, impatient.

"*Bonjour*," I said. "*Avez-vous des cercuils*?"

"Non!"

"Oh. Then can you make me one?"

"Not before next week."

"Next week? But the man is already — Well, can you tell me where I can get one today?"

"No." He added disgustedly, "Too many people die." He hung up. I thought, that ought to motivate him to keep more coffins in stock.

Then I remembered the Tahitian who makes canoes, just down the road. He was an Adams, a descendant of the Bounty, and, although the family had become almost pure Tahitian again, he was proud of his British ancestor and was a good friend of the British and American residents. Perhaps he would make a coffin for me. I slipped out the side door of the hotel and walked to Adams' place, a short distance by way of the beach.

Adams' house was typically Tahitian. The high pitched roof was thatched with layers of plaited coconut fronds and the walls were of the intricately woven bamboo. Hibiscus and *tiare* bushes dotted the garden that was simply cleanly swept, white sand. Adams was in his workshed, a wide thatched roof that protected him from the sun and rain. He was chopping into a freshly peeled tree trunk, destined to be an outrigger canoe, when I arrived. He put down his ax to greet me and listened courteously while I told him about my problem.

Yes, Renée," he said, "I can make you a coffin before

noon. But you say that you have to ship the body away. It is not easy. You had better call Dave Cave."

"Dave Cave? The American who has the construction company?"

"Yes. He has a welding shop. He knows what you have to do. You can call him from here." He took me into the house. Adams' wife, smiling hospitably, dropped her broom and tightened the knot of her *pareu* as she hurried forward to show me where the telephone was. I dialed.

"Hello Dave? It's Renée, at the Hotel Maeva —"

"Renée — yes, of course. What can I do for you?"

"An American has died at the hotel." I told him about my problem. "— and Adams said you could tell me what must be done to ship the body to the United States."

"Good thing you called right away," Dave said. "It's pretty complicated. First you must get two coffins."

"Two coffins?"

"I'll tell you why in a moment. Then you'll need a shipping box, a steel case, and that is what I'll make for you. It'll be expensive."

"Mrs. Dudley says it doesn't matter how much it costs. You must be sure and send the bill to the hotel before she leaves in the morning."

"She isn't going to wait until her husband's body goes?"

"Oh, I didn't tell you. She wants us to try and get him on the same flight she is to leave on tomorrow."

"Tomorrow? The ten o'clock flight? That's awfully close. Well, the shipping case can be finished by four this afternoon and I'll reserve the space on the plane. You have to get the shipping permit — but get the *Certificat d'Etat Civil* first. The death certificate is not enough."

"I know. Someone from the hotel has gone to get it."

"Where is the body now?"

"In his bed. We're waiting until most of the guests are out of the hotel."

"You'll have to get him to the morgue, you know."

"The morgue?" I hadn't thought of that.

"It's behind the hospital. Better give them a call." He went on. "Now you need a zinc coffin. I'll have one made and sent over to the morgue by noon. That coffin will be sealed and placed inside the wooden coffin, which will be put inside the steel case. Is that clear?"

"Yes, quite," I said. "Dave, I really appreciate your help—"

"All right. Now this is what you have to do meanwhile. You must get a hold of some charcoal, lime, and sawdust."

"Uh what—?"

"Crushed charcoal, quicklime, and sawdust. It's to be put in the zinc box with the body."

"Oh!" I understood. For the same reason why people were buried so fast in the tropics. No such thing as embalming either. "I can get the sawdust at a lumber mill and there's charcoal at the marketplace, but where do I get quicklime?"

"From the Sanitation Department. Get a couple of bags of each and take everything to the morgue. And call the Gendarmerie. They witness the sealing of the coffin."

"Okay. I think I can remember all that."

"Good girl. You'll manage it, I'm sure."

Thanking the Adamses for their kindness, I returned to the hotel. I called the hospital for an ambulance. The attendant asked, "You say the man is dead?"

"Yes. You'll have to take him to the morgue."

"No, lady, we don't. Not when he is already dead. You get the hearse for that."

"But it isn't for a funeral. He has to go to the morgue."

"I heard you. Get the hearse."

Damn. It meant dealing with that dismal woman, Madame Berthaut, who was here this morning. As undertaker, all she actually did was drive the hearse from the bereaved house to the church and cemetery. If it was just a question of transportation — I thought of Manunui, the porter, and his old pick-up, but then shook my head. It wouldn't do.

I called Madame Berthaut. Needless to say, she was delighted to be of service. She asked, "What time will the coffin be sealed?"

"It can't be sealed until it gets to the morgue."

"Oh. Is that all right with the gendarmes?"

"Well, I should think so. Everything to be put in the coffin will be at the morgue."

"You had better call the Gendarmerie and ask their permission," said Madame Berthaut.

I called the Gendarmerie and explained about my problem. The agent listened attentively and then asked, "Do you have the *Certificat d'Etat Civil*?"

"*Naturellement*," I said. Gaston must have gotten it by now.

"And you want to transport the body from the hotel to the morgue."

"Yes."

"The coffin to be sealed at the morgue."

"Yes."

"You can't do that."

"I can't?"

"It is the law. You cannot take a body in an unsealed coffin through the streets of town."

"Oh." I thought quickly. "Well then, it can be sealed here."

"But you are sending the body away."

"Yes."

"In another coffin."

"Yes." I explained it all over again and concluded with what I thought was a logical solution. "— and so we can just open the coffin to put the body in the other one and seal it again."

"No," he said, "you can't do that."

"Why not?"

"Once it is sealed it cannot be opened. It is the law."

Hm, I thought to myself, we'll see about that.

"And, Mademoiselle," the agent continued, "I advise you not to have us come and seal the coffin there and then go to the morgue and do it again. This is a small town and we will know what you are up to."

I hung up. Damn. How to do it? Pierre! My former neighbor and good friend, the Chief of the Commissariat de Police. Why hadn't I thought of him before? I quickly found his number and rang. "Pierre? It's Renée Denis. Remember? I lived next door —"

"*Oui, certainement*! What can I do for you?"

I explained my problem again. "But that is not a problem," he said. "Just a moment." In a couple of minutes he was back on the line. "*Voila*! It is very simple. You shall put the body into the wooden coffin in the hotel. You will close it but not seal it. To satisfy the Gendarmerie we will have two policemen standing there watching you. And they will accompany the body in the hearse to the morgue. They will wait there until the body is prepared in the other coffin and then they will see it sealed. Is that all right?"

"Oh, yes, Pierre! Thank you. You are very kind."

"*C'est bien*. Now don't be a stranger. Come and see us."

"I will, I will. *Au revoir*!" How fortunate and nice to have good friends.

It was now ten o'clock. Where was Gaston with the *Certificat d'Etat Civil*? I crossed over to his office. No one there. I went out to the reception desk.

"Oh, there you are," one of the girls cried, having forgotten to bring me the message. "Monsieur Gaston called. He said to tell you the *tavana* went fishing at Moorea because the weather is so fine. He won't be back until tonight."

"Tonight!" I sank into a chair. Then the body can't possibly leave tomorrow. Without the *Certificat d'Etat Civil* we can't get the shipping permit. Maybe the police will not even let the coffin be sealed without it.

Well, I still had to get the body to the morgue. Adams said he would have the wooden coffin ready at noon. I had two hours to get the stuff to put into the zinc coffin. I went to my office and called the Department of Sanitation.

"*Allo*."

"*Bonjour*. This the Hotel Maeva Beach calling." People were usually impressed. "We have a death in the hotel, and we need —"

"This is the Department of Sanitation."

"Yes, I know. It seems that you have quicklime, and—"

"Quicklime? You'll have to talk to Dr. Perrin about that." She put me on hold while she buzzed his office.

Damn. Cantankerous old Dr. Perrin. He was head of the department and I had thought that for a little lime—

"*Allo*?"

"Dr. Perrin? This is Renée Denis, at the Hotel Maeva Beach. We have had an unfortunate occurrence—" and in as professional a manner as possible I explained what had happened. "So you see," I concluded, "we need a small amount of quicklime."

"We don't sell quicklime here," Dr. Perrin said sharply. "We are not a store."

"Oh, I know you don't sell it. I would buy it if that were possible, but—"

"Do you think we keep a supply on hand for the tourists who die in Tahiti?" he demanded. "This is the Department of Sanitation." He made it sound like the House of Lords. "Try Taravao. They may have some at the dairy. Good-bye."

Taravao! That was the other side of the island. It would take several hours to go there and back. I called Dave Cave again. "You know Dr. Perrin at the Sanitation Department?"

"Uh-oh. Yes. Why?"

"That buzzard won't give me any lime!"

"Hm. Maybe the island is in short supply. Well then, you have to know the right people."

"Who's that?"

"In this case, General Durand's wife. You are in luck. It so happens my wife is entertaining Madame Durand and other illustrious guests in honor of the opening of the library's English Room. Madame is very fond of Americans."

"I don't understand."

"The army has quicklime. The army never runs short

of supplies. I shall call Madame Durand, who will call General Durand, and he will be delighted to give you some lime."

"Are you sure?"

"I am sure. Now you go get the sawdust and charcoal."

I hung up. At that moment more help arrived in the form of a timid little man who appeared at my door. French, not Tahitian, my mind registered, because of his balding head, glasses and white shirt. "Yes?" He came in hesitantly and introduced himself.

"I am Monsieur Berthaut. My wife, Madame Berthaut, the undertaker, has told me to come and help you." I gazed at him, thinking, surely the perfect husband for a woman undertaker.

"That is very kind, Monsieur Berthaut. I really don't know —" But perhaps he could help. "Monsieur, I have much to do. Could you possibly go to town for me?"

"Certainly," he said eagerly. "I have a car." He waved towards the window. There was a Deux Chevaux, the humpbacked tin car that looks as though it is about to plow onto its nose.

"That's fine. Now what we need is some sawdust and charcoal."

"Sawdust and charcoal? Is that all?" He was disappointed. I explained what it was for. He nodded and then asked, "Do you have all your papers?"

We didn't have any. "I wish we did," I sighed. "The *tavana* has gone fishing so we don't have the *Certificat d'Etat Civil.*"

"Then how can you ship the body tomorrow morning?"

"It looks as though we can't."

Monsieur Berthaut said, "You could have him die in Papeete."

"What?"

"The American. You could have him die in Papeete instead of here in Punaauia."

"What are you saying?"

"I have a friend, a doctor. He can make you another death certificate saying that the American died in Papeete. We take it to the Registry Office at the town hall and they give us a *Certificat d'Etat Civil.*" I stared at him. This meek little man proposing an intrigue. He shrugged his shoulders apologetically. The phone rang.

"Mademoiselle Denis? This is Dr. Perrin." He sounded so nice!

"Yes, Dr. Perrin?"

"I beg you to understand that there has been a shortage of quicklime in Tahiti. We have to be very careful. But I have just received a call from the adjoint of General Durand. The army will be glad to replenish my supply of quicklime if I should run out. It is generous of them. So now I can give you some quicklime."

"You are very kind," I said in my sweetest tones. "Thank you, Dr. Perrin. I shall come personally to get it. *Au revoir.*"

I turned to my new assistant. "Monsieur Berthaut, things are looking up. However — I shall have to give your idea some thought. Now let us go immediately to town. You get the sawdust and charcoal, I'll get the lime, and we'll meet at the morgue."

I drove to the Department of Sanitation. As I gave my name at the desk Dr. Perrin, portly, immaculately dressed in tropical whites, came through the door on the other side. "Ah, Mademoiselle Denis," he greeted me. He spoke to the clerk at the desk. "Have a boy bring up two sacks of lime and put them in the lady's car." He took me by the arm, saying, "It will take a moment. Come into my office while you wait."

"Thank you. I do appreciate your help," I murmured as he pulled out a chair for me. He sat down in another chair facing me quite closely.

"Now tell me what happened," he said. Again I repeated my story. He listened, very sympathetic. Something nudged me — why not try Monsieur Berthaut's idea on him?

"I have a friend," I said casually, "who says he knows a doctor who could help me. He would be agreeable to making out another death certificate for Mr. Dudley so we could get a *Certificat d'Etat Civil* in Papeete."

"Hm," said Dr. Perrin. "You mean, (he put his hand on my knee) saying that he died in Papeete?"

"Well, yes." I smiled as though it was just an amusing idea.

"I see." He leaned back in his chair and studied me. "It wouldn't make any difference now, would it? As Mr. Dudley is already dead. And you are performing an act of kindness in trying to get his body shipped out on the plane tomorrow." I nodded. After a moment a broad smile it his face. He stood up then and said heartily, "Well, let's say I don't know anything about it, my dear — and I hope you succeed!" He led me to the door. "Glad to be of service, my dear. Call me at any time. *Au revior* !"

Grinning to myself, I ran lightly down the steps to my car.

"Mademoiselle! Mademoiselle!" It was Monsieur Berthaut. His car skidded on the gravel as he braked, while holding up the flap window to peer out at me.

"Monsieur Berthaut — what is the matter?"

"It is the charcoal, Mademoiselle. It comes only in lumps. What shall we do?"

Suddenly feeling hysterical I collapsed against the car, unable to keep from laughing. Monsieur Berthaut stared at me through his little window, shocked by my behavior. Getting myself together, serious again, I said, "Monsieur Berthaut, we must smash the charcoal. But first, you and I will go to see your friend, the doctor. Leave your car here. We'll take mine."

He quickly parked the Deux Chevaux and hopped into my car. "To the left and into the center of town," he directed me. The young doctor, newly arrived in Tahiti, was fascinated by our story which of course I had to tell all over again. Within twenty minutes we had our new death certificate saying that Mr. Dudley had died at a certain address in Papeete. We rushed to the town hall — and arrived too late. The offices had just closed for the noon break. "We can get it this afternoon," Monsieur Berthaut said. "Anyway we must hurry to the hotel. Madame Berthaut and the police will soon arrive there."

"Yes, that's so, but we must also prepare the charcoal. Monsieur Berthaut, can you take it to Dave Cave's welding shop where they will certainly have large hammers? His men will be glad to help."

Monsieur Berthaut nodded, looking distressed, but I had too much on my mind to wonder why. As I left him at his car he said gallantly, "I shall manage."

I drove back to the hotel. Gaston had returned and he shook his head regretfully when he saw me. "Sorry about the *Certificat d'Etat Civil.*"

"It doesn't matter now," I assured him. "We're getting another." I started to tell him about our tricky solution, when Gaston cried, "*Mon Dieu*! Look at that!"

I whirled to look out the window behind me as Gaston banged upon the bell on his desk. A porter rushed up and Gaston cried, "Go down there and get those men into the service entrance!" Shaking his head in exasperation he said, "I forgot to tell them to be sure and come back by way of the side road."

Three swarthy Tahitians were coming up from the beach, as sunbathers and guests on the terrace sat up and stared. The men were carrying the solidly built coffin on their shoulders, for all the world as though the body was already in it. The lone man in back was complaining as the other two hesitated about which way to enter the building. So much for our discretion. All the guests watched, fascinated, as the coffin was directed into the hotel.

When the hearse arrived — ancient, a dull purple, with black plumes looking like wretched crows sitting on each corner of the high roof, Madame Berthaut drove it around to the side entrance and waited, the engine running noisily. The policemen had arrived and were taken up to Mr. Dudley's room, where they witnessed the body being placed inside the coffin which was temporarily sealed. The coffin was quickly carried through the hall, down the service elevator, and out to the hearse. The policemen climbed in with the coffin and Madame Berthaut took off in what I thought was unseemly haste. I ran to my car and caught up with her at the morgue; she explained that if she drove slowly the hearse had a tendency to stall.

The zinc coffin had been delivered as well as the sawdust, and I had the quicklime, but Monsieur Berthaut had not yet arrived with the charcoal. I looked across the road to the Sanitation Department and was surprised to see his car still in the parking lot. I went over and found Monsieur Berthaut behind a hedge, kneeling on the cement walk, a large stone in his hand. He raised his face; it was streaked with sweat and charcoal dust. He was smashing the charcoal on the sidewalk.

"The welding shop was closed. The noon break, you know."

"Oh, Monsieur Berthaut!" The little man grew six feet tall in my estimation. I helped him pour the charcoal back into the bags and we returned to the morgue.

We all watched solemnly as the attendants transferred the body to the zinc coffin. They put in the charcoal and sawdust, the powdered quicklime going in last. When they tipped the bags over the edge of the coffin, the lime poured out in a dense white cloud that quickly filled the room. Choking and sneezing, the attendants, policemen and all of us, stumbled blindly out the door.

Finally the air cleared but we were covered with white lime. Brushing at our clothes futilely, we filed back into the morgue and watched the sealing of the coffins. I was waiting apprehensively for the officers to ask me for the papers. But they only glanced at each other and then nodded to me, saying "*C'est bien*" as they backed out the door, still brushing at their uniforms.

As soon as they turned the corner I rushed out to the car, Monsieur Berthaut at my heels. We raced to the town hall. An elegant French-colonial building of the last century, with gingerbread railings along its wide, airy porches, its charm was only exterior. Inside, the offices were a rabbit warren of cubbyholes, dark paneled walls absorbing any light that filtered in through tall, narrow windows. The registry office was lined with volumes of records, their labels dating back a hundred years.

Monsieur Berthaut led me to the desk of Madame Lety, the registry clerk. Madame Lety, a bird-like woman of seventy years of age, peered up at me through tiny glasses. "Yes?" she asked.

"There has been a death, Madame," I said respectfully, "and we have come for the *Certificat d'Etat Civil.*"

"You have the *Acte de Décès*, the death certificate?" I gave it to her.

Madame Lety had a large, thick book on the desk in front of her. Opening it at the beginning, she began to work her way to the back, several pages at a time. Each page was divided into two parts, each part headed by the name of a deceased, followed by a long paragraph, all written in a fine, spidery script. Madame Lety finally reached the last entry. She pulled the death certificate towards her and studied it.

"Where is this address?" she asked suddenly.

"Why — in Papeete," I said.

"I see that, but in which *quartier*?"

Monsieur Berthat leaped forward. "*Quartier* !"

"Hmm. I don't know that house." Madame Lety probably knew the town inch by inch.

"It's a new house," I said helpfully. "See? Americans."

"Mm." She shot a look at me. "Were you there?"

"What?"

"When he died. Were you there?"

"Oh! Yes. I was."

"Write your name here. And signature. As witness."

"Yes, certainly. Here?"

Madame Lety studied my handwriting. Finally she began to inscribe in her book, pursing her lips tightly as she copied out the foreign name of the deceased in that spidery script. Then she started on the paragraph. What on earth was she writing about for so long?

I fidgeted. I still had to get the shipping permit from one of those offices that had no break at noon so they closed at four o'clock. I glanced at my watch again. Madame Lety looked up sharply. "You don't need to wait," she said.

"What?"

"You don't need to wait," she repeated. "The director isn't here. Your copy won't be valid without his signature. He won't be back today."

"Oh! But the body must go out on the plane tomorrow at ten o'clock!"

Seeing my distress she softened a little. "I can call the director and inform him of the emergency. He'll sign the certificate first thing in the morning. Be here at eight. I promise you'll have it then."

What else could I do? I thanked her with as much grace as I could manage and gloomily left the office. There I was with two death certificates and no *Certificat d'Etat Civil*. And running out of time.

But — wait now. From the town hall I could get to the Bureau de Control by eight-fifteen. Ten minutes to get the shipping permit and twenty minutes to the airport. I was supposed to be there by eight-thirty.

Monsieur Berthaut watched me. He knew exactly what I was thinking. "You can do it, Mademoiselle," he said with unfailing confidence. "So you'll be a little late. They know you, don't they?"

True. The hotel and the airline were the same company. Everyone at the airport knew me and my reputation for efficiency. I rallied again to his support. "You are right. I can certainly try. I promised to do my best."

I took Monsieur Berthaut back to his car. "My good friend, you have been of inestimable assistance. *Je vous remercie, Monsieur.*" The little man glowed with pleasure. I last saw him as he climbed into his Deux Chevaux, his eyes moist with emotion.

I returned to Punaauia in time to attend the church service at the little chapel near the hotel. Mrs. Dudley trustingly accepted my assurance that everything was being taken care of. It was night when I got back home, exhausted. I set the alarm and fell into bed.

The next morning at eight o'clock I was at the town hall. A clerk opened the door and I ran to Madame Lety's desk. She was waiting for me — the morning paper was spread out before her. The headline caught my attention before I was aware of Madame Lety's shrill cry. TOURISTE MORT A L'HOTEL MAEVA BEACH.

"He died in Punaauia! Not in Papeete! You lied to me! You deliberately let me write it in The Book. Never before has there been a single mistake, and now —" Madame Lety fumbled for the page as her glasses slipped down her nose. I fled.

The original plan. Thank God I had not yet given the real death certificate to Mrs. Dudley. I dashed to my car and raced back to Punaauia. Surely, after playing hooky all day yesterday the *tavana* would be in his office today. I made it to the green store in twenty minute flat, jumped out and ran down to the back room. The *tavana* had just arrived and was lighting a punk to drive the mosquitoes away.

Contrary to the bureaucracy in Papeete, official business in the districts was taken care of with little concern for protocol or precision. The *tavana*, a big, amiable

Tahitian, dispatched certificates with a minimum of detail. He scrounged around in his desk, pulled out a form and a copy, and in less time than it took to get it together he scrawled out my *Certificat d'Etat Civil.*

"Thank you," I cried, already out the door. "I think I'll make it!" I was in my car before I looked at my watch. Eight-forty. Oh, no. I slumped in the seat. They had probably canceled the shipment already. I had to admit I was licked.

No! I could still do it. I jumped out of the car again and dashed into the store, grabbed the telephone and called the airport. The operator got the extension for me immediately. "Freight Department. *Allo*?"

"*Allo...Oui*, it's Renée Denis, calling about the big case—"

"*Finalement*! We have been waiting for you. Did you get the permit?"

"Yes, I've got it. I'm very sorry to be late but I am on my way to the airport now."

"Ah. How soon will you be here?"

"It will take about half an hour. You know the traffic at this time—"

"Yes, yes. All right, we'll put the case on the plane and you get here as soon as possible." Before he hung up I heard him shouting to the men to load up.

I went to the counter to pay for my call. Fresh croissants were piled upon a platter. I hadn't had breakfast. I bought one and went out to the car. Munching the croissant, I drove slowly to the airport, four miles away. But I didn't go down to the terminal. I stopped the car at the top of the hill from where I could see the plane.

There it was, the huge steel box at the door of the freight compartment, and, behind it, thirty or more carts of baggage. I watched them all go into the plane, checking my watch now and then. At nine-fifteen exactly, I started the car and sped down the hill to the terminal and around the building to the freight department. I braked fast, flung the door open and ran across the tarmac.

"Here! Here it is for the coffin!" I cried, rushing up to the harried comptroller on the loading platform.

All right! Good for you," he said as, checking a list on his clipboard, he stopped long enough to reach for my paper. Just as I hoped, he didn't even glance at it. He slipped it in with the other papers under the list on his clipboard. The freight agent in the office looked out the window.

"*C'est le permit*?" he called. I held my breath.

"*Oui*, I've got it," the comptroller replied with a wave.

I went to my car and drove off to the side, parking where I could still see the platform. The comptroller finished counting and put his board down. He called out, "Close her up!" and the huge doors of the freight compartment slid shut. A few minutes later the engines started up. I heaved a sigh of relief.

Of course I still had to get the shipping permit for the freight agent before he found out he had the wrong paper. I drove back to town with the *Certificat d'Etat Civil.* The folded paper I had given was the phony death certificate.

I went to the Office de Control. A modern office, it looked very efficient. Rows of filing cabinets, desks in perfect order, smart shirts identifying the personnel assured a perfect control. I presented my *Certificat* to the clerk at the counter. If requesting a permit to ship a body was at all unusual he gave no sign of it.

"You want a shipping permit?" he asked. I nodded.

"The body is going to the United States?" Indeed it was.

"When are you sending it?"

"Today," I mumbled.

"Today? There is only one flight." He looked at his watch. "That left at ten o'clock. Where is the body now?"

"Where?" I fumbled for words. "Uh, you see, there was a hurry. Actually — the body is on the plane."

"On the plane? But it is after ten. What do you mean?" Then, realizing, he exclaimed, "Are you saying that the body has gone?"

I looked at him helplessly.

"You simply cannot do that!" he cried. "It is impossible!"

Perhaps. But I did it.

20

International Tour Conductors

"Can you take an Orient tour leaving June 7th?" It was a telex from Travelworld, one of the top tour agencies in Los Angeles. I had written to the owner and manager, Ron Harris, a few weeks before, and he had answered that he would be pleased to hire me as a tour escort for the next season, starting in September. I gave my resignation immediately to the manager of the Hotel Maeva, as three months notice was required to give him time to find a replacement. That was six weeks ago; June 7th was in one week.

I was in the middle of improvements on my house to rent it, and, although I had started training someone to replace me at the hotel, my job was so diversified it also would take time. But I didn't want to lose this chance to leave. The manager was the same who had been so unconcerned about the death of Mr. Dudley, and other problems we had had with tourists. He was interested mainly in the social life of Papeete; I could not work with him as I had with Ivan, the first manager.

"Do you think he would let me leave now?" I asked the manager's secretary who brought the telex from Travelworld to my office.

"If Françoise and Mareva can handle things well enough until your replacement gets here, I think he'll let you leave. He is rather annoyed when you keep telling him about things that should be done —"

"Well, he's been here six months, and he still says he just got here, give him time and he'll take care of it. I think he'll be glad to let me go." He did (and I was smug when I returned from the trip, and was met by someone the manager had sent to the airport to ask if I would come back to work).

I telexed my reply to Travelworld, then went to get the first of my cholera and typhoid shots. Within a couple of hours my arm began to swell and throb. A brother-in-law and a neighbor came to take over the painting and yard work. There was already someone who wanted to rent the house and he was delighted to get it sooner than he expected. After all day at the hotel, initiating two girls into the duties of public relations, advertising and entertainment, I returned home with my aching arm, to pack and carry my belongings to the family house. The fifth day I went for the second shots, and took Tane to Moorea, where he would stay with Nini and Viri until I came back at the end of the holidays.

Arriving in Los Angeles, I went straight to the Travelworld offices to find out where I was going. Several friends worked for Travelworld, tour escorts as well as Ron Harris, who had come to the hotel several times, but it was Sunday so there was no one but the person with whom I went over my program. First to Japan —Tokyo, Kyoto, Myanoshita in the mountains, and down through the Inland Sea to Beppu. Temples, shrines, gardens and the Bullet train. Then Taiwan, Bankok, Kuala Lumpur in Malaya, Singapore and Hong Kong. I didn't know that Taiwan was the former Formosa. "You are taking twenty-five people to Asia and you don't know where you are going," my instructor remarked. I protested there had been no time for me to study. "Not to worry," she reassured me, "the guides in the Orient are excellent." It was a good thing because I lost somebody the second day, at Nikko, where one tour member followed another group, just as in *If It's Tuesday It Must Be Belgium*. I took the rest of my group back on the train to Tokyo, while my wonderful guide found the man in the swarming crowds and got him back that night.

The day of the tour departure I went to the airport two hours before boarding time, and waited at the counter for my tour members. I felt terrible, nauseated, and my arm was swollen to almost twice its normal size, as the doctor had warned me would happen, taking the second shots within a few days instead of two weeks. Paregoric every few hours was making me terribly sleepy. I sat on a bench, watching for people carrying Travelworld bags. The Wells couple arrived first. I stood up quickly, and was horrified to see them becoming surrounded by darkness. I was going to faint. Hanging onto the airline counter, in what I hoped appeared to be a manner of securing my lists, I managed to stay on my feet and the darkness receded. The Wellses were wonderful; although they had no idea how ill I was, they helped me watch for other tour members and check them in. Somehow everybody was accounted for and we boarded the plane, whereupon I slept for at least ten hours. Feeling much better upon arrival, I was soon able to enjoy the trip as much as the tour members.

It was quite a change for me, being accustomed to the cheerful wave of the hand, or exuberant embracing everywhere between casual acquaintances and friends, in Tahiti. In Japan I observed the ritual of bowing that formalizes every meeting, arranged or by chance, and every departure. Bernard Rudofsky wrote in his book, *Kimono Mind*, "When it comes to salutatory attention, the Japanese outdo any and all nations. Although they do not necessarily bare their heads in greeting, they stop in their tracks and, as a mere preamble for bowing, unwind their shawls and take off their overcoats, even on the street in the cold of winter. To the uninitiated it looks like preparations for a brawl. Bowing deep and copiously is performed with gusto in such unlikely places as public conveyances, railroad stations, and department stores. Since there is little elbow room, greeters risk being carried away by the torrent of passersby, yet, even with their bows knocked askew, serenity prevails and no stampede, downpour, or blizzard will interfere with their maneuvers. The vigor of their bowing ebbs ever so slightly, to forestall their being caught upright while the opponent is still bent low."

When we arrived at the International Hotel in Bangkok, I went to the reception desk to collect the keys for my group. Suddenly someone shrieked my name from across the lobby. I turned and was dismayed to see Joan, an Australian girl who had lived for a while in Tahiti and worked as a guide. I hardly knew her but disliked her because she met tour groups at the airport smoking a cigarette, wearing miniskirts that barely covered her panties, and a wig with loops of two inch curls on top of her head. I don't know why her boss permitted it but he may have found her attractive. Astonishingly, she became a Travelworld escort.

Running across the lobby, she cried, "*Quel plaisir de te revoir*!" although English was the mother language for both of us. She threw her arms around me. I mumbled something, and turned back to the desk, but she said quickly, "Renée, the office told me you were coming and asked me to see you. Will you come up to my room when you are free?" Of course I had to, and went to her room an hour or so later.

"Renée, I know you don't like me," Joan said, which is embarrassing to hear under any circumstances, "and maybe in Tahiti I was pretty wild, but I like you and want to help you." She then proceeded to take me through the rest of my tour, giving me all the useful bits of information that make an escort really helpful to clients and knowledgeable about each destination. She also generously told me about "the sweetest guide in Kuala Lumpur; you must ask for him, he is just great in bed." By the end of the afternoon I completely changed my opinion of Joan. She was a very warm and amusing person. Although the office received some letters about her dress style, she was only advised to be more moderate.

Joan escorted a tour to Tahiti later, when I was public relations director one season at the Intercontinental, and one member of her group was a blind woman traveling by herself. Joan took care of her with just the right attention and humor ("Here, chicken, take a wing," she'd say, offering her arm, to the delight of the woman), without neglecting the other tour members in any way. I don't know any other escort who could have done it with such ease that everyone in the group was always pleased to see them both.

Another escort, Jeremiah, was apt to be brash and excitable but always stayed cool and courteous with his tour members. This was not so with some escorts, like the one who loudly complained to the airline officials because we were delayed in leaving Katmandu, which is surrounded by mountains that the planes cannot fly over if the winds are too strong. Jeremiah and I had side-by-side world tour programs, so we knew each other well enough after a few days so he could relax and be himself when we were away from our people. He was like a brother, coming to blow off steam with me when he was exasperated by an exigent client, or figuring out a way to escape from our groups for a few hours.

It was Jeremiah who initiated the trick on Carl, the new German escort, whose maiden trip was a tour of French Polynesia. I was working in the hotel then, and the office told him to see me about everything he needed to know. He met the Front Desk manager and, both of them being young and handsome, they went out on the town. I hardly saw Carl during his tour group's stay — but I heard the speech he gave his clients, in which he told them how much he expected of them in tips.

We were five or six Travelworld escorts in Singapore at the same time, and we took Carl to Boogie Street, closed off to traffic at night and becoming one long, outside café. Prostitutes, fabulously dressed or nearly undressed, had Carl panting with excitement. Jeremiah did not hesitate to invite one statuesque beauty to our table. Carl was enchanted, and sure enough, after a short while the two of them left for the hotel. Half an hour later Carl came back, furious. "God damn it!" he yelled at Jeremiah. "You knew he was a man!" Of course. All the prostitutes were men, and we hadn't stopped laughing since he left with his sweetie.

In Delhi, Jeremiah became enchanted by a very tall, blond American woman wearing Indian pants and jacket like Nehru, because, I learned later, her suitcases were lost and she couldn't find anything else to fit. He managed to get a date with her for dinner that night. It was about midnight when I was woken up by a pounding on my door. Quickly alert, because a tour member could be ill and needed emergency care, I opened the door. It was Jeremiah, who charged in and flung himself on my bed. "She wouldn't sleep with me!" he exclaimed.

"Did you really expect her to?" I laughed, and he soon laughed too.

"No, but I thought, the way she was dressed, she was a bit crazy and who knows? She might have. I spent a lot of money on that dinner."

She happened to fly to Beirut on the same plane as my group, and we sat next to each other. We liked each other immediately and never stopped talking the whole flight. She told me she had been married to Armour, of Armour ham products, and showed me a box of fabulous jewels that she always carried in her hand luggage. "I have a large apartment in New York and would love to have you come and visit." Unfortunately I lost the telephone number she gave me. Jeremiah remarked, "How come she likes your company instead of mine?"

From Egypt I took Jeremiah's tour group as well as mine into Israel because he was Israeli and couldn't risk going into his country and be inducted into the army. In his group there was a black mother with her physically quite developed thirteen-year-old daughter. When we landed at Ben Gurion International airport, we were shocked to see men with guns surrounding the plane (I learned later it was usual in those days of 1972). We were told to stay in our seats, while two burly men came slowly up the aisle, looking us over. As they reached the black couple, the daughter raised her hands as though holding a gun and made a noise like a machine gun, as a child would do. The toughs didn't think it was funny; they hauled her out of her seat. Her mother screamed, they pushed her back, and I jumped up. "I am the tour escort! I am responsible for her!" I cried, and they let me go with them. They pulled the terrified girl under the plane. A woman searched her and, persuaded she wasn't dangerous, they let her go.

When I went to Hong Kong the first time, and my people were settled in the hotel, I went to the Travelworld office and asked for assistance to find a cousin of mine. The attractive Chinese woman at the desk said, "I would be pleased to help you. What is her name?"

"She is the author, Han Suyin. Remember *A Many Splendored Thing?* She is first cousin to my father, who is Belgian —" I wondered why the woman looked very surprised at first, and then smiled delightedly.

"You have come to the right person," she exclaimed. "Han Suyin was the first wife of my husband. They are still good friends. Come back this afternoon; I will ask my husband where she is and have him call her for you. What a nice surprise!"

It was a remarkable coincidence, but unfortunately I had arrived in Hong Kong too late. Walking back along the street after lunch, I saw the headlines in the newspapers, "HAN SUYIN CROSSED THE RED CHINESE BORDER TODAY." I wrote to the address that Han Suyin's former husband gave me, but it was several weeks before I received the reply in Tahiti. It was a postcard picturing "A Revolutionary Modern Ballet" Further explanation was given: "The despot landlord, Huang Shih-jen, and his flunky come to carry away Hsi-erh by force as payment for the family debts. Yang Pai-lao and his daughter resist bravely." The postcard was issued by the Foreign Language Press, Peking, China. Han Suyin wrote, "Your letter reached me here and I hasten to answer it. I no longer stay in Hong Kong but live in Europe part of the time, the rest in India and China. (I knew she had lived in England, where she and my father were very close.) Yes, I did hear of your father's death and I am very sorry. Your father was a very fine and great person. My publisher can always find me and perhaps we shall meet. Affectionately, Han Suyin."

We returned by way of Tokyo to Los Angeles, because of Travelworld's agreement with Japan Airlines, which was an excellent company, but the extra routing made an exhausting trip. Going east, the accelerated sequences of night and day made it impossible to sleep much. It took me two days to recover, holed up in Hollywood, in the then unnoticed picturesque Hotel Roosevelt of the 1920s, next door to the tour office building. Then I was off again, this time on a tour of the South Pacific. The first stop was Tahiti, where my children were awed to see me arrive with two dozen adults that I was fussing over more than I had ever fussed over them. I kept the children with me at night in the hotel and took them on the day tour of the island; they sat perfectly still in the back of the bus and never said a word. Going along a quiet stretch of the road, the chauffeur said, "Bring the kids up here to sing." Leilani came forward and hesitantly took the microphone, but, being the granddaughter of the lead singer in church, she sang perfectly naturally in her clear, lovely voice. Tane, hearing the unexpected sound of her voice over the microphone, couldn't resist trying it too. The tour members were enchanted, begging to hear more of the popular Tahitian songs. When they and the driver sang, in their soft Tahitian language, a traditional English hymn, a couple of ladies had tears in their eyes.

Peter said he had noticed me at the hotel but "being a rather timid person," he said, he dared not speak to me. I laughed because he certainly was at ease with his people when our two groups met at the airport to take the flight to Tonga. He reminded me of a little boy as he slyly glanced at me, hoping I would notice him as he cheerfully and efficiently herded his chattering people onto the plane. There were not many flights between the islands so tour groups often traveled together for several days; escorts always checked the reservations and arrival lists in hotels to see who may be the other escorts there at the same time.

We had just spent four days in Apia, Western Samoa, enjoying the old, Somerset Maugham atmosphere of Aggie Grey's hotel. I loved Western Samoa. The island was as Tahiti must have been fifty years before, at the time of Gauguin. But there were no French colonists, and if there were any English we did not see them. The people appeared to be pure Polynesian, perennially cheerful and friendly. As we passed through the villages early

Sunday morning, many people were still in bed, as we could see because their *fare* walls, simply panels of woven palm fronds, were raised to the roofs. There was little traffic in the island, and the people appeared to be delighted to see us pass, as they sat up in their beds and waved.

We were to spend two days in Tonga, a much smaller, coral island. We were covering a large part of the South Pacific in seven weeks. From French Polynesia and American Samoa (where brown-skinned Tane, on a later trip with me, after I told him he would be in his first American country, said, "But Americans are just like us!"), we would go on to New Zealand, Australia, Papua-New Guinea, the Solomon Islands (Guadalcanal), and New Hebrides. The trip was fast and tiring, but well-planned.

Tongatapu was small enough to see everything in two days. Fortunately there was an excellent little brochure in the hotel which I quickly scanned upon arrival, because our local guide never showed up. I had said I had been to Tonga before — it is better, if possible, to assure tour members the escort is familiar with every place they are going. Having already read the history and geographical highlights of the island, and as the people and vegetation were the same as in other Polynesian islands, I easily managed as guide.

Peter, a New Zealander, about 40, had introduced himself to me at the airport as we had boarded the plane for Tonga. We were sure to see each other during the next few days and he hoped I would join him for dinner one evening. I accepted, liking his comfortably friendly face. He told me a lot about New Guinea and the Solomons, of which I knew little but their histories. It was the first impressions that were important for travelers, and I wanted to be prepared with knowledgeable answers if my people asked about mosquitoes, and would it be cold at the high altitude of Garoka.

I hardly saw Peter in Tonga; their schedule of activities did not coincide with ours. The last evening, our group went to a beautiful cove for a Tongan party. We went early to have a swim first, and I sat on the beach as the sun went down, feeling a little tired and wishing I could spend two days there with nothing to do. Someone sat down beside me and I turned, surprised to see Peter. "Yup. I hope you don't mind?" he said. "I asked to join the party." I laughed, very pleased to see him, as he grinned and handed me an ice-cold beer. We joined the others and had a delicious feast of roast pig and *fafa* just like home. Afterwards we all went into the cave, a natural theatre, carved a long time ago by the waves pounding the coral cliff, the floor of clean white sand. Dancers appeared with torches and put on a beautiful show, ending with a wild dance including all the party.

I rounded up my excited, happy tour members at the end of the evening. Peter and I had gotten separated while I was putting them on the bus to return to the hotel; I was sure he would be at the bar, hoping I would join him. But I hesitated to go, and went up to my room along with my people, calling goodnight to each other.

The next morning we were up early to take the flight to Fiji. I saw Peter in the lobby, also checking out. He raised a quizzical eyebrow and asked, "See you on the plane?"

He joined us sooner than expected because it just so happened his bus to the airport was too small for all his group, so he and a couple of other people came in our larger bus for the short ride. Peter was already a hit with my group — tour members find it romantic when their escort catches somebody's eye. When we were on the plane and one of my ladies found herself next to me, she hopped up and insisted upon changing seats with Peter.

The plane was small, the seats were narrow, and our arms were against each other on the partition between us. As we talked I realized it was pleasant to be touching him. I glanced at him to see if he was aware. He caught me; he was aware. He grinned at me in his delightful way but said nothing.

I wanted to buy watches for the children in Suva, Fiji being duty free. Peter said he would take me to a shop where he could get the best prices. I got my people settled in the hotel for the evening while Peter rented a car. We drove downtown to the shop where we discussed the merits of different watches, and I saw one that I decided to buy for myself as well. It wasn't expensive but it was very nice. Peter had said to let him bargain and pay, and I would reimburse him afterwards. When I gave him the money, he took only enough for the children's watches. I started to argue but he kissed me lightly on the cheek. "I want to give you your watch. Perhaps you will remember me when you look at it."

We drove up a long and tortuous road to a small settlement in the hills, to visit a family Peter knew. The mother was Fijian, the father Indian. They were very happy because we went to see them at their home. Peter had told me, "The Indians have a hard time because they are resented by the Fijians, nor are they popular with the English. Emigrating long since from India, they now make up the largest percentage of the population. They have most of the shops and they can be irritating, especially when they try to entice you in off the streets, but they are very different when you know them."

We drove back down to Suva, to walk along the pier and see the sailboats. Peter, a true New Zealander, had a ketch and sailing was his passion. We had dinner in a picturesque restaurant on the pier, and danced to the romantic music of the small band. Later we went to the Grand Pacific Hotel, a landmark in the Pacific because of its gracious old colonial lobby. The bar was equally elegant but there were few people and the atmosphere was a bit sad. Perhaps that's why I told him about my brief marriage to Teva, and he told me about his frivolous wife who both resented and took advantage of his traveling. Not many tour escorts were married, or had a good relationship with their wives or husbands. Some couples, both were escorts, and I imagine that brief romantic encounters with other persons were overlooked. Peter tried to stay married because of his two children, but he finally divorced.

It was late when we drove back to the hotel and walked through the empty lobby to my room. I realized Peter was deeply hurt by his wife. She was a good mother but she wanted only money and security from him. He had so much love to give, so much kindness and humor. How could she not care for him? His love filled me with warmth and happiness, a sense of closeness and security, if only for a few days.

He came to spend the night with me at Yanuca Island, driving three hours from Suva. I expected him at five o'clock but he arrived at two. I sent my people off on the sugar cane train, and we made love all afternoon. That night I said I wouldn't be able to sleep; I wasn't accustomed to sleeping with someone. I had a dream — we were on the narrow sugar cane train and Peter was standing next to me. All of my tour members were on the train, they could see us and I kept telling Peter not to stand so close. Poor Peter — I may not have wanted to push him off the train but I almost pushed him, startled and protesting, out of bed.

There are large, tall trees just beyond the porches of the Fijian Hotel on Yanuca Island. I always had a room on the third floor, the highest, where the branches of the trees spread over the roof. In the morning the trees were filled with bell birds, bursting with song as the sun rose. It was the most exotic sound I ever heard. I went back to Yanuca a few months later. I stayed with my tour people as much as possible so I would not realize how much I missed him. But in the morning, when I woke up to the sound of the bell birds, I cried for Peter.

21

Tetiaroa, Marlon Brando's Islands

The sparkling clear green of the lagoon and the mysterious deep blue of the ocean surrounding the ring of white sand islands make a spectacular picture of the atoll. Tetiaroa is not one island; it is a small coral atoll of thirteen islands, if one counts the tiny bird sanctuary. The atoll is unique because there is only one narrow pass through the reef, so the lagoon is protected from large predatory fish. Because of this there is a rich variety of coral fish and crustaceans in the lagoon. Marine biologists come to explore the aquatic life of Tetiaroa, and fishermen throw out their lines both inside and outside the reef. Several large coral blocks, part of a *marae*, is all that remains of the epoch, one hundred and fifty years ago, when members of the ruling Pomare family and their entourage came for rest and recreation, and the women fattened themselves to become more alluring. In those days there were often as many as a hundred canoes lined upon the beaches, as servitors came from Tahiti, forty-four kilometers distance, to supply the royal court with meat, produce and household goods to maintain a luxurious life at the royal retreat. The islands were neglected after the Europeans arrived in Polynesia, and finally abandoned altogether.

According to the archives, in 1902 an English dentist, J.W. Williams, came to Tahiti, where there was a desperate need for dental care. The last king, Pomare V, required the good doctor's services but he did not have money to pay his bill. Instead, he gave the Englishman the atoll of Tetiaroa. Going over to see the property he had acquired, the new owner was dismayed to see that the coconut palms, which would give him the only income he could expect from the islands, were ravaged by rats. Half-chewed green coconuts littered the ground; this was before the days of banding the trees with metal strips to keep the rats from climbing them. The Englishman went back to Tahiti and put an ad in the local paper for people to bring him cats. He was given enough cats to fill several cages; he stopped feeding them for a couple of days and then took the howling cats on the boat to Tetiaroa. Let loose upon the islands, it was not long before they had destroyed the rats, thus saving the coconuts. Williams and his family eventually moved to Tetiaroa and Williams died there in 1937. His widow stayed on, although she was almost blind. She had ropes strung between the coconut trees and so was able to follow them and move about. When she was eighty-one she decided to move to the United States, so she wanted to sell the property and be paid in dollars.

Marlon Brando, discovering the beauty and peaceful

life of the South Pacific when he came to play the part of Fletcher Christian in the *Mutiny on the Bounty*, wanted to buy an island. That was almost impossible for a foreigner, but apparently because the atoll of Tetiaroa was already owned by a foreigner, authorization to purchase was given by the French government. In 1965, Brando bought half of the islands for the sum of $200,000, and finally all of the atoll for a total of about $600,000.

Marlon Brando dreamed of projects to take advantage of the natural resources of his new territory, principally aquaculture, raising sea turtles and other indigenous marine life that would eventually become available in the markets, and raising lobsters as well as crayfish in the protected lagoon. His ideas were met with enthusiasm by the Economics Research Associates, Los Angeles, and laboratories working on hatching lobsters, but the experiments had not yet progressed to the extremely difficult process of raising lobsters. Brando learned that they were being raised successfully in France. Brando's advisors apparently presumed the lobsters would acclimate somehow in Hawaii, where they were to make a stopover of several months. The establishment of the lobster farm in Tetiaroa was estimated at $600,000, and the yearly exploitation would be at least $125,000. A hundred thousand lobsters delivered to the international market would bring in perhaps $300,000 a year. With the aquaculture production of vegetables to fatten the profits, the Economics Research Associates were optimistic. So in 1973, $30,000 worth of baby lobsters were purchased in France — but they all died in Hawaii.

The turtle project started out with greater success, there were hundreds of the jewel-like baby turtles in fenced areas, but the fences broke, all the babies swam off into the lagoon and were eaten. As for the aquaculture, it never got off the ground.

The government told Marlon Brando he would have to develop a tourist attraction if he wanted assistance to commercially develop Tetiaroa. He agreed to build a small hotel and also have day-tours for visitors taking the short flight to go and see his islands.

Ivan Falesitch, first manager of the Hotel Maeva with whom I worked so well, came to tell me he was now business manager for Tetiaroa. I had not seen him since we both left the Hotel Maeva, and I was surprised to see how he had changed from the image of executive to laid-back, casual man. He wore a beard, neatly trimmed, polo shirt and chino pants, and slouched comfortably in a chair on my terrace.

"How about going to Tetiaroa to help me organize a hotel and visitors' program on the islands?" he asked.

"Sounds interesting, but I'm still doing international tours."

"You're not going to be a tour director for the rest of your life, and this is the kind of thing you are good at. Would you go to Tetiaroa and see what you think can be done? When are you leaving on your next tour?"

"Not until September. I just got back from the last one."

"Then you have four months. A lot can be done in four months. At least go to Tetiaroa and see what it's all about."

I flew over to Tetiaroa the next morning in a small plane loaded with construction material and food. There was a slight haze over Tahiti and Moorea which lay to the west as we headed out across the ocean, but in a few minutes I saw the line of the reef, flat against the horizon, and then broken in a rough necklace of islands encircling the pale green lagoon of Tetiaroa.

We flew over the atoll to come in on the far side. The arrival was scary; the runway, from the reef-edged island rim to the lagoon where it seemed to drop into the water, looked much too short. "Going into the wind to land it's okay," the pilot said. We touched the ground, coasted along the grass strip through the palms and came up short of the beach. A couple of Tahitian men arrived as we opened the doors, and along a path through the palms I saw a young woman in a bright *pareu*. It was a close friend of Marlon Brando who was staying on the island as guardian and radio operator.

Reiko was a Japanese-American who had been a dancer in the States; I never understood why she came to stay in Tetiaroa alone. She had come the first time four years before, was alone for several months, then met a young fisherman who came to the island and stayed with her for a year. She left when it was apparent that no work would be started at that time, and returned five months ago, when construction finally got under way. There were workers in barracks on the other side of the small island but Reiko saw few people except the pilots.

I stepped out of the plane and realized it was very windy. Intending to stay two or three hours, I turned to the pilot to ask if he could still pick me up later.

"The wind will get worse, Renée. I doubt I can come back to get you today. You had better look around as much as you can while the plane is being unloaded." I nodded and went to meet Reiko, whom I liked right away. We went directly to see the *fare* that was to be the main building of the hotel. It was huge. Beautifully constructed but so large, like an airport hangar, that a hundred people would be lost in it. However, the open space could be broken up and made more intimate, as there would never be many people there at one time.

I couldn't help but be enchanted by the islands, and, as Ivan had told me about the projects that Marlon still hoped to establish, I became intrigued by the challenge. I turned to Reiko.

"Would it be all right if I spend the night?"

"Oh, please do. It's a good idea to stay so you'll have time to see everything. The plane can come and pick you up tomorrow." We went back and told the pilot, who agreed to return the next day.

Reiko took me to the old house, long since abandoned but with a re-thatched roof, where she lived. A radio was installed so she could communicate with Tahiti and Hollywood, taking messages, ordering for the construction crew, the people on "Turtle Island" and the projected "Lobster Island."

There was a well nearby and several pails lined up on a bench. "The water is potable but it smells stagnant so we use rain water for drinking and cooking," Reiko said. "I pull up those pails early so that by afternoon the odor is gone and I can use the water for laundry. Same thing for the shower. There is a generator but it is so noisy I don't leave it on. There's a kerosene refrigerator. The food is all in those garbage cans because there are still some rats and several cats. No mosquitoes." The wind was a nuisance, one seemed to be in a bubble of sound, but it had its advantages.

Marlon Brando had put the architect, a Canadian who spoke French, in charge of the projects he wanted developed at Tetiaroa. Forrest had no experience in tourism and had never been to Tahiti. I had seen the original drawings of the hotel — the roof of the main building had corners like a Chinese pagoda. Of course, after he arrived in Tahiti, Forrest saw the Tahitian construction of thatch roofs and woven bamboo walls, and changed his plans accordingly. The huge roof of the main building was simple enough to build, as that was practically all there was to it, but the cottages looked terrible. Forrest was away for a week and I could see the Tahitians were goldbricking. "Why didn't you use bamboo for the walls?" I asked the Tahitian foreman.

"Forrest likes the look of the woven palm fronds inside. He says it's more native," the foreman told me.

"I like it too, inside, but why is it so messy outside? You don't usually leave the outside untrimmed, nor have it run together at the corners."

"He doesn't know that. He tells us what to do so we do it."

I recognized the attitude immediately. They did not like Forrest.

The style of the hotel project was similar to Club Med, individual *fares* but group tables in the dining room, and all activities included in the AP (American Plan, which has become more the foreign plan for American travelers). Besides the tours to the *marae*, visits to the other islands to see the turtles, lobsters and aquaculture projects, there would be fishing, snorkeling in the lagoon for shells, canoeing and sailing. People would be discouraged from getting too close to bird island. Young birds would not budge from the bushes, but eventually the adult birds wheeling overhead would find other islands where they would not be disturbed.

Back at the main building, I studied how the space could be broken up. Reiko helped me figure out a working plan of personnel duties and guest activities. That night I woke up with my head full of ideas. I couldn't sleep well, even with a sleeping bag to supplement a sagging rope bedframe. I went outside, spread a *peue* on the beach, and watched the moon set in the early hours of the morning, making a false dawn and tempting me to get up. I was eager to get back and talk to Ivan. When the plane came for me, Reiko jumped in too — she had not been off the island for weeks, so she took advantage to spent two days with me in Punaauia. Ivan came and we talked for hours. He was sorry to see I wasn't very optimistic.

"It isn't that I don't think it is a great idea," I said. "You told me that Marlon didn't have any intention to have a hotel; it is necessary only to get assistance from the government, so he wants it to be as economical as possible. The way the *fares* are laid out is so impractical that we will need more personnel than should be necessary. Women personnel means houses for their families. I know Marlon wants a population established there eventually, he even talked about having a church or at least a meeting house, but he has to build houses for the personnel now. When the first tourists come they won't be satisfied to hear what we are intending to do; the activities, with personnel to guide and assist, must be ready opening day. Remember how disappointed people were in the Maeva when the swimming pool was closed, even though the ocean was a few yards away? They felt cheated —"

"Yes, of course, I know all that," Ivan interrupted. "I notice you are saying 'we,' so are you going in with us? That's what I want to know."

"You are planning to open in two or three months. I think it isn't possible. I remember how difficult it was to install all the plumbing and evacuation system, for instance, at the Club in Moorea. You have to find more workers to get it all finished in time. Those men are not working fast."

"What's the trouble? It's their own kind of construction, and I was amazed to see how quickly Tahitians build a *fare*."

"Yes, when they're left alone to build it. They don't need more than the basic dimensions and what you want in the way of windows and doors. It seems that Forrest interferes too much. He's the boss, so they do as he says."

"Why don't they tell him when it's wrong?"

"That's typically Tahitian. If they like a boss they won't hesitate to speak up, and if they don't like him they won't say anything."

"Do you know a construction man the Tahitians like, someone who knows the way they build, and will be a go-between for Forrest?"

"I know a Tahitian who doesn't hesitate to say what he thinks. He worked for me at the Maeva and was so efficient and dependable, I stole him away when I worked at the Intercontinental one season. Tahitians are very loyal. The hotel directors from France usually change every two years, so Tahitians are loyal to the boss who stays."

"Do you think you could get this man?"

"I hear he's working on a construction site in Moorea now. We can go over and talk to him."

We flew to Moorea and rented a Mehari, a plastic sort of jeep. The gas tank was almost empty and the French owner of the dozen or so cars had a hard time getting the cap off.

"Do you have to lock it here?" Ivan asked.

"Yes," the man said. "Tahitians will steal the gas." I was astonished to hear it.

"Since when do Tahitians steal gas?" (I never heard of it happening to this day when many Tahitians own cars, although easier thefts were becoming common.)

"Stealing started here in Moorea about five years ago," the Frenchman said, "ever since we got television." I could understand his blaming television even for stealing. The Tahitians were changing fast as new ideas, particularly through sophisticated (for the Polynesians who had rarely seen pictures more modern than westerns) and increasingly violent films, were being introduced to the islands.

Would Tahitians now agree to take their families to Tetiaroa where Marlon did not want television, modern houses or bars? Would they be happy going back to the simple way of life of the untouched islands? Marlon wanted to establish a village, build the *fares* and eventually give property to each family. He would provide boats for ocean fishing, tools and whatever they needed to start cottage industries such as shell work, and weaving hats, baskets, and *pandanus* mats from the trees that grew thickly on the islands. Men would be paid to collect copra and work for the various industries that were developing, and women could work in the hotel. I knew many people in Moorea, so we started searching there for families who would like to move to Tetiaroa.

We drove to the construction site where Tioni, the man I wanted, was in charge of the carpenters and masons. It was just eleven o'clock, lunchtime, and Tioni took us to a restaurant that catered to the workers. We ordered *maa tinito* (Chinese food, in Tahitian, though it isn't particularly Chinese), a popular dish of red beans, pork and vegetables. Ivan hesitated. "I've never eaten it," he said.

"Never eaten *maa tinito*? How could you have lived here two years and never eaten *maa tinito*?"

"I lived in a hotel, Renée. I hardly know anything about the Tahitians. Why do you think I need you?" We ordered *maa tinito* for him and he liked it.

Ivan took to Tioni right away. Intelligent, experienced in construction, Tioni didn't hesitate to give his opinion when Ivan asked for it. He liked the idea of going to Tetiaroa, said he would give his resignation in to the present job, get other workers, and help us find families. I knew I could count on him. Ivan was impressed and delighted. "I was also impressed by that man's powerful build," he said to me later. "Some of those Tahitians seem about to burst out of their skins, the muscles of their arms and legs bulge like weight lifters."

I was amazed by the activity that was going on in Moorea. When I was living at the Club we were practically isolated on the far side of the island from the airport, the administration center in the village of Afareaitu, and the two other hotels, the Bali Hai and Eimeo in Cook's Bay. Now, around the Club, souvenir shops, car rentals, restaurants and small hotels were invading the coconut groves. Fortunately the beaches were not tampered with and they were never crowded.

Ivan had never made the tour of Moorea so we went back to the airport by way of the two bays, taking the road that goes through the valley and up into the mountains from Opunohu to Cook's Bay. We stopped at the *marae* restored by Sinoto, the Japanese archaeologist, with its courtyard invaded by magnificent *mape* trees. Their weird, buttress-like roots sometimes raised the trunk of the trees above the ground, leaving a space in the center underneath. The site was eerie, silent and beautiful. We drove past the plantations of the agricultural school, to the Belvedere, from where we could look far down to both bays, and up to the mountains around us. They were steep and jagged; clouds poured over the edges like smoke out of a caldron. We were so far above the ocean and its eternal low roar of the waves breaking on the reef, it was like being in another country.

I was glad to help Ivan find workers, eventual personnel and families, but I felt I didn't want to stay at Tetiaroa without my children, at least not before there was a school. When I saw Ivan a couple of weeks later I asked how the construction was going.

"Much faster, thanks to Tioni, but Forrest was pissed when they had to do the walls of the *fares* over again. Some of the coconut logs had to be replaced too. Tioni bawled the workers out for using logs that were soft inside. They said Forrest told them to use those logs because they were straight. Of course they didn't tell him they were too green. Tioni was disgusted. He doesn't like Forrest either but he says, "You take a job, you do it right."

"Have you found candidates for manager of the hotel?"

"There is a couple, a Canadian and his Tahitian wife, who are managing a small hotel in Moorea, and they would like a change." The *popaa* (foreigners) switched from hotel to hotel, job to job, more often than the Tahitians. "There are two others I was told about." He looked at me casually. "One of them, the way he was described — Tahitian, a good-looking entertainer living in the States, has an American wife — I thought he might be your husband."

"My ex-husband, and, yes, living in the States," I said. Ivan left shortly after but the germ of what he had said must have lodged in my subconscious. It wasn't until that evening I started thinking about Teva.

When I saw him one time in Florida, Teva had said he would return at the end of that year. He didn't come, nor did he write to say he was delayed for some reason, or changed his mind. I was sure he wanted to return but possibly he didn't have enough money saved for the tickets, or Nikki was giving him a hard time about letting their son go.

I realized that Teva would be excellent for the position of *tavana* — chief — as Marlon wanted the manager called. The business manager would handle the office work, reservations and finances, in Tahiti. The *tavana* would be responsible for the personnel, guests' activities, and entertainment in the evenings. Teva was admired by other Tahitians; he could count on their cooperation. He was excellent at creating a convivial atmosphere, encouraging the personnel to play the guitar and dance, and getting the guests to join in. I called Ivan to tell him what I was thinking.

"He would be great!" Ivan exclaimed. "And you, would you agree to go to Tetiaroa if Teva is there?"

"We're divorced, Ivan. We're on friendly terms, I know he wants to come back to Tahiti, but I don't know how he feels about me. Working together could be very awkward."

"You would like it, though, wouldn't you?"

"Of course I'd like to have my family all together, but suppose he has another wife now?"

I wrote to Teva. He cabled back; he wanted the job. He still wanted to come back to Tahiti. Obviously he had not been able to return yet because of financial or other reasons. I had not said anything about the idea to Teva's parents as I did not want them to be disappointed again, but now I rushed over to the *magasin* to tell them. Teriivahine cried with happiness, and Teriitane said, "I am not surprised. I knew he was going to come back at the first opportunity."

"Perhaps you know him better than I do," I admitted, to please him, "but I still don't understand why he doesn't write."

"That is the way Tahitians are," Teriitane said emphatically. He was making excuses for Teva but he knew how often Teva wrote to me at first. "Why write all the time when we know the important things? We would be begging him to hurry, without understanding his problems. He would have written when he could come back."

It was hard for me to accept that, with my American mind, wanting to know why about everything, and be reassured. "Well, now he can," I said happily.

Ivan was very pleased. He was certain we would make a good team, although I kept reminding him that that possibility was not to be depended upon. I cabled Teva to pick up the tickets for him and his son. Perhaps Nikki wouldn't let him take the child but I wanted Teva to know that he was welcome. Teva cabled back to say he would come in two or three weeks. I was disappointed but realized that of course he had to give notice to leave the job he had there.

I went over to Tetiaroa again, taking Tane with me. Marlon Brando came to meet us at the plane. He was dressed in an old shirt and torn shorts, and his hair straggled almost to his shoulders.

"You are Renée," he said, having been told I was coming. "Ivan thinks a lot of you. You've sent us some good men. You'll have a *fare* this time, although there's not much in it but the beds."

We walked over to the *fare*, I left my bag, and we went on to the main building. "It will all come together, you'll see," Marlon assured me. "By the way, Forrest is leaving." That should make things easier, I thought.

At lunch, Marlon continued talking. Tane, nine years old, had run off to explore the island as soon as we landed, and was now drawing at the table as he often did. Marlon noticed and picked up the paper. There were several sketches, including one of a hand. "Did you do this?" he asked Tane. Tane nodded. "Just now?" Tane nodded again. "Look at this," Marlon said to me. "Do you know how difficult it is to draw a hand? This boy has talent. I would like to send him to a good teacher."

A few minutes later, when the attention was turned off him, Tane asked me, "When do we meet Marlon Brando?"

We didn't see much of Marlon during the week we were in Tetiaroa that time. He stayed mostly in his *fare*; he wasn't feeling well. Reiko was not there either. Everyone was new, including a business manager who, I was sorry to learn, would be replacing Ivan in a couple of months. Ivan was going back to Europe.

So there was this typical, pale-faced New York accountant type, very serious and quiet, his wife who was half American Indian, and their two pale-faced, serious sons. The boys had inherited nothing from their mother's side. They followed Tane around all day, amazed. Tane climbed trees, took the boys out fishing, and built a fire on the beach. The boys were astonished. Tane baited fish hooks for them and took the fish off their lines. They wouldn't eat the fish, and Tane was disgusted when they couldn't even keep their hotdogs from falling into the fire. We didn't see much of their mother; she stayed in her *fare* most of the time. When we did see her it was obvious she was not at all pleased, despite her Indian blood, to be here in this savage land. The family left the day after the boys came to dinner alone. It was while we were eating that we heard a commotion on the pathway coming from the *fares*, and suddenly the father came rushing out through the bushes, yelling. Behind him came his wife, yelling too, and brandishing a knife. The men jumped up and subdued her. Her husband then agreed they should go back to New York as she wanted.

And there was the interior decorator, a six-foot black woman from Mozambique and New York (When was Marlon in New York collecting these people, I wondered). She was very aggressive and insisted upon going over to Tahiti on every flight, bringing back whatever caught her fancy. She decided she would buy the food for our meals too, but admitted her French was not adequate when she ordered a *boîte* (sounded like a box) of coffee, thinking she would get a case to last a week, and only one can was delivered. It did not discourage her from interfering with everyone's work, and even Marlon got involved. He didn't help matters; he didn't say clearly who was responsible for what. Marlon decided his housemaid was responsible for the first aid kit and everyone must ask her for the key. We seldom needed it, but the idea that we would have to ask the maid for the key was the last straw for the decorator. She flounced off the island. I think Marlon enjoyed creating awkward situations to see how people would react, as though he might have a role in a similar situation some day.

Marlon's check for the work I had done for Tetiaroa bounced. It was honored later but construction was stopped until more funds became available.

22

Grandpère and Tahitian Medicine

Grandpère was ill again; he had to go to the Clinique for treatment. He was having to go more and more often. He had high blood pressure and diabetes, which was common in French Polynesia. Grandpère could not stay on a strict diet for more than a week or so. He didn't want to go to the Clinique before Teva returned but when I told him about the delay at Tetiaroa he agreed to go.

I drove Grandmère to the Clinique twice a day to visit Grandpère. After chatting a few minutes with them I took out the folder of short stories I was working on and sat at a table in the corner, while they talked for an hour or two. We often met people whom Grandmère knew, also visiting patients, and she asked them to come and see Grandpère. They came in, whole families, kissing all around, myself included, after Grandmère's quick introduction which sometimes included Teva, and sometimes implied I was their daughter *faaa'mu*. They installed themselves comfortably on the beds and chairs they rounded up from outside, and gossiped. They looked at me curiously sometimes, wondering what I could be writing about. Grandpère and Grandmère couldn't tell them but I could see they were proud of me. Sometimes I took the children to the Clinique after school; one could visit at any time. There were always people standing in the halls and out on the balconies. Children waited patiently and silently. Nurses came through with trays of medication and meals. The rooms were very clean and of course there were sheets and pillows, but patients brought more pillows, brightly colored, and *tifefes*, the appliquéd or patchwork bedspreads. The homey atmosphere certainly reduced much of the trauma of a stay in the Clinique.

I drove Grandmère in the big Chevrolet pick-up to do her shopping for the *magasin* when Grandpère was in the Clinique. We went to the marketplace in Papeete, where *les trucks* were lined up along the street and cars could barely squeeze by. Pick-ups and large cars had to wait if they couldn't get through, and everyone behind was very patient while purchases were being loaded. Some took advantage to run into one of the Chinese stores for ice cream or *mapes*, the Tahitian "chestnuts" strung on a stick, that taste more like potatoes than chestnuts. Others start to chat with friends selling flowers or passing the time sitting on the steps. They forget there may be a dozen vehicles backed up behind until someone finally gets impatient.

After four days Grandpère was able to go home again. I hoped he would rest, but it was doubtful he would get much sleep with every customer of the *magasin* wanting to visit with him, and the men gathering to drink beer and play *petanques*, the metal balls connecting with a sharp report, the men cheering. When I went to see him, Grandpère was usually lying on a bench in back of the house, stretched out on his side and looking rather like a beached whale. He said, "I feel better except for my foot."

"Did you tell the doctor about stepping on the fishbone?"

"Oh, that was nothing, a little puncture. It's just the "maladie." It is always the same. The pain starts in my head, goes down through my body and into my legs, and then out my feet."

The doctor stopped by to see Grandpère. This time Grandpère showed the doctor his foot. I learned about it later. "*Il n'est pas là* — he isn't here," Adèle, the adopted daughter who lives with them, told me. "He's gone to the Clinique."

"That's good," I said. "He'll get that sore from the fishbone taken care of. Will he be back soon?"

"Maybe he won't be coming back tonight. They have to *couper* — cut — his foot." It was a common expression, and I was not surprised that the sore had probably turned into a boil that needed to be lanced. "Grandpère didn't want to go the Clinique," Adèle went on. "He and Grandmère argued all morning and by noon they weren't talking to each other. Then Grandpère's foot began to bleed and that frightened him, so he went."

The Chevrolet arrived at that moment, driven by a son-in-law. Grandmère got out, looking drawn and tired. She told us, "Grandpère has to stay at the Clinique. He is hurting a lot. Perhaps tomorrow or the next day they will cut his foot."

It must be a bad boil or carbuncle by now, I thought. None of them understood, or they didn't want to understand, what the doctor was saying.

I took Grandmère to the Clinique the next morning. Grandpère was sleeping, looking very gray and exhausted. The Chinese man in the other bed said he had slept badly, but then Grandpère woke up and looked quite cheerful when he saw us. I asked him what medication he was getting, and he answered, "So many shots I am black and blue, but they're doing me good. I feel much better. My foot is going to be all right."

Grandmère smiled happily. "To think they wanted to cut off his foot!"

Alarmed, I went to see the doctor. "How is Teriitane's foot?" I asked. He shook his head — I recognized the usual expression of exasperation when the French try to explain to the Tahitians exactly what are their "maladies." He told me what was wrong and what they would have to do. I stared at him, shocked.

"*Mais oui, ma fille*," the doctor said, "gangrene has set in. We must amputate. With the greatest luck it will be only his big toe, but it is more likely we have to remove all the toes or half the foot. We are filling him with medication to increase the blood circulation before we operate. We'll probably do it tomorrow."

I didn't want to say anything to Grandpère or Grandmère. They were probably told the first day, but with their usual Tahitian optimism they refused to believe it. They were sure that the medication was going to be sufficient and he would be cured without an operation.

The next morning I was on my way to the *magasin* when I met Adèle who was coming to get me. "He is here!" she cried, and at first I thought she meant that Teva had arrived at last, and thank goodness, before the operation. But she said, "No, it's Grandpère! He's at the *magasin*. He left the Clinique and took a taxi home."

The doctor had told him again last night that they were going to amputate part of his foot in the morning. I cried, "Oh Lord, can't he realize that it's better to lose a foot than risk his life!" I didn't go to the *magasin*; I went straight to the post office and cabled Teva. "Your father very ill. Foot must be amputated. He will not let doctors operate. Please come immediately." Only Teva might be able to persuade Grandpère to let them operate.

I went to the *magasin* and found Grandmère sweeping the porch, looking as though everything was as usual. She greeted me with a smile. "Grandpère came home. He's going to be all right. Adèle has gone to the *tahua*, the healer, to get *ra'au tahiti* and that will cure his foot." *Ra'au tahiti* — Tahitian medicine.

Grandpère woke up when he heard us talking. He looked much better. "Yes," he assured me cheerfully. "That will cure it. Those idiots at the Clinique want to cut off my foot! They said only part of it, but then it would be more and more, until they cut off my leg."

"Grandpère, that could happen if you delay any longer—"

"I never should have gone there in the first place. Don't worry, Adèle has gone to the *tahua* to get *ra'au tahiti* and my foot will be cured," he said emphatically. Seeing my expression, he teased me because I didn't know anything about Tahitian medicine.

I knew about one treatment that was terrifying, and I had told Teriitane never to have my children treated with Tahitian medicine. When I came back from a trip I was furious to learn he had taken the children to the *tahua* to treat them for an illness. I never learned what illness. Fortunately they'd had anti-tetanus shots. Both of them were treated for the *arero maa.*

Arero maa means double tongue. The Tahitians mean the underside of the tongue. "When a child has a fever," the *tahua* says, "the 'second tongue' can swell like the gills of a fish and suffocate the child, so it must be cut." He uses a sharp blade of bamboo. "Now the fever can come out with the blood." There is of course great risk of infection and tetanus poisoning. I had heard a neighbor's child had died in an epileptic seizure, but actually had died in a tetanus convulsion. His mother told me her family would not let her take the child to the hospital, although treatment was free.

"They took my baby to the *tahua* and had the *arero maa* cut. For a week my baby's mouth hurt and he couldn't eat. Then he went into a fit and died," she sobbed. "And now my brothers say I caused his death by putting a spell on him to keep the *tahua* from touching my baby." I heard other stories of casting spells, and real witchcraft in the outer islands, but it was not common in Tahiti.

Certain Tahitian remedies were excellent. Tane's persistent heat rash was cleared up in a couple of days after he was bathed in water in which Nini had soaked special leaves that she collected in the valley. Nini mixed up an excellent remedy for coughs and made poultices for swollen ankles. I went to the house one day and found the family keeping company with an old man who was up to his neck in herbed water, to ease his sore back. For lack of a tub they had him soaking in a canoe. I took an elderly cousin to the *tahua* because she said only bee stings would ease her aching joints. As we left her house her sister came out and decided to go for the treatment too, just in case. The two of them were stung several times by the bees applied by the *tahua* and I couldn't help laughing to hear them both crying, *"Ai! Ai! Ai!"*

Later I read an article about the research initiated by Paul Henri Pétard. A pharmaceutical doctor at the hospital in Papeete from 1937 to 1945, he studied the island plants and Tahitian medicine, and published two books on the subject. He says, "Tahitian medicine can be astonishingly effective and the cures as successful as modern medicine for many illnesses. They can have little effect on illnesses introduced to the islands, but they are excellent for rheumatisms, infections, otitis, neuralgia, sprains, and women's illnesses, including cancerous tumors. I observed the treatment of a tumor as large as my fist; in two and a half months it disappeared completely.

"In the old days, the knowledge of *ra'au tahiti* was passed secretly from father to son and it had a magical connotation, so when the missionaries succeeded in dominating the people, the *tahua* received a severe blow. There are still many *tahuas* and many of the remedies are known to everyone, but incomplete knowledge of a treatment might completely defeat its efficacy. Unfortunately some medicinal plants can no longer be found. The preparation of many concoctions may take a long time and the treatment may be tedious. Seeds, bark and leaves may have to be ground to a powder or paste with a mortar and pestle. The concoctions must never touch metal so modern machines cannot be used to process them.

"A remedy for fish poisoning is to cut up the rind of watermelon and boil it in sea water for fifteen minutes. Drink two or three glasses of the liquid. The poison of the *nohu*, the stone fish, can be fatal and requires a stronger medication. Grind together the bark of *mape* and *atae* trees and apply the mixture to the wound; the cure is rapid. (I saw a man who, having stepped on the spines of a *nohu*, was given a shot of morphine and was still twisting in agony). For rheumatisms, grind a handful of the beans that grow near the ocean, and make a paquet of the paste with fibres of *mo'u*, a fine, tough grass. Soak the paquet in coconut oil, drain it and then plunge it into boiling water. Twice a day, rub the painful parts of the body with the warm paquet, and cover with a warm cloth."

These treatments and remedies were developed over centuries of research and practice, just as modern medicine developed. Also, particularly in civilizations close to nature, some people have the gift, just as pianists and artists do in their fields, to divine the nature of a body's illness and how to cure it. Fortunately, *ra'au tahiti* is being taken seriously by more doctors like Paul Henri Pétard, and this wonderful domain of medicine will not be lost.

I had no phone because I traveled so much and lived in whichever house was available when I returned. There had never been a phone at the *magasin*, so I had given the number of my neighbor and close friend, Cecile, to Teva and the UTA airline, to reply to my cable. Cecile, a beautiful Tahitian-Chinese woman who had also worked at Club Med those first years, had married and lived next door to me at the Pointe des Pecheurs. Tane and her son, two years younger, were like brothers, and Hugo often stayed with me when his parents took a holiday. I had left Tane at Cecile's house, and went over to get him. "They're out fishing," Cecile told me, so I went down to the beach. The boys were out on the reef beyond the pass. Tane was a strong and prudent swimmer but Hugo was still at a clumsy age, and his mother certainly didn't know he had crossed the pass. I waded into the lagoon and called until Tane heard me. It was several minutes until he got all of their fish strung on his fishing pole and they started walking back along the reef. They plunged into the pass; Tane yelled at Hugo to stay closer to him but Hugo was swimming mostly underwater to avoid the surface current. I swam out to them although Hugo was really in no danger because he knew to just swim steadily until he reached the shore farther down. We went up to the house. "Relax and stay for dinner," Cecile said. "There'll be no answer from Teva until tomorrow. Taina's coming and she'll take your mind off your worries."

Taina was a tall, beautiful, cinnamon-colored *demie*, half Tahitian, half French. Her hair was the same color as her skin, and in a brown bikini she looked like a living statue. She had married an American tourist and was astonished when she went to the States and learned he was a wealthy man, immersed most of the day in business and the nights in social obligations. She was not happy although she tried for several months to adapt to a life that most girls would envy. She came back to Tahiti often, staying until her husband came down after her again. She had been a nurse and knew Teriitane so I asked her what she thought. "How long will he live if the gangrene goes up into his leg? You know how stubborn he is. He might suffer for weeks—"

"Not many weeks," Taina said sadly. "He would not live that long."

"Don't talk like that. Teva will convince him to have the operation and he'll be fine," Cecile said, and changed the subject. "Taina, you said you'd come this morning. What happened?"

"My cousin came over last night and we went out to *bringuer* (to paint the town red)," she recounted delightedly. " It was four o'clock this morning when we got to the marketplace to sober up by eating *maoa* and *taioro* (grated coconut meat); it didn't work so we still felt great. You should have seen Mama's face when we set up the croquet set at five o'clock, still in our long dresses, and played a game. So then I slept all day." She told us about her little boy. "One time I was trying a face pack, you know? and he was watching me while I smeared the stuff all over my face. He asked me why, and I told him it would make me more beautiful. When I took it off a while later, he said, "*Raté*, é — failed, didn't it?"

That night I slept badly and at six I went over to see Grandpère. Adèle had just finished re-bandaging his foot.

"My foot is already better," Grandpère said. "It's the *ra'au tahiti* that's working. Here." He handed me a baby food jar with some tea-colored matter in it. It looked like wet sawdust. It was sawdust of the *aito*, the ironwood tree. "That will draw all the poison out," he told me. Grandmère and Adèle were as confident as he was.

I sniffed the jar. "It smells good," I commented.

I had coffee and *pain buerre*, bread and butter, with the family and then went to Cecile's house. Her maid was raking leaves in front when I arrived. I said, "*Bonjour*," and went in, calling to Cecile. No answer. I looked out to the beach but she wasn't there. I returned to the front and asked the maid where she was. "Cecile has gone to town," she told me.

Surprised she hadn't said so sooner, I asked, "Did anyone call for me?"

"Someone called but I didn't catch the name. She said for you to call back. The number is by the phone." (Cecile said later that the maid was told to wait specifically for that phone call, and for me to come by.) I ran back into the house, hoping she had the number right. It was the airline company; Teva was arriving the next day.

I raced to the *magasin*; Grandmère saw me coming and shouted to Grandpére, Renée is coming back already — she must have an answer!"

"Yes, he's coming," I cried as soon as I stopped. "Tomorrow morning at six-fifteen!" Grandmère hugged me, laughing and crying, and Grandpère beamed with satisfaction.

Grandmère started planning. "We must go to the market today — no tough yard chickens for the party—" I said I would take her when I went to get the children at noon, as it was Friday. They would all come to clean the house *raro* (the family house) for Teva. We left Grandmère at the *magasin* and went on to the house, ate *poisson cru* (marinated fish) and sweet potatoes Adèle had prepared for us, and went to work. Shortly after, three men rode over on vespas, carrying a chain saw. They were the first to show up at the *magasin* to drink beer and pass the time of day under the *purau* trees.

"Teriitane said for us to come over and cut the old tree back there for the *himaa* (ground oven). He wants to have *maa tahiti* (Tahitian food) tomorrow. You are to take us into the valley to get a pig." They cut and piled the wood, and then, fortified with some bottles of local Hinano beer, installed themselves in the back of the Chevrolet. I drove into the valley, along the river bed. It had rained so little that year the river was not more than a stream wending its way through the wide bed of stones.

Tua, the old man, called, "Stop here." Here? We were nowhere yet. "Yes. Look over there."

Across the river bed, the thirty foot cliff showed how high the river rose in the rainy season. Through the trees at the edge of a cliff I saw a *fare* and then a couple of pigs, visible at that distance because they were huge. "500 pounds," Tua said. "We need one only sixty pounds."

"You're going to get one there? How will you get there?"

The bank this side was not very high. The men crossed the river bed and shouted as they reached the foot of the cliff. A man promptly showed up on top, let fall a rope attached to a tree, dropped over the edge and made his way down. He shook hands with Tua and the others, and they all sat down to discuss the price of pigs. I found a boulder to sit on too, and waited. I couldn't hear the men and the valley was very quiet because we were well in from the ocean. I could see far along the river bed where it cut through the mountains and, high above, the sharp points of the volcanic crown, the Diadème, stood out against the stark blue sky. Finally a truck came along so I started the pickup to move it out of the way; the men heard and remembered I was waiting. All of them, including sixty-year old Tua, climbed up the rope. It took them a while to choose a pig, drag him over to the cliff, tie him up and let him down on the rope, squealing so loudly I could hear the echoes all over the valley. The men hoisted the pig on a pole, got him back across the river bed and up into the Chevrolet.

Back at the *magasin*, we discussed how we could all go to the airport. I would take Grandmère, who wanted to go to the market first and buy fish and taro; Etienne, a son-in-law, would drive Grandpère and the children in the Chevrolet. Before daylight the next morning Grandmère and I went to the market, then to the airport. We stopped at the flower *fare* to buy leis, and as we came out we saw the Chevrolet arriving; Grandpère was driving! Grandmère was furious but Grandpère was thoroughly pleased with himself. He leaned heavily on the canes he had made, and sank gratefully into a wheelchair. The children and I left the grandparents at the arrival gate, and went up to the terrace where we could see the passengers get off the plane. And there was Teva.

23

Teva, Tetiaroa and Grandpère

I could see Teva's expression as he approached the terminal. The children shouted to him, waving and laughing. He looked up and smiled. I called down, "Teva, he's feeling better. He's here." Teva told me later how much he appreciated that; he didn't worry so much going through immigration. He came through the arrival gate and went directly to his parents, who had tears of happiness in their eyes. We all kissed him and covered his shoulders with leis, Leilani and Tane stepping forward shyly because they hardly remembered him. The girls told Teva how Grandpère had driven the pick-up to the airport. Teva asked Grandpère how was his foot. "Much better — the *ra'au tahiti* is going to cure it," he answered emphatically. Teva glanced at me and saw my expression. He understood.

The children and I went into the baggage area to get his suitcase, but there was a delay and finally Teva came in too. He asked me exactly what had happened with Grandpère. I told him the whole story and finished, "— and now he is going to show the doctors they were wrong, that his foot will be cured with *ra'au tahiti*."

Teva said, "You understand now why I stayed away from him for ten years?"

I turned away, hiding my anger. I wanted to say, "You stayed away from me for ten years! You left me with him, without any warning of what he was like!" Teva didn't notice I was angry.

"I came on the first flight after I received your cable. I had to promise my boss I would go back to Florida in two weeks."

Then he had no intention of staying, despite his first cable saying he would accept the position at Tetiaroa. What about his father? What if he should die? Who would take care of the family? And the *magasin* that supports the family? Who would know what to do about — everything?

"It is not fair!" I cried. "He is your father, your family! I have stayed by them for ten years. Now it's your turn!"

Teva was silent for a moment. Then he said, "You are right." We stood there, watching the children search for his suitcase. He asked, "What about Tetiaroa?"

I shrugged my shoulders. "I don't know. I haven't seen Ivan for the past week. He may have come by but I wasn't at home much. I didn't want to see him anyway because you didn't write to confirm when you were arriving, or if there was a delay."

"But you knew I was coming as soon as I could?" I nodded; there was no point in saying anything. He asked, "Is Marlon Brando here now?"

"No, he's been gone since he learned he was nominated for the Academy Award."

"You wrote that no decision would be made until he comes back. Then there is still a chance for us?"

"Yes, certainly," My spirits lifted. "You get Grandpère home, and I'll call Ivan and tell him you're here."

Teva drove the pick-up to the house with his parents and the children. Teriitane was tired but happy to see the tables were set up; the pig, root vegetables and wild bananas were in the ground oven, and there was plenty of beer in the ice chests. Many people came, those of the family who were invited, and neighbors who had learned that Teva had arrived. Everyone joined in drinking to Grandpère's health and Teva's homecoming. It was late afternoon when the feast was over and most of the people had left. Grandpère had long since gone to bed, and by evening he was rested enough to talk two hours with Teva. Teva told me about it the next morning.

"It was bad. I tried to tell him he should go back to the clinic and have the operation done while it would still be only part of his foot. He absolutely refused. He is sure the *ra'au tahiti* will cure it. I shouldn't have said those old women's remedies were no good except for rashes and sprained ankles. My father got angry and said I don't know anything.

"I changed the subject and asked about the *magasin*. Why do they live there when they have the family house, where the children should be? It is bad for the girls to be raised in what is practically a bar, with a lot of men drinking and hanging around under the *purau* trees. My father said they could not leave the *magasin*; there was no one they could trust. I tried to explain how they could make a business deal with some one else to operate the store and pay them rent plus a percentage of the proceeds. He said no, that he had kept the *magasin* for me, that he wanted me to stay and work with him. So I told him I would never work with him, though I would help to get someone else to run it. I said I am not going to be a storekeeper for the rest of my life. He was furious, saying that anyway I am not capable of running it alone; he must stay at the *magasin* in case anything happened. 'What has ever happened?' I asked. I was furious too. My parents had never seen me like that and my mother was crying.

"I didn't mean to be disrespectful or make them angry, but my father never gives me a chance to talk, he doesn't listen, and even being gone for ten years doesn't convince him that I can manage anything by myself. I want to stay now and help them, but he still wants to manage my life."

"Teva, I know. You asked me last time why I divorced you. He told me that under Napoleonic law, if the husband is not here, the wife must be obedient to his father. He wanted me to live like a nun, even though you were not writing and we didn't know if you would ever come back."

"I am sorry about that. But you see, nothing has changed."

Monday morning we took the children to school and went to wait for Ivan at one of Vaima's sidewalk tables. Friends came by, stopping and shouting with delight when they saw Teva. Right away they started talking about the businesses they were getting into, about politics, the progress in Tahiti and how much there was to do. Teva, already stunned by the number of cars in Papeete and construction everywhere, was dismayed by the change in the people. His friends urged me to convince him to stay. Everyone seemed to take it for granted that we were together, despite the divorce.

Ivan arrived and was favorably impressed by Teva. He had another appointment so he couldn't stay then. He asked if Teva could go with him to meet Forrest that evening, as Forrest was leaving the next day for Los Angeles and would be giving Marlon the latest news of Tetiaroa. Ivan wanted Marlon's okay to start hiring personnel for the projected opening, even if Marlon had not yet met them, and of course the *tavana* was the most important. Teva was impressed with Ivan too, and looked forward to the meeting.

We started back to Punaauia. Teva asked, "Do you have anything you must do today? Would you go out to Papeari with me to see a friend? We could have lunch and go on around the island." I had nothing special to do. It was a beautiful day and a wonderful drive. We stopped at Atimaono where the projected golf course was being built. I did not know that Teva was a golfer; in fact, an excellent golfer, and already becoming known in the States. We ate at a Chinese restaurant that catered to the small population at Taravao; the meal was cheap and delicious. There were not many changes beyond fifteen kilometers in either direction from Papeete, and Teva was glad to see Tahiti as it always was. We drove slowly, talked and talked, and had a wonderful day.

The next day Teva told me about the meeting with Forrest, who apparently had been retained by Marlon after all. Forrest also liked Teva immediately, and agreed he would be the right man for the job. But not right away. The construction was under way again, although many of the workers had left for other jobs because they could not wait without pay. He was assured now of the funds to complete the hotel. As Teva said he had to return to Florida for a while, he could stop and meet Marlon in Los Angeles, and the contract could be signed then. That sounded good, even though I was apprehensive about Teva's return to Florida. He never explained what he was involved in but one problem seemed to be, having gone in with other members of the troupe to buy a house, he had to sell his share. I hoped that that was all, that he would be able to take his son and come back quickly.

Teva had to pick up the *tahua* to come and see his father. The medicine he had prescribed was not as effective as it was the first days, and the *tahua* said they had better apply bees too. Teva must go to get him at five o'clock in the morning because that was when the bee stings were the most potent. Teva, listening to all this, became angry again but of course he went to get the *tahua*. After he took the old man back home he again tried to reason with his father. Teriitane would not listen. He said he would prove not only to the doctors but his disbelieving son, he would be cured with Tahitian medicine.

I went to see Teriitane. "We've hardly seen you these days," he said. "We miss you, Grandmère and I." I thought it was an opportunity to talk to him about his foot.

"Grandpère, I think it is poor circulation that is keeping the *ra'au tahiti* from being as strong as it was the first days. Remember, the doctor was giving you medicine to increase your blood circulation? Perhaps you should have the doctor come and give you shots for that. Why not try both medicines at once?"

"No, you can't mix them. The *ra'au* will work alone. It isn't the circulation; it's the gout. It will get better. I've always taken *ra'au tahiti* for that." It was true he always recovered quickly whenever he had an attack of gout.

Teva came over to see me every day. He simply couldn't stay long with his father. One evening he and Tane went fishing and Teva cut his foot on the coral. It was bleeding copiously and we applied pressure for half an hour before it stopped. Teva said ruefully, "Just like a tourist. I have no more strong Tahitian blood. Too much hamburger."

One morning he borrowed Adèle's vespa and went to see his cousin up the road. The gendarmes stopped him, demanding, "Where is your helmet?"

"I'm just going up the road. Is it the law to wear a helmet now?"

They thought he was trying to be smart. "You're Tahitian. Where do you live?"

"Here — my father has the *magasin* Tiare Anani."

"Oh, I know," the Tahitian gendarme said, "You're the one who's been gone for ten years."

I went again to see Grandpère. Teva was there, sitting alone behind the house, although there were a dozen of his old friends under the *purau* trees. I remarked upon the number of vespas lined up in front of the *magasin*, like horses in front of a saloon in a western film. Teva managed a smile but he was very depressed.

"It's really bad. I think of the girls growing up here and it makes me furious." He glanced at the house. "I can't stay more than a few minutes in there now. We're both ready to fight as soon as I go in. I can't do anything to help him."

I went into the house. Grandpère was sitting up and looking much better. "You see? The bee stings and the *ra'au* are working." He had a smaller bandage on his foot, and his ankle and foot were not at all so discolored as they had been.

"Is it the same *ra'au*?"

"The *tahua* grated some of these and added that to the *ra'au*," Grandmère said, showing me some tiny green coconuts.

Something was working. I began to wonder.

I told the doctor about the improvement that Grandpère's foot appeared to have made. He was surprised and interested, but smiled when I described the treatment. He said tactfully, "I would have to see his foot, of course, before I could advise you in any way, and he refuses to let me even see it. He has a remarkable resistance, that is evident, and maybe this stuff is slowing down the progress of the disease. There is no way I can tell you how long he will live without proper treatment."

I tried to talk to Grandpère about Teva. He said, "No, Teva cannot take over the *magasin*. He knows even less now than he did before. Those ideas he has may be all right for America but not for here. He is an American now, not a Tahitian."

Teva decided to leave then, to be able to return that much sooner. He saw Ivan, who told him again to stop in Los Angeles and see Marlon. Marlon had just refused the Academy Award to protest against the treatment of the Indians by the American government. I wondered if he had so much on his mind he had forgotten Tetiaroa.

I was painting the cupboard in the garage when I heard the rattletrap noise of the Deux Chevaux, the canary yellow "office car" that Ivan was knocking around in lately. I had not seen him since Teva had left, and what with Grandpère, and chauffeuring Grandmère and the children, I was not at home much. Ivan climbed out of the car and ambled over.

"Hi," he said. He seemed to become more laid back every day, his excellent English Americanized with slang expressions.

"Hi," I answered, scraping excess paint off the brush and dropping it into a can of water.

"Are you going to rent the house again? Moving back to the other house?"

"Yes, but I'll be leaving in a couple of weeks."

"Leaving? I thought you were going to cancel your tours."

"It isn't a tour. That program starts in September. I'm going to take the children on a charter trip to Mexico. My mother has never seen them."

"That'll be nice. When will you be back?"

"I don't know. As a tour conductor, I get privileges. There are several one week charters so we can come back on a later date."

"What about Tetiaroa?"

"Yes, what about it? It seems to be put off indefinitely."

"No, not at all. We'll open the hotel July 1st. Aren't you coming with Teva? Haven't you heard from him?"

"No. I thought you might have, through the office." I would not say I was hurt because Teva had not written to me. Perhaps it was all in my mind that he seemed to still care about me. "I have the impression Marlon has lost interest in Tetiaroa."

"Not at all! The delay has been in hiring personnel because he wants to meet everyone first, and the Academy Awards thing keeps holding him up. Look, I'll call Los Angeles and get a definite answer for you. Perhaps not right away — Marlon's gone off to South Dakota to get an Indian to accept his award."

"Maybe he'll get the idea that Sitting Bull or Rain-in-the-Face could be chief of Tetiaroa," I said.

"Come on, now. I'll let you know as soon as possible, at least what he says about Teva."

I went back to my painting and tried to forget about Tetiaroa and Teva. I was not going to risk losing my job as tour escort. I was surprised that Travelworld still wanted me, after my refusal to make the second South American tour because it was badly organized. That had really surprised me; I had chosen Travelworld because their tours were the best. I was a less desirable tour conductor because of my insistence that my first tour start in Tahiti and the last would bring me back. But I didn't mind if I didn't have immediate connecting tours because then I flew down to Guadalajara to stay a few days with Mother. I could not cancel now if I still wanted to travel.

The next day I picked up the children, shopped for groceries and came home. Tane went first into the kitchen *fare*, that was never closed. He came back out, saying. "Mommy, what does the plate on the table say?"

"Plate? Say?" There was nothing else, no papers, on the table. Ivan had come by and left me a message, choosing a plate painted with roses, that Grandmère and the children had given me for Mother's Day. The message, written with a marker, said, "Called Los Angeles this morning. Teva has the job. See you later. Ivan"

The children were wildly excited. The girls told me they had asked Teva if they could go to Tetiaroa too, and he said of course. They would be away from the *magasin* at last. And I hoped he meant we would all be together.

Grandpère's foot slowly but surely healed. I insisted

the doctors from the Clinique should come and see for themselves. They had to admit that the *ra'au* stopped the gangrene and cured his foot. Grandpère was proud and delighted to have proved it to them.

Marlon was not in Los Angeles when Teva stopped there on his way to Florida. Teva didn't write. Forrest was still in charge of whatever was going on in Tetiaroa, and Ivan left for Europe. I didn't cancel my tours, and I continued traveling for another two years. All the projects for Tetiaroa were abandoned for lack of funds, except for the hotel. It was finally finished and opened to have considerable success, as it would in any case of course, because it was on Marlon Brando's island.

24

Mexico

Marisa was waiting at the gate when I arrived in Los Angeles. I hadn't seen her since she went back to Italy to study at the Academia di Belli Arti in Florence. She completely astonished me. Although she was wearing an old, tight pair of jeans, a tank shirt, and a denim hat in need of mending (she said she would never go out like that in Italy), she was utterly sophisticated and beautiful. She was very slim, tall, graceful, and completely self-assured. Her hair was wavy, the color a mixture of sand tints to sun, and her skin was pale gold. A little make up accentuated her blue-gray eyes. We hugged each other fiercely, but I kept pushing her away so I could look at her. We were laughing and trying to say everything at once.

An airline representative came up, taking a poll of passengers, and inquired about the trip. Marisa said quickly, "Yes, I just arrived from Tahiti. I make the trip often—" and started to tell such a fantastic story that we soon broke up again. This girl could handle any situation, I thought, for fun or for real. But she met her match as she was about to drive her father's Mercedes out of the parking lot, and stopped when she saw another car backing out at the same time. Marisa blew the horn frantically when she realized the car was going to hit hers, but too late. The driver backed right into her car. Marisa leaped out and dashed up to the driver's side — the woman, with a face like a bulldog, was about to go off without a glance behind her.

Marisa cried, "Lady, you just hit my car!"

The bulldog looked at Marisa coolly. "I know, but I didn't hurt it."

"How do you know? You were going to drive off without looking!"

"I hardly touched your car." The woman changed into forward. "In fact, little girl, I don't think I touched it at all." Marisa was left there, gaping, but she had the presence of mind to catch the license number.

We bought our tickets to fly to Guadalajara a couple of days later. The plane rattled so much it seemed it would surely fall apart before we arrived, but we were so glad to be together again that we giggled and flirted with the cabin crew all the way. Mother came to meet us at the airport but she somehow missed the announcement that the flight had arrived. After some time we found her wandering around in the building, waiting. She had in fact noticed Marisa while I was in a booth trying to telephone her, and thought to herself, how beautiful that girl is. She was enchanted to discover the girl was her own granddaughter.

We drove home, chattering like magpies. Mother didn't live in Guadalajara itself. Her house was in a village called Ciudad Granja, beyond the outskirts of the city. We turned off the *circumnavigación* and bumped off the shoulder of the highway down into the back yard of a church. I gasped, "Mother, you missed the road! There's a sign up ahead."

"No," Mother answered breezily. "That's where the road was, but it washed out in the rain last year. This way is better." We circled the church and drove along the main street of the village. The street was wide enough but as it wasn't paved the ruts were deep, so the traffic in both directions stayed on whichever side was more navigable. There were few cars in the village, but there were horses, sometimes very good ones, ridden by real Mexican cowboys, and donkeys ridden by children. We never saw a woman on horseback. The women were sturdy peasants, wrapped up in shawls that cradled the latest baby of several in those devote Catholic families.

Mother's house was hacienda style, with a courtyard in the middle, and another courtyard for the stable. She had two horses; they were for her husband who loved to ride but he had died the year before. Bill wanted to be cremated, so Mother had to sneak his body out of the village to Guadalajara. She returned with his ashes concealed in a shoe box. The next day she told the yard man to saddle the horses. Mother never rode so the man was surprised when she mounted one and told him to mount the other, to lead her up the slope of a nearby hill. Near the summit she told him to stop. Mother rode over to the other side where she scattered Bill's ashes onto the ground. The good Catholic villagers always wondered, what happened to Bill's body?

Marisa and I were in heaven, in Mexico with two horses. The cowboys watched us stolidly when we rode by, but we always said, "*Buenos días,*" and after a while they smiled when they answered. The women always smiled; Mother had "adopted" many children.

Mother returned to the States a few years later so I went down to Mexico on my own, sometimes by car, going by way of Arizona, Nogales, and Mazatlan. One trip I went as far as Puerto Angel, well beyond Acapulco, and came back by way of Oaxaca and Mexico City. I was alone but had no fear of Mexican men because they respect women (the way men no longer do in the States). Only once was I frightened, when I misunderstood a soldier's signal at one of the many control points, thinking he had waved me through as they generally did. His sharp whistle brought me up short. He came up to the car looking angry and mean. He jerked the car door open, ordering me to stand aside while he thoroughly searched the car. A group of men nearby, with guns, stood up to watch. Finally the soldier stepped back and said I could go. I got in and started the car but then he cried, "*Alt!*" Terrified now, I was frozen in my seat. He came to the window, leaned over and asked politely, "Do you have some toothpaste you can give me?"

I drove on to the town of San Miguel de Allende, having heard about a group of hopeful writers there who met each week to critique and assist each other. The family who owned the Posada de las Monjas made me feel very welcome and I soon had friends, Mexicans as well as several of the writers. I had studied Spanish for a year before making a South American tour, and was proud to have translated competently for the Peruvian guide in

Machu Picchu. I continued to read Spanish novels so as not to lose the language. I felt so at home in Mexico and the colonial town of San Miguel was so beautiful, I stayed six months.

The next time, I flew to Guadalajara and took the bus to San Miguel de Allende just to see the wide open spaces and brick farmhouses of the countryside of Guanajuato, that glorious state of Mexico, with its high blue skies and sparkling air. Horses and multi-colored goats grazed on the unfenced land. In the towns, vendors sold fruit and cold drinks to the passengers through the windows of the bus. About a hundred miles out of Guadalajara, another driver took over our bus. With a big smile for us, the passengers, he got into his seat, smartly changed gears and took off like a maniac. I was sitting at the front; seeing animals and peasants flee for their lives, I suggested he should slow down. He grinned cheerfully, assuring me he was an excellent driver. We came to a small village with narrow streets, and approached an intersection where the facing street came in at an angle. Before we had stopped a truck came at us, no doubt driven at the same speed as our bus. Both drivers deftly swerved away from each other but too late; whatever hit our windshield smashed it. We were showered with tiny cubes of glass — fortunately it was safety glass and no one was hurt. The driver, ever smiling, pulled up at the side of the road, told us not to worry as there was a bus station in the village and he would run to get us another bus. Somehow he quickly did so and assisted us to change buses, as gallantly as though he were hosting a party.

Arriving in San Miguel, I was greeted with open arms by "my family" at the Posada de las Monjas. After the *mozo* had taken my suitcases up to my favorite room with its terrace overlooking the valley, I was eager to walk up to the square in the center of this beautiful colonial town.

I walked up the familiar sloping sidewalk, stepping on the cracks because the large square stones were polished by the passing of many feet, and they were slippery. There were two streets to cross on the way to the square, one with traffic coming up the hill, the other with traffic going down. I remembered to calculate the speed of the cars that would turn left and stepped between with perfect timing, like a toreador before a bull. Arriving at the Jardín, as this square was called, I strolled along with other visitors, Mexicans and Indian vendors.

The sun was setting and its rays lit the inside of the filigreed walls of the cathedral, where four large bells ring the hours. A few birds flew in from somewhere beyond the cathedral as the sun set on the horizon. The birds know then they must come to roost; they came in increasing numbers until the sky was filled with them. They flew directly into the densely leafed trees of the Jardín, calling and warbling as they settled in, making so much noise that people had to raise their voices. Birds and voices were drowned out when the bells began to ring. The cathedral was lit up then, its pink-gold spires silhouetted against the darkening sky. The noise was almost deafening so I retreated to the far side of the square. Two little Indian girls came by, giggling, one propelling the other. They had several small bags of pistachio nuts which they offered to people sitting on the benches. It was apparently a game for them because, laughing when their wares were refused, they cheerfully turned to the next person. I bought a bag of pistachios from the delighted pair, and was about to leave when an adorable little Mexican girl in school uniform stopped me and asked if I spoke English. Pleased to be mistaken for a compatriot, I told her I did, so she said, "My brother was supposed to pick me up but he hasn't come yet. Would you please ask that American lady who always sits on the bench there, to take care of me until he comes?"

I walked on to stroll through the arcades, stopped at one of the stands to eat some tacos, and then started back to the hotel. Waiting at the first corner that was always very busy, I noticed the policeman in the center of the cross streets, directing the traffic. A broad-shouldered, gray-haired Mexican, I thought he was very attractive. He turned at that moment and saw me. *"Buenas tardes,"* he said with a large smile.

"Buenas tardes," I answered, smiling also, as he stopped the traffic and I crossed to the other side.

I often walked up to the Jardín twice a day, on my way to the market, the bank, or just to stroll around the square. Every time I crossed the street at his corner, the policeman and I exchanged *"Buenos días"* or *"Buenas tardes."* I said to a friend, Mary, who was with me one day,

"You see that good-looking policeman? He likes me."

Mary scoffed, "Just because he smiles and says *'Buenos días'?* He says that to everyone."

"No, he doesn't. I haven't heard him say *'Buenos días'* to anyone else."

"All right, you'll see tomorrow if he does with me." The next morning Mary called, *"Buenos días"* instead of me, and he answered, *"Buenos días"* with his big smile.

"You see?" Mary said to me.

"That's because he sees you're my friend, and it's politic to be polite to the friend of someone one he admires," I laughed.

"Admires! You're pretty sure of yourself."

"All right, let's see if he smiles and answers when you're not with me."

The next day she went to the corner alone and called, *"Buenos días"* as she crossed the street. The policeman turned and answered politely, but he barely smiled. "I guess you're right," Mary said, and I was satisfied.

I was surprised one evening as I stood at the corner waiting for the cars to pass, when the policeman came over to me quickly, caught my hand in a strong grip, and said, I thought, *"Aquí, a las diez."*

He returned to the center of the traffic while I stood there, confused. Without thinking about what he might be suggesting, I wondered if I had heard right. *A las diez?* It had to be the next morning because it was past ten now. I crossed the street in front of him, and asked, *"Mañana a las diez?"*

With his big smile, and catching my hand again, he said, *"Si, a las diez."*

I went back to the hotel and crowed to Mary, "I've got a cita!"

"You've got what?"

"A *cita.* A date."

"A date with whom?"

"My handsome policeman!" I told her what happened.

"What kind of a date is it at ten o'clock in the morning?"

"I don't know. That's what I want to find out."

"You're crazy. He's certainly married. All Mexicans marry before they're twenty."

"I know that, but he is old enough to be a widower, possibly. Anyway, I'll find out tomorrow."

The next morning I dressed with care and strolled up to the corner, a little late because I didn't want to risk waiting for him. He was not there. I went into the doorway of the pharmacy behind me and was glad to see another friend from the hotel, with whom I could chat a while. Then Mary came, looking around. "He isn't here?"

"No, and I'm leaving now."

"You're not going to wait? It's only quarter after. We're in Mexico, you know."

"You're right, but, look over there where that Indian usually comes to sell pottery. Have you ever seen a well-dressed 'Americana' sitting on a doorstep in San Miguel?"

"No. What do you suppose she's doing there?"

"Hah! I think my handsome policeman made a date with three women, and he has gone with the first one who showed up."

"You're crazy," she laughed, as I pretended to be outraged.

We went back to the hotel. An hour later, Mary asked, "Are you sure he said to meet you at the corner?"

"What? Why, I don't know. I was not sure whether he meant at ten in the evening or ten the next morning, and I presumed he said *'Aquí.'* "

"Everyone meets in the Jardín. You know that."

"I didn't think of it. Well, it's too late now."

That evening we walked up to the Bellas Artes. The policeman was at the corner directing heavy traffic, but he saw me. He gave me his big smile and waved. *"Buenas tardes,"* he called. *"Buenas tardes,"* I answered, smiling too.

The next day a maid came to tell me that a Mexican woman wanted to see me. I immediately suspected who it might be and prepared myself for the worst. In the courtyard was a plainly but neatly dressed, attractive older woman who watched me descending the steps. Then, to my surprise, she smiled, and put out her hand to take mine. "You are a lady, just as I thought. And I am Señora Garcia, the wife of your friend, the policeman. Can we talk a moment? Oh, I see you didn't know his name." I was so surprised and charmed by this lovely woman that I didn't know what to say. She looked concerned and asked, "Don't you speak Spanish?"

"Yes," I finally answered, smiling too. "Well enough, I think. Please, let us sit down." Then I started to apologize but she interrupted me.

"Oh, no! Men do these things! My husband wanted to meet you but he decided against it, as he is a good husband and a loving father." She went on to tell me all about their eight daughters and their home. We talked for perhaps an hour, perfectly at ease with each other, as I told her about my children and we compared the ways each of us lived. Then she asked, "Would you like to come and visit us?" I accepted with great pleasure but said I was leaving the next day. I promised to visit the next time I returned to San Miguel. We hugged each other good-bye, feeling as if we were the closest of friends.

25

Bora Bora

I was in Mexico when I received a cable, three weeks in arriving, that there had been five hurricanes in Tahiti, and the roofs of my two cottages had all but blown away. The last hurricane in Tahiti was in 1906. Although there are heavy storms that hit the coral islands, since then Tahiti was always spared. This time the hurricanes were caused by the rise in temperature of El Niño, the principal Pacific Ocean current along the west coast of North and South America. That heated the air, which created hurricane conditions in the center of the South Pacific. Houses were not built in anticipation of high winds, but for heavy rains. The tightly woven or layered palm thatch roofs came well down over the walls and windows of Tahitian *fares*.

I returned as fast as possible, to find my houses deserted, of course, by the tenants. It took a month to get them scrubbed, repainted and the roofs re-thatched. Fortunately both houses rented again immediately. I was afraid I wouldn't find someone who would put up with the neighbors on one side. Half of their house was blown away — the front half — so one could see the beds pushed under the remains of the tin roof, with the stove, table and TV out in the yard. The occupants were two men, three women, five or six children, and two Chinese fairies. They had no bathroom and used another neighbor's. I called the Sanitation Department, then learned the family had practically no income. I bought a toilet, there was enough wood around to build a bathroom for the household, and I hired one of the men to finish the work on my houses and garden.

I was exhausted by the time my tenants moved in so I decided to go to Bora Bora for a few days of rest. Being almost broke, I took the cargo boat that went to Huahine, Raiatea, Tahaa, and Bora Bora, a trip of about twenty-four hours. The six cabins were taken so I camped on deck along with a hundred other passengers, thinking there was no problem because it was not the rainy season.

The evening we left was pleasant, but as the sun set the sky suddenly became thick with black clouds. We were hit by a splattering of rain, then by sheets of rain that swept periodically across the decks half the night. The crew tried putting up canvas tarpaulins but they were of little use in the wind and driving rain. We took them down and made tents, several people huddling under each. I shared one with three young Tahitian men who had sort of adopted me; we managed to get some sleep although we were soaking wet. One guy got rather horny, and was hit on by the others for being disrespectful. The rain let up by two o'clock and a full moon poured its pale yellow light upon us. At Tahaa a church group came on board. The women were dressed in severe navy skirts and white blouses but their hats were crowned with flowers. They sat upon their flower-printed mattresses, taken with them to camp in a meeting house in Bora Bora. To the soft strumming of two or three guitars they sang through the early morning hours until dawn. It was magical.

I stayed one night at the Oa Oa Hotel, island-style, colorful and cheap, and the next two nights on the boat of a delightful American couple anchored off the pier. In the bar I met an old friend, Alex. He had lived in the islands for thirty years, had been an excellent hotel manager, but was now a total alcoholic. He told me about a position that was coming up at the hotel where he was working. That evening the owner of the hotel called and asked me to go over for dinner. Thus I met Baron George.

Baron Jerzy Hubert Von Dangel. "Dangle?" the uninitiated asked with a smirk. No, he pronounced it the French way, "Von Don-jel," with the accent on the jel. Anyway, nobody called him that; even if he wasn't one, he was called Baron George. He had been just about everything else; air force pilot, fashionable shop owner in Caracas, Venezuela, and doctor, he says. Born in a castle in Poland, he says, immigrating when a child to Australia, he came to Tahiti in the early seventies and fell in love with a Chinese-Tahitian girl. Although he didn't marry her he managed to stay in French Polynesia. He tried exporting pearls and Tahitian craftwork, smoking fish, and playing bit parts in films — *Hurricane*, *Beyond the Reef*, and others. In his forties, tall, balding, he was often asked if he was Kurt Jergens' brother. He reminded me of Gustav, not only his looks but also his personality. All ego and talk, he was charismatic and absolutely charming when entertaining guests. Unlike Gustav, though, if no one was around he was crude. He became aware of my distaste and then spoke with more "class" in front of me, instead of talking like an Australian dockworker as he did with his pals.

I had asked Alec what was the name of this baron's hotel. "Bloody Mary's? How unattractive!"

"On the contrary," Alec said, " the Americans love it. It's the restaurant they come for. As for the hotel, six of the ten *fares* were blown away in the hurricanes. The others are in fair shape and I'm living in one, so that doesn't leave much of an hotel." George had built the restaurant with the Tahitian who later became his assistant cook. It started out with just a roof, a section walled off to be the kitchen, and a long grill. Tables were made of split coconut logs, seats were log stumps. That was when *Hurricane* was being filmed; apart from the hotel Di Laurentis had built to house the actors and film crew, and the rather sophisticated Hotel Bora Bora, there was no other restaurant that catered to tourists. Although fish obviously is the specialty of any restaurant in the South Pacific, Baron George hit upon an excellent idea. Fish was all he served, except grilled chicken for the unfortunate people who didn't like fish. The catch of the day was spread out upon a large table decorated with leaves and flowers. The clients chose the fish they wanted, from mahi mahi to shrimp to Pacific lobsters, or crayfish. Baron George presided over the grill while pretty waitresses garnished the huge plates with rice, salad, and a choice of George's famous sauces. There was a wide selection of wines, a well-furnished bar, and that was it. Not cheap, but it was crowded every night.

The roof had been expanded recently; there was space for about twenty-five tables, still made of split logs with the uncomfortable coconut stump seats. The floor was clean white sand. There were no walls and plants crowded in around the restaurant. Occasionally a large land crab sidled in from the garden. Bora Bora had thousands of land crabs and the ugly crustaceans with their bulbous, weaving stick eyes, did not endear themselves to the tourists.

That night, George absolutely charmed me into staying to be a hostess, and the next day I was promoted to manager. Alec knew he was on his way out, which was why he recommended me. I had never had anything to do with a restaurant but it seemed simple enough. Being broke and without a job, I decided to give it a try.

Between Alec, who had not paid the merchants for six months, and George, who helped himself from the till or made out checks — the bank account in his name was the account for the restaurant — from $6 for a bottle of suntan oil to $1300 for his new TV, the bookkeeping was total chaos. With an average of fifty people a night, at an average of twenty-five dollars each, money should have been no problem. But George took off on trips to Las Vegas and elsewhere at the slightest whim, taking a girlfriend with him, of course. He had half a dozen women, and four or five children. That was his pleasure, fine, but I tried to have a separate account for the restaurant and that made him angry. It was sweet that he didn't want to fire any of the help, but five of the fifteen were too many. He owed three years of social security, over twelve thousand dollars a year, as well as fifty thousand he had borrowed and gone through in about fifty days. I spent a week finding accounts that were destroyed by the hurricanes or simply thrown away. I went to see all the suppliers and promised to pay a past month's bill along with the present one — that project lasted two months. The cook was ordering far more than we needed; it didn't take long to discover how he was siphoning food off to his friends and pocketing the money. George would do nothing about it because Tihoti was his friend who helped him build the place. I cut the orders but the savings were a drop in the bucket. After three months I managed to get twenty thousand in the bank and was going to distribute it to the creditors who were counting on me to catch up, when George appeared in the office with his suitcases. He had taken all the money out of the account. I wailed, "George, you can't do that!" He threw down his suitcases, all but stamped his feet with rage, and cried, "You're fired." Nobody was surprised; I was the thirteenth manager in four years.

The fourteenth manager was the shrewd half-Chinese nephew of one of George's Tahitian lady friends. A most affable and generous person, he stayed on for several months, lived like a prince, and really milked George.

I lived in Bora Bora for another year. I was surprised one day when George came to my house and asked me to help him. He hadn't fired Pascal, he wasn't leaving the island, so I didn't see what I could do for him. He told me he was going to be forced into bankruptcy or he had to sell the restaurant (the land was leased, and also in arrears) to pay his most pressing debts, about $80,000, immediately. He would sell the business to me if he could stay on the way he was, bestowing charm and public relations from his dais at the grill. I was very interested, and flew to Papeete to consult with my bank.

For all of Bloody Mary's fame in Hollywood as well as magazine articles acclaiming it across the United States, it was unknown in Tahiti. Baron George was known, though, and the bank thought I was crazy, obviously, when I couldn't come up with a year's accounts. I could have managed without the bank, paying George a couple of thousand or so dollars a month, and paying off the debts in time. But George wanted a lot of cash right away. He stalled until an American showed up with an offer that pleased him. Negotiations dragged on for months, and as the conditions of the lease were not being met, there was a chance of the owner getting the land back, with the buildings, by default. That would have been all I needed, as the owner wanted to lease it to me. I stayed on in Bora Bora until finally the American came up with half the money, took over the place and moved Baron George out.

I saw George a couple of years later in Tahiti. He greeted me like a long lost sister. "Oh, Renée, I should have sold you Bloody Mary's. That bastard ended up paying me little more than that first cash payment!"

"George, that's terrible! How did it happen?" But I knew in the vague way he replied, that probably, like with the accounts, he had paid no attention and let himself get rooked.

The last I heard, George was a preacher with psychic healing powers.

26

Rupé Rupé Ranch

The group of riders galloped up the bank and into the ranch. Ramay, my Marquisan outrider, turned his horse back to the beach to be sure that the stragglers arrived safely, as all the horses liked to race up the slope.

"Renée, we have to do something about that dog," he said as I joined him.

"What dog?"

"The one that almost bit Royale yesterday, when that so-called cowboy took off in front of the Capitaine Cook place."

"Didn't you warn him?" The dogs always got excited and chased after the horses if we galloped on that stretch of beach.

"Yes, and I told him again today, but the guy is a show-off. He thought it was funny to see the dogs trying to nip at his horse's heels, but that new dog was playing no game. She was lying in wait this time, and she went for the throat."

"What! Tell me about her when I get back." We helped the riders off their horses. I drove them back to Club Med and returned to the ranch. Ramay was washing down the horses.

"Okay now, tell me more about this dog. Had you never seen her before?"

"Not before yesterday, and then she was with some people so she seemed all right. She is no ordinary Tahitian dog, though. She is a big pit bull."

"Then she belongs to somebody, and must have broken her chain. I'll see if I can find out who owns her."

I drove over to the Capitaine Cook, a small complex of apartments for hotel personnel. There were rarely more than two or three people on the beach, so there was plenty of room for us to canter by. I walked over to the beach and immediately saw the dog. She was a brindle, and maybe half pit bull. I approached her slowly, ready to retreat if she growled. She raised her head, turned and saw me, and, wagging her tail, came right to me.

Dogs can smile, and she was smiling. I rubbed her head, scolding her, and she was ecstatic. I had brought a chain in case I could approach her, and she stood still for me to attach it around her neck. I really didn't need it because she followed right at my heels, and at the car she jumped in and sat down obediently. I went from house to house in the neighborhood, but nobody recognized the dog. I took her home to the ranch and attached her chain to the porch railing. She lay down on the top step and put her muzzle on her front paws, so that only her eyes moved to follow me in and out of the house. She was perfectly satisfied to have found a home and a mistress.

I called the Tahitian SPCA for advice about what to do. They said, "Nobody has called us about a lost dog like that. Maybe she was left behind by some people coming back to Tahiti."

"I can't imagine anyone abandoning such a good dog."

"Pit bulls can sometimes cause a lot of trouble."

"I doubt this one would, unless the owners happened to acquire horses."

"I'll put a notice in the paper and, meanwhile, can you keep the dog?"

"Yes. I'll have to keep her tied up but I guess it'll be all right for a short while."

Each passing day I became more fond of Brindy, as I called her. In the evening, when the horses were in the corral, quietly feeding, I took Brindy off her chain for a run on the beach. As she paid no attention to the horses when we passed by the corral, I started teaching her to become familiar with them, letting the dog and horses sniff each other. I said to Ramay, "How nice it would be if she could learn to go out with us on the rides, and perhaps keep the other dogs from running out on the beach and startling the horses."

Ramay was doubtful. "Could be she'd enjoy a good fight any time she can. Pit bulls are like that."

"Not this one. She is as friendly with other dogs as she is with people."

I called the SPCA but they had received no response to their notice, and said that now she could be adopted. "She isn't spayed? Then we'll have it done, if you can manage to bring her to Tahiti."

Ramay took us to the boat, as the Suzuki had to stay at the ranch for transporting riders, and in town we took

Le truck. Brindy made herself as small as possible between the bench and my feet, looking apologetically at the other passengers who sat as far away as they could. At the vet's, she was put to sleep without any trouble, and I went back to get her the next day. I had shopping to do and took her with me to the marketplace. I couldn't resist buying a smart collar and leash, and also an expensive muzzle that I thought would be wise to put on her when she was loose in the ranch. On the street Brindy stayed close beside me, but everyone moved out of the way, murmuring, "Pit bull!"

The next morning, while the horses were quietly eating, I put the muzzle on Brindy and let her off the chain. I thought she would be subdued by the wide bandages taped tightly around her middle, but she took off like a shot after an early risen chicken. The chicken, squawking desperately, half stumbled, half flew across the ranch and into the neighbor's yard, Brindy hard after her. The neighbor had ducks, and all hell broke loose. Brindy didn't pay any attention to the ducks; she wanted that chicken, and soon they came back again, with the neighbor right behind. When I finally got Brindy's attention and she flopped down beside me, I got the scolding from the neighbor.

"Are you going to keep that dog? Why do you want a dog now? You don't need a dog. They're nothing but trouble, and obviously she's one of the worst!" The man loved his ducks.

I promised I would put a fence up between the ranch and his property. I bought chicken wire fencing for $100 and had it installed. The next day Brindy sailed over it in pursuit of another chicken, and pushed under it on her way back. No amount of scolding would keep her from chasing chickens. She'd see one, look at me guiltily, and take after it. I smacked her with a rolled newspaper. She withstood the punishment and showed me how sorry she was by crawling under the porch for half an hour. Then off she'd go after another chicken.

I seemed to be making progress with her acceptance of the horses. She paid no attention when they were being brushed and saddled, and only watched intently when the horses and riders left the ranch. She even got used to Ramay and me riding around the ranch and herding the horses down to the beach. I thought that she had now understood that horses were off limits.

I always tied Brindy to the porch when I left the ranch, but one day I didn't snap the chain onto her collar properly. I was at Club Med, checking accounts for the rides, when Ramay called me. "Brindy got loose and attacked one of the horses!"

I raced back to the ranch, where Brindy had returned and was hiding under the porch. The horse had two bad bites on his shoulder and Ramay was pouring antiseptic on them.

"How did it happen, Ramay?"

"Brindy was probably looking for you and she went down to the beach. She saw the horses galloping, and I guess her instincts were too strong to resist."

The horse's wounds would heal completely, but I could not take the risk of it happening again. I took Brindy to the SPCA.

Fortunately, they had a request for a guard dog out in the country. I drove to the house to find out if there were any horses in the neighborhood. There were none, but there were chickens. I had to warn the family about Brindy and chickens. The father said, "So she chases chickens? That's wonderful! I never could keep the pesky things from crapping up the yard where the children play!" I knew that Brindy would be very happy there.

A ranch? Yes, I bought a ranch and it was the dream of my life. Owners and managers of the riding stable on Moorea had changed several times since Club Med sold the horses we rode those first years. I never learned what happened to my favorite, Tonga. New Zealand horses were more prone to illnesses than the Marquisans who had become acclimated to the humid, tropical weather and had bred for many generations. They became smaller, stockier and tougher than the original Arabians the Spaniards introduced to the Marquesas Islands. Most important for us, they had inherited the hard hoofs of their Arabian ancestors. There was no farrier on Moorea so one had to shoe one's own horses if necessary. It was simpler to have only Marquisan horses, but at the riding stable, or ranch, as the French called a riding stable, there had to be some big horses as well for tall and heavy riders. Horses had been imported from New Zealand by the land wealthy French-Tahitians, mainly for the races that were popular in Papeete. These bred with the local horses, and in Tahiti and Moorea there were many part Marquisan, part New Zealand horses. I had had to sell my horse, Akahiki, when I started traveling as a tour escort, and he became a race horse at the track in Papeete, competing in the traditional last race of the day's program. Marquisan horses are ridden without saddles by Tahitians wearing only a *pareu* tied up around their thighs. The riders wear a crown of flowers and the horses a lei around their necks. The race is wild and dangerous, at such breakneck speed that one or two riders are often thrown from their mounts on the curves. It is the most exciting and colorful race of the meet. We had a few New Zealand horses at the ranch, but the Marquisan horses were our favorites.

Unfortunately, my Tahitian children did not inherit my love of horses. Akahiki had been teased by children before I took him from Moorea and he never got over that. He would turn to kick if any child approached, so of course Tane and Leilani didn't like him. Tane's favorite activity has always been the classic Polynesian sport of surfing, and Lani is a dancer.

What happened to everyone during those several years since Teva left us for good? Grandpère lived in reasonably good health for a few more years. After he died, Grandmère, inconsolable, lived on a short while and both are buried in the beautiful cemetery at the Pointe des Pecheurs, at the edge of the sea. Teva finally came back to Tahiti and to the perfect job for him. He became the pro at the golf course of Atimaono. The sport became very popular in Tahiti, mainly due to Teva. He became an idol of the French population and was more comfortable with them than with the Tahitians. I was not in Tahiti when he returned, and when I came back I realized we had grown worlds apart. He finally moved to France.

When our son, Tane, was fifteen years old and had had four years of English in school, I told him we must speak only English together. "Languages are so important, particularly English, if only as a kind of insurance. You can always get a job in tourism, or as a translator, in your own country or another," I told him. But it is very difficult to switch to another language when only one language has been spoken in the family. I should have spoken English to my children when they were little, but as Tane spoke only Tahitian in Moorea, it was essential for him to become fluent in French because of school. Now it was high time to start speaking English, but when I insisted, Tane didn't speak at all. Exasperated, I decided, "All right, we'll go to the States and you'll attend school there. Then you'll have to speak English." To my surprise, Tane agreed.

We spent two years in California. I chose Newport Beach so that Tane could still go surfing. We had an apartment down on the Balboa peninsula, the top half of a duplex, half a block from the ocean. There were wide windows and a large terrace that was filled with plants, and the rent, then, was only $350 a month. I had two part-time jobs, as I had no training for any serious occupation, nor did I want a job that would keep me indoors. One job was maintaining plants, trimming and watering plants in offices, restaurants and private homes. The other job was taking folders of advertising materials from one newspaper office to another and back. The delivery car happened to be a Mercedes and I drove from Newport Beach to the airport, along the coast road bordering the beaches, often through glorious sunsets.

Tane went surfing in the morning when it was barely daylight and the fog was sometimes thick as pea soup. An hour later he came back, shivering cold in his wetsuit, showered, jumped on his bike and went to school at Newport Harbor High. He did very well in English and Spanish because of the emphasis put on English grammar in his classes in Tahiti, and Spanish was similar to French. He was popular in school but was not interested in the girls. He had met his first love in Tahiti shortly before we left, and he remained loyal to her. He spoke English and was understood easily by his friends, but he had such a strong accent I could hardly understand him. Although school was easy for Tane and he had good times with his surfing buddies, he wanted to return to Tahiti.

Marlon Brando called me and asked if Tane could go to Tetiaroa with his son, Tehotu, friend and surfing pal for five years by then. Marlon had engaged an American teacher, Jim Sanders, to tutor Tehotu, but Tehotu refused to stay on the island unless Tane was there too. I met Jim, we talked about the program he was planning for the two boys, and I agreed that it was a great opportunity for Tane. He would learn about so much more than in any school in the States or Tahiti. Tarita, Tehotu's mother, was living on the island but the boys had their own *fare* and usually ate their meals in the hotel kitchen. Jim held classes in the one room, thatched "schoolhouse" or out under the palm trees. Whatever interested the boys, from alternative energy to evolution, Jim incorporated in their studies and Marlon had sent whatever books and materials they needed. They installed a solar water heater, cultivated a garden after chemically correcting the sandy soil, learned how to mix cement to make a smooth, solid floor, and, with a specialist, Tane built a wind generator. Jim's wife had a class of several younger Tahitian children, including Tehotu's sister, Cheyenne. Marlon's plans to have facilities for scientists, as well as a college specifically for students of marine life and aquaculture, unfortunately fell through when the funds ran out.

I spent a month at Tetiaroa while the boys were there, recovering from a car accident I was in on the 405 Freeway. I had planned to go down to Tahiti for Christmas but the medical costs wiped out my savings until the insurance companies paid up. Marlon heard about the accident and telephoned me. "This is my Christmas present to you and Tane," he said, and sent me the ticket to make the trip "to stay at Tetiaroa as long as you like." It was not the first present of a ticket he gave us. Tane had worked in Newport Beach, cleaning sailboats, to buy his ticket to go home for vacation but he didn't have enough money yet. A few days before the school term ended he received a small package, a present from Marlon. He opened it and was surprised to see it was a tie, a very ordinary brown tie. I was as perplexed as Tane until he took the tie out and saw beneath it, four hundred dollars to buy his ticket home.

Recently I came across a letter I had written to my brothers, and remembered how spoiled Tehotu and Tane were despite, or because of, their wonderful opportunities. The boys and I were to fly to Tetiaroa on Monday morning.

"Tane said the plane was to leave at 8:30. I called the Brando house at 8:00 to ask if they wanted me to stop and pick them up, but there was no answer. I supposed they had left already but when I arrived at the airport they were not there, nor did they show up by 8:30. There were several workers, who took turns spending weekends in Tahiti, waiting to leave with us. I asked the pilot if he had to leave without the boys but he said, 'No, we have to wait. This flight as well as the Friday flight is chartered for the two boys.' Tehotu insisted upon going to Tahiti every weekend. The flights cost over $600 a week. At ten to nine the boys showed up, red-eyed and obviously exhausted. I didn't ask where they had spent the night. Since Tane had returned to Tahiti, and his first girlfriend had moved to France, he had become aware of other girls, older ones, I learned. It was these young women who were blatantly eager for the company of the two handsome boys, although Tane was only seventeen and Tehotu sixteen. When Tane told me they were going to nightclubs, I demanded to know how he managed his share of expenses with his allowance. 'We don't pay,' he said. 'They're school teachers, and they always pay.'"

The school program ended when Tarita learned that Jim had ordered the video films of Cosmos, and had taught the boys about evolution. She was very religious, believing totally in the dogmas of the Bible, and was horrified by the sacrilegious theories that Jim was exposing to her son. Her protests evolved into a fight with Marlon, and she demanded that Jim be dismissed. Unfortunately, Marlon, tiring of the issue, gave in to Tarita. It was bad luck that she had heard about the films at all. Although she had a modern house in Tahiti, and could travel around the world if she wanted, she preferred to stay at Tetiaroa, and just happened to hear of the classes about evolution.

Tane was deeply disappointed by Marlon's sudden abandonment of the school program, and all the excellent projects he had planned for Tetiaroa. Tehotu was sent to Hawaii, where he was on his own and expected to be responsible enough to attend school. I heard that that idea was not a success and he returned to Tahiti. Meanwhile Tane had become resentful and wild. He resisted every attempt I made to get him interested in continuing his education or getting a job. He finally agreed to go again to the States if I found a program specializing in alternative energy that he could follow. I went to see Jim in Los Gatos and we discussed what was offered in Santa Cruz and Berkeley, but realized he would have to grasp very technical English. Unless he studied seriously he could never master the subject on his own. Finally, a true Tahitian, he wouldn't leave the islands. He went to Moorea to live with his family *faaa'mu* and became a G.O. at Club Med. Later he started canoe racing, organized the Moorea team and went on to win trophies, participating in the competitions in Honolulu and New Zealand. Mama Nini gave him some land and he started building what became a series of houses, basically Tahitian, but in his own inimitable style. Still a surfer, he eventually converted the family house in Punaauia, and opened a shop called "Tahiti Surf n Style."

I stayed in the States another year, bought a motorhome, drove down to Mexico for several months and then across the United States to New England for the summer. The village of Thompson had not changed; it is a National Heritage village and the classic white houses around the common will remain always the same. I went to see the farmhouse where my brothers and I had spent the happiest years of our childhood. The house had not been changed at all but the barn had been turned into a garage for rebuilding vintage cars. The couple who live there were thrilled to meet me because they knew all about the Armand Denis-Leila Roosevelt Expeditions. They had found some photographs left behind in the attic, and had copies of all the newspaper articles about our parents. They asked about the studio and were delighted to learn that it was where Father had edited and finished the film, *Dark Rapture.* The playroom had been turned into an office, but the other rooms were the same, and I was filled with nostalgia for our wonderful years there. Mother had lived in New York after her divorce with Father, then she and her second husband, Bill Westley, went back to Thompson and opened a portrait studio and photographic service. Later, they moved to the village of Ciudad Granja, in Mexico, and after Bill died Mother stayed on there in her hacienda, where we all went often to visit her.

Father of course continued to travel and made his base in Nairobi, Kenya. He married again, and Armand and Michaela Denis became famous throughout the United Kingdom when they started the ten year series of adventures, *On Safari*, on BBC Television.

My brothers continued traveling also. After the African trip, David, Armand and Father traveled across the jungles of South America. David and Armand returned to take over the operation of the Chimpanzee Foundation in Florida. Father wanted to return too, but his wife refused to live there. Father unhappily gave up on the Foundation and it was finally closed. Armand went to Guatemala, later he spent several months in Haiti with André, and then worked for Cordis in Florida, developing and perfecting cardiac pacemakers. Last year he came to spend several weeks with me at the ranch in Moorea. David went to Germany with the U.S. Occupational Forces, and then used his G.I. Bill to study engineering at the University of Florida. After several years with Esso, he joined Scubapro, Gustav's company, as Vice-president and Director of Research and Development for his diving products. He left Scubapro, worked for Fluor and spent eight years in Saudi Arabia. David and Armand got together again in California, and purchased two houses, workshops, and a store to supply camper's needs — all historical buildings built in 1930 — in the middle of a wilderness campground, by Crystal Lake in the San Gabriel Mountains. Both of them could fix anything from the fire engine to frozen water pipes; it was necessary to be able to do so when the closest service was more than twenty-seven miles and forty-five minutes down a serpentine road sometimes littered with stones knocked down by wild animals or earthquake tremors. This family liked challenges.

My daughter, Marisa, returned to Italy where she studied art and theater production. She became well known among the artists centered in St. Paul-de-Vence, France, where she lived with her husband and two children, Theodore and Basile. She opened a theater in Sophia Antipolis, Le Theatre du Paradoxe, for which she wrote plays, painted the decors, and helped her actors make their costumes. The actors were children learning English; playing parts, no longer themselves, took away their self-consciousness. The theater was a great success.

Leilani, her "birthmark" long since disappeared, grew pretty enough to win third and second prizes in two beauty contests. Although she often came and stayed with Tane and me, home was always with her grandparents, and she remains totally Tahitienne. She participated in sports and, with her older sisters, traveled to islands across the Pacific with the Tahitian volley ball team, and later with the women's canoe racing team. She has a daughter, Rauhéré, but refuses to marry. Tane also has not married but has a child, named for his great grandfather, Teriitane, and Tane is raising his son himself.

One evening we were all together in Punaauia, and watching *Ushuaia*, the TV program of the French adventurer, Nicolas Hugot. When he announced "the most dramatic and fantastic film sequence I have ever seen," I just knew what was coming. Sure enough, it was the sequence of the Burmese peasant priestess and the snake god, the cobra. There had been no boy children born in the village for the past two years. The woman's daughter was pregnant again. The grandmother, slim and graceful, went to the hills to plead with the god cobra. She approached the narrow cave where the cobra lived and began to softly call him. Suddenly the huge snake slid out, more than fifteen feet long. When he raised himself he towered over the crouching woman, who had stepped back only slightly. She continued to talk to him and slowly rose, undulating slightly. He struck at her but, as she reached forward to break his lunge and deftly parted her knees, he spat poison only on the skirt of her sarong. He reared back up and she talked softly until he lowered his head, partly hypnotized by her voice and the swaying motion of her body. Then she leaned over and kissed the cobra on the top of his head. He struck again, and again she avoided the blow. She had to kiss him three times. Her skirt was splashed with poison as he tried to reach her, but she managed to subdue him and kiss him two more times. Then she quietly told the cobra to go back in his cave, which he did.

My children could hardly believe what they were seeing, and when the program showed Armand Denis and Leila Roosevelt, the producers of the film, they were thrilled. "They were your parents? Our grandparents?"

I had tried to tell them about our family, but in Tahiti it was hard to imagine such remarkable people and their extraordinary adventures.

The End

To Live in Paradise

A Great Gift for Anyone

To order additional copies
Phone (800) 773-7782 or (707) 964-9520
or
Fax a copy of this order form to (707) 964-7531
or
Mail a copy of this order form to:

Lost Coast Press
155 Cypress Street
Fort Bragg, CA 95437

Price per book: $15.95
Add $3.00 shipping & handling for the first copy, $1.00 for each additional copy (California residents please add $1.20 sales tax per copy). Please inquire for quantity discounts.

Payment method:
☐ Check ☐ VISA ☐ MasterCard

Number ________________________ Expires __________

Cardholder's Signature ______________________________

Address ____

City ____________________ State ____ Zip ____________

Ship to:

Name ____

Address ____

City ____________________ State ____ Zip ____________

Include card signifying that this book is a gift from:

____

(To send additional gift copies, photocopy this form.)

To Live in Paradise

A Great Gift for Anyone

To order additional copies

Phone (800) 773-7782 or (707) 964-9520

or

Fax a copy of this order form to (707) 964-7531

or

Mail a copy of this order form to:

Lost Coast Press
155 Cypress Street
Fort Bragg, CA 95437

Price per book: $15.95

Add $3.00 shipping & handling for the first copy, $1.00 for each additional copy (California residents please add $1.20 sales tax per copy). Please inquire for quantity discounts.

Payment method:
☐ Check ☐ VISA ☐ MasterCard

Number ______________________________ Expires ____________

Cardholder's Signature ______________________________________

Address ____

City _______________________ State ____ Zip _______________

Ship to:

Name ____

Address ____

City _______________________ State ____ Zip _______________

Include card signifying that this book is a gift from:

____

(To send additional gift copies, photocopy this form.)